Lecture Notes in Computer Science 6960

Commenced Publication in 1973
Founding and Former Series Editors:
Gerhard Goos, Juris Hartmanis, and Jan van Leeuwen

Yiannis Cotronis   Anthony Danalis
Dimitrios S. Nikolopoulos   Jack Dongarra (Eds.)

Recent Advances
in the Message Passing
Interface

18th European MPI Users' Group Meeting, EuroMPI 2011
Santorini, Greece, September 18-21, 2011
Proceedings

 Springer

Volume Editors

Yiannis Cotronis
University of Athens
Greece
E-mail: cotronis@di.uoa.gr

Anthony Danalis
University of Tennessee
Knoxville, TN, USA
E-mail: adanalis@eecs.utk.edu

Dimitrios S. Nikolopoulos
University of Crete
Heraklion, Greece
E-mail: dsn@ics.forth.gr

Jack Dongarra
University of Tennessee
Knoxville, TN, USA
E-mail: dongarra@eecs.utk.edu

ISSN 0302-9743   e-ISSN 1611-3349
ISBN 978-3-642-24448-3   ISBN 978-3-642-24449-0 (eBook)
DOI 10.1007/978-3-642-24449-0
Springer Heidelberg Dordrecht London New York

Library of Congress Control Number: 2011936883

CR Subject Classification (1998): C.2.4, F.2, D.2, C.2, H.4, D.4

LNCS Sublibrary: SL 2 – Programming and Software Engineering

Typesetting: Camera-ready by author, data conversion by Scientific Publishing Services, Chennai, India

Printed on acid-free paper

Springer is part of Springer Science+Business Media (www.springer.com)

Preface

Parallel computing has entered a new era. Multicore processors on desktop computers make parallel computing a fundamental skill required for all computer scientists. High-end systems have surpassed the Petaflop barrier, and significant efforts are devoted to the development of the next generation of hardware and software technologies toward Exascale systems. Processor architectures, high-speed interconnects and programming models are bound to go through dramatic changes. The Message Passing Interface (MPI) is today the most successful and widespread programming model for parallel computing. An open question is whether MPI will evolve to meet the performance and productivity demands of Exascale systems.

EuroMPI is the successor of the EuroPVM/MPI series, and is a key conference for this community, established as the premier international forum for researchers, users and vendors to present their latest advances in MPI and message passing systems in general. The 18th European MPI users' group meeting (EuroMPI 2011) was held in Santorini during September 18–21, 2011. The conference was organized by the University of Athens and the Innovative Computing Laboratory at the University of Tennessee. Previous conferences were held in Stuttgart (2010), Espoo (2009), Dublin (2008), Paris (2007), Bonn (2006), Sorrento (2005), Budapest (2004), Venice (2003), Linz (2002), Santorini (2001), Balatonfured (2000), Barcelona (1999), Liverpool (1998), Krakow (1997), Munich (1996), Lyon (1995) and Rome (1994).

The EuroMPI 2011 program provided a balanced and interesting view on current developments and trends in message passing. The main topics were communication, I/O, networking, and implementation issues and improvements; algorithms and tools; interaction with hardware; applications and performance evaluation; fault tolerance. We received 66 paper submissions out of which we selected 28 papers for presentation, and 10 posters with short presentations. Each paper had three or four reviews, contributing to the high quality of accepted papers. Two papers were selected as best contributions and were presented at plenary sessions: Tobias Hilbrich, Matthias S. Mueller, Martin Schulz and Bronis R. de Supinski, "Order Preserving Event Aggregation in TBONs" and Adam Moody, Dong Ahn and Bronis de Supinski, "Exascale Algorithms for Generalized MPI_Comm_split." The conference included the special session on Improving MPI User and Developer Interaction (IMUDI 2011) organized by Dries Kimpe and Jason Cope.

The Program Committee invited four outstanding researchers to present lectures on aspects of high-performance computing and message passing systems: Pete Beckman on "Exascale System Software Challenges – From Bare Metal to Applications," George Bosilca on "Will MPI Remain Relevant?," Michael Resch on "Experience of a PRACE Center" and Sudip Dosanjh on "Achieving Exascale Computing Through Hardware/Software Co-design."

The Program and General Chairs would like to sincerely thank everybody who contributed to EuroMPI 2011, the authors, the reviewers, the participants and our sponsors Microsoft, ParTec, the Innovative Computing Laboratory at the University of Tennessee, Mellanox and the University of Athens.

September 2011

Yiannis Cotronis
Anthony Danalis
Dimitris Nikolopoulos
Jack J. Dongarra

1st Special Session on Improving MPI User and Developer Interaction IMUDI 2011

While for many researchers MPI itself remains an active research topic, for many others it has become an invaluable tool to extract useful science from some of the most powerful machines available. Unfortunately these MPI application developers – and their highly valued experience and use cases – don't always find their way to the EuroMPI conference. The 1st Special Session on Improving MPI User and Developer Interaction (IMUDI 2011) aims to improve the balance by actively reaching out to the application developer communities. By evaluating the MPI standard from the perspective of the MPI end-user (application and library developers) we hope to provide application developers the opportunity to highlight MPI issues that might not be immediately obvious to the developers of the various MPI implementations, while at the same time enabling the MPI developers to solicit feedback regarding future MPI development, such as the MPI-3 standardization effort.

For this year's session, we invited Torsten Hoefler (University of Illinois at Urbana-Champaign) to give a keynote address on the software development challenges associated with parallel programming libraries using the MPI standard. We peer-reviewed and selected three papers from the six papers submitted to the IMUDI 2011 session. These papers cover several topics that address software development challenges associated with the MPI standard: MPI interfaces for interpreted languages, using C++ metaprogramming to simplify message-passing programming, and group-collective MPI communicator creation. We believe that the discussion of these topics in the IMUDI 2011 session will bring together MPI developers and MPI end-users, and help MPI users and implementors understand the challenges in developing MPI-based software and how to effectively use MPI in parallel software products.

We are grateful for the support and help provided by our colleagues for this event. While we cannot list them all, we especially thank the EuroMPI 2011 conference organizers, including Jack Dongarra (University of Tennessee - Knoxville), Yiannis Cotronis (University of Athens), Anthony Danalis (University of Tennessee - Knoxville), and Dimitris Nikolopoulos (University of Crete), for their invaluable feedback. We also thank the members of the IMUDI 2011 program committee for reviewing the session papers and their help in organizing the session. The program committee for this year's session included George Bosilca (The University of Tennessee - Knoxville), Christopher Carothers (Rensselaer Polytechnic Institute), Terry Jones (Oak Ridge National Laboratory), Wei-Keng Liao (Northwestern University), Shawn Kim (Pennsylvania State University), Andreas Knüpfer (Technische Universität Dresden), Quincey Koziol

(The HDF5 Group), Jeff Squyres (Cisco Systems, Inc.), Jesper Träff (University of Vienna), and Venkatram Vishwanath (Argonne National Laboratory). We also thank William Gropp (University of Illinois at Urbana-Champaign) and Rajeev Thakur (Argonne National Laboratory) for their support.

September 2011   Dries Kimpe
                                                        Jason Cope

Organization

Program Committee

Richard Barrett	Sandia National Laboratories, USA
Gil Bloch	Mellanox Technologies, USA
George Bosilca	Innovative Computing Laboratory, University of Tennessee, USA
Aurelien Bouteiller	University of Tennessee Knoxville, USA
Ron Brightwell	Sandia National Laboratories, USA
Yiannis Cotronis	National and Kapodistrian University of Athens, Greece
Erik D'Hollander	Ghent University, Belgium
Anthony Danalis	University of Tennessee, USA
Bronis de Supinski	Lawrence Livermore National Laboratory, USA
Jean-Christophe Desplat	Irish Centre for High-End Computing (ICHEC), Ireland
Frederic Desprez	INRIA, France
Jack Dongarra	University of Tennessee, USA
Edgar Gabriel	University of Houston, USA
Francisco Javier García Blas	University Carlos III Madrid, Spain
Markus Geimer	Juelich Supercomputing Centre, Germany
Al Geist	Oak Ridge National Laboratory, USA
Brice Goglin	INRIA, France
Ganesh Gopalakrishnan	University of Utah, USA
Sergei Gorlatch	University of Münster, Germany
Andrzej Goscinski	Deakin University, Australia
Richard Graham	Oak Ridge National Laboratory, USA
William Gropp	University of Illinois at Urbana-Champaign, USA
Thomas Herault	University of Tennessee, USA
Torsten Hoefler	University of Illinois at Urbana-Champaign, USA
Joshua Hursey	Oak Ridge National Laboratory, USA
Yutaka Ishikawa	Graduate School of Information Science and Technology / Information Technology Center, The University of Tokyo, Japan
Tahar Kechadi	University College Dublin, Ireland
Rainer Keller	HLRS, Germany
Stefan Lankes	Chair for Operating Systems, RWTH Aachen University, Germany
Alexey Lastovetsky	University College Dublin, Ireland

Pierre Lemarinier	University of Rennes, France
Ewing Lusk	Argonne National Laboratory, USA
Tomas Margalef	Universitat Autonoma de Barcelona, Spain
Jean-François Mehaut	University of Grenoble, INRIA, France
Bernd Mohr	Juelich Supercomputing Center, Germany
Raymond Namyst	INRIA, France
Christoph Niethammer	Universität Stuttgart, Germany
Dimitrios Nikolopoulos	FORTH-ICS and University of Crete, Greece
Rolf Rabenseifner	HLRS, University of Stuttgart, Germany
Michael Resch	HLRS, University of Stuttgart, Germany
Casiano Rodriguez-Leon	Universidad de La Laguna, Spain
Robert Ross	Argonne National Laboratory, USA
Frank Schmitz	Karlsruhe Institut of Technologie (KIT), Germany
Martin Schulz	Lawrence Livermore National Laboratory, USA
Jeff Squyres	Cisco, USA
Rajeev Thakur	Argonne National Laboratory, USA
Vinod Tipparaju	Oak Ridge National Laboratory, USA
Carsten Trinitis	Technische Universität München, Germany
Jesper Larsson Träff	University of Vienna, Department of Scientific Computing, Austria
Jan Westerholm	Åbo Akademi University, Finland
Roland Wismüller	University of Siegen, Germany

Additional Reviewers

Abdul Aziz, Izzatdin	Meilnder, Dominik
Aspns, Mats	Miceli, Renato
Brock, Michael	Miranda-Valladares, Gara
Broquedis, François	Moreaud, Stéphanie
Church, Philip	Palmer, Robert
Clauss, Carsten	Reble, Pablo
Denis, Alexandre	Roth, Philip
Duarte, Angelo	Rytter, Wojciech
Girotto, Ivan	Segredo, Eduardo
Hadjidoukas, Panagiotis	Segura, Carlos
Hunold, Sascha	Spiga, Filipo
Kegel, Philipp	Steuwer, Michel
Laros, Jim	Tao, Jie
Li, Dong	Tarakji, Ayman
Mckinstry, Alastair	Wong, Adam

Table of Contents

Experience of a PRACE Center

Michael Resch

HLRS, Stuttgart, Germany

Introduction

High performance computing continues to reach higher levels of performance and will do so over the next 10 years. However, the costs of operating such large scale systems are also growing. This is mainly due to the fact that the power consumption has increased over time and keeps growing. While a top 10 system 15 years ago would be in the range of 100 KW the most recent list shows systems in the range of 1-2 MW and more. This factor of 10 makes costs for power and cooling a significant issue. The second financial issue is the increase in investment cost for such systems. 15 years ago a mere 10 Mio € was enough to be in the top 10 worldwide. Today about 40-50 Mio € are necessary to achieve that same position.

Bundling of resources has been the answer to this increase in costs. In the US national centers evolved very early. The US and Germany followed. In Japan national large scale projects were set up. The national efforts in Europe seemed to be not competitive enough for many. So an initiative was funded by the European Commission called PRACE (Partnership for Advanced Computing in Europe). Its aim is to create a common HPC environment in Europe and by doing so bundle all resources to make Europe a relevant player in HPC again.

PRACE

In order to compete with the US and Japan PRACE needs to create a distributed Research Infrastructure. No single site in Europe can host all the necessary systems because of floor space, power, and cooling demands. PRACE is therefore aiming to create a persistent pan-European High Performance Computing RI with all necessary related services. Four nations (France, Germany, Italy and Spain) have agreed to provide 400 million Euro to implement supercomputers with a combined computing power in the multi Petaflop/s range over the next 5 years. This funding is complemented by up to 70 million Euros from the European Commission which is supporting the preparation and implementation of this infrastructure.

In order to support scientists all over Europe users will be supported by experts in porting, scaling, and optimizing applications. Training programs are supposed to accompany the PRACE offering teaching scientists and students how to best exploit the large scale systems. A scientific steering committee will control a peer review process through which access to the resources of PRACE will be granted based on scientific excellence.

Y. Cotronis et al. (Eds.): EuroMPI 2011, LNCS 6960, pp. 1–4, 2011.

Strategic Impact

Any large scale HPC center in Europe is affected by PRACE. In order to be relevant also as a national center one has to make sure to become part of PRACE. Integration into PRACE would require a clear a common strategy though. This means that Europe would have to work out which fields re relevant in HPC, which user communities should be supported, how many large scale systems would be required, how these large scale systems could be integrated with medium sized systems in regional centers, and how all of this would translate to specific focuses in individual centers across Europe. In other words, one would have to leave the question of "big iron" aside and would have to answer the question "what do we want to achieve with HPC". For a large scale center it is necessary to find its own role within such a pan-European strategy. The better roles are defined within PRACE the easier it is going to be to optimize European HPC centers.

Financial Impact

PRACE was a reaction to the growing financial costs of high performance computing. So in the first place one would expect it to solve the financial problems of the community. However, the issue of funding is not an easy one. So far national governments have devised their own strategies for funding high performance computing. In some cases like in the UK funding was given by the national government alone putting high performance computing under the control of a national funding agency. In other cases like Germany a mixed funding model was employed that would integrate local state funding with federal funding schemes. European funding comes into such existing systems and will certainly in the long run change them.

A positive scenario would put European funding on top of the existing funding. In case national centers would benefit and would see European money as a sound basis for medium term financial planning. A worst case scenario would see a reluctance of national governments to subsidize a European activity, pointing at the existing European funding and leaving national centers in a financial quagmire. As of today it is unclear which way the European commission and different national governments are headed and how this is going to work out for national centers.

Organizational Impact

As of today national centers are working based on national organizational strategies. There are established rules for accessing the systems. There are establishes rules for providing service to the users. Again we see different model. Some, like the UK, tend to separate services from cycle provisioning. Others, like Germany, keep expertise both for hardware and application support in single national centers. Access to systems is in most cases based on scientific merit. At the national level regional preferences have become widely irrelevant and animosities between different disciplines of science are reconciled by scientific committees that help keeping a well worked out balance.

With a European strategy a new level of organization is introduced. Committees have so far been set up in PRACE. It remains to be seen how rules and regulations work out. It is open how well the scientific decisions are accepted both with a national background and with competing scientific ambitions of the users.

For a national European high performance computing center this opens another can of worms. New rules have to be integrated into existing governance. New users and ne reviewers have to be dealt with. This could be a fresh breeze for organizations that have been set up 15 to 20 years ago. It could also put some of the existing regional and national systems out of balance. It remains to be seen which way PRACE is headed and how this is going to work out for individual centers.

The Special Case of Industrial Usage

All over Europe there are some centers that not only supply CPU cycles and/or support to scientific users. These centers have well established concepts of how to support industrial users. The most well-known industrial co-operations are established in the center in Bologna (CINECA) Edinburgh (EPCC) and Stuttgart (HLRS).

For these centers PRACE has created yet another challenge. Providing access to industry one usually has to focus on a variety of issues:

- Security: Industry does have an issue with data security as well as with secure access to systems.
- Reliability: Reliability and availability for industry is a highly relevant issue while science usually can accept downtimes as long as they are not close to a paper deadline.
- Network access: The typical researcher has access to the internet through public research networks. The bandwidth usually is high enough to connect to any European HPC center. For industry network connectivity is much more difficult to get. And beyond bandwidth there is an issue again with network security.
- License availability: For research licensed software usually comes at an acceptably low price. For industry licensing is a big issue and costs for licenses usually by far exceed costs for CPU cycles. Providing industry with access to HPC systems requires therefore a solution of the licensing problem that simply does not exist for researchers.

Furthermore there are a number of financial issues:

- Cost: While researches take system access for granted and do not ask for financial cost, prices are extremely sensitive for industrial customers. At the same time public funding agencies are not at all happy with tax payers' money being used to subsidize industry. It is therefore mandatory to calculate the total cost of any CPU hour sold to industry in order to keep competing nations like the US or Japan from filing complaints against European funding agencies.
- Tax: As long as HPC is kept in the realm of science and research tax usually is not an issue. It becomes an issue once industrial usage or even exchange of CPU cycles becomes the norm. German tax regulations have it that giving

away CPU cycles to a company – be it a non-profit or a profit organization – obliges the provider to pay tax for the assumed value of these CPU cycles. Tax regulations may change over time but should not be ignored when talking about system costs that are in the range of 100 Mio. €.

- Regulations: Funding for HPC is provided by research funding organizations based on the assumption that the money is spent for research purpose. If systems are used for industrial production simulation away has to be found to reimburse the funding agency. As an alternative the money earned can be reinvested to compensate for the loss of CPU time of scientists by industrial usage.

Summary

European large scale computing centers are headed for an interesting future. With China, Japan and the US competing for the fastest available computer system they might be left behind in terms of performance. The negative impact on their users could push them to seek access to systems in one of the countries mentioned. This would put European centers out of business within three to five years.

European funding is a chance to counter these efforts. The European Commission has decided to make PRACE its method of choice to support high performance computing in Europe. This introduction of a new player in the field opens a variety of questions. These questions will have to be answered if Europe wants to be competitive in high performance computing in the future.

For any national center it is vital to find a role within PRACE. At the same time it is necessary to carry over well established and worked out procedures to turn PRACE from a project into a stable funding and operational infrastructure that has a positive impact on an existing, well-established, and very successful HPC eco-system.

Achieving Exascale Computing
through Hardware/Software Co-design

Sudip Dosanjh, Richard Barrett, Mike Heroux, and Arun Rodrigues

Extreme Scale Computing
Sandia National Laboratories
Albuquerque, NM 87185
Sudip@sandia.gov

Several recent studies discuss potential Exascale architectures, identify key technical challenges and describe research that is beginning to address several of these challenges [1,2]. Co-design is a key element of the U.S. Department of Energy's strategy to achieve Exascale computing [3]. Architectures research is needed but will not, by itself, meet the energy, memory, parallelism, locality and resilience hurdles facing the HPC community — system software and algorithmic innovation is needed as well. Since both architectures and software are expected to evolve significantly there is a potential to use the co-design methodology that has been developed by the embedded computing community. A new co-design methodology for high performance computing is needed.

Hardware/software co-simulation efforts that will be central to co-design are underway [4-7]. One example is the Structural Simulation Toolkit (SST) which is being developed and employed by Sandia National Laboratories, Oak Ridge National Laboratory, the Georgia Institute of Technology, the University of Maryland, the University of Texas, New Mexico State University, Micron, Intel, Cray Inc., Advanced Micro Devices, Hewlett Packard and Mellanox [4,5]. SST is an enabling tool for several co-design centers being formed and is being applied to understand how current and future algorithms will perform on Exascale architectures and is being used to perform tradeoff studies to analyze the impact of architectural changes on application performance. Co-design requires that the benefit of these architectural changes be considered in relation to their cost (R&D investment, silicon area and energy). It will be critical to leverage existing industry roadmaps whenever possible. An example SST calculation is show in the figure below.

Research is also assessing the potential use of mini-applications as a tool for enabling co-design [1]. The goal is to try to capture the key computational kernel in a mini-app that consists of approximately 1,000 lines of code. There has been tremendous interest in the use of mini-apps by both the research community and by computer companies. It is difficult for a microprocessor company to understand applications that consist of millions of lines of codes. There can also be export restrictions on the application code or key libraries, making it difficult to engage the worldwide research community. Mini-apps also need to evolve and can be a mechanism for code-teams to learn what will and will not work on Exascale systems. Research is focused on understanding how well mini-apps can represent a full application by comparing the performance of the application and the mini-app on

Y. Cotronis et al. (Eds.): EuroMPI 2011, LNCS 6960, pp. 5–7, 2011.
© Springer-Verlag Berlin Heidelberg 2011

different microprocessors and scaling up the number of cores. Preliminary results indicate that mini-apps can be used to predict the relative performance of applications on different microprocessors and can also predict scalability (at least to tens of cores on a microprocessor as is shown in the table below). This is potentially an important result because it would represent an O(1,000) reduction in complexity for the co-design process.

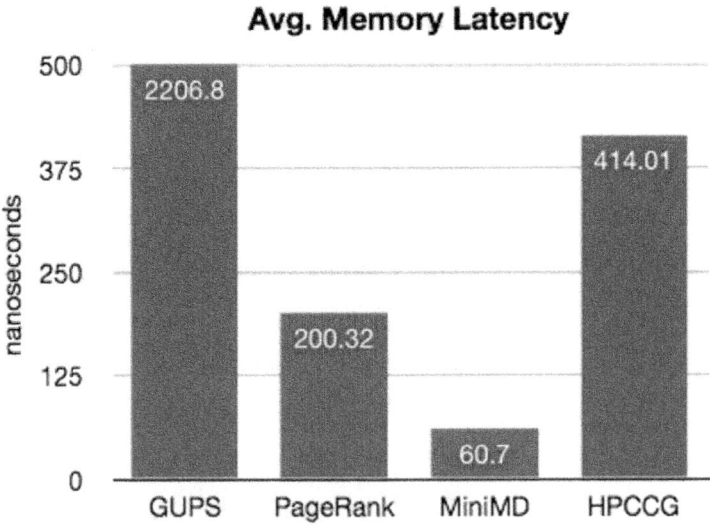

Fig. 1. Average memory latency in nanoseconds calculated by SST for a variety of mini-applications on a current generation memory subsystem

Table 1. Parallel efficiency for the linear solve in a electrical device code (Charon) compared to the results from a mini-application

Cores	Miniapp	Charon – linear solve without preconditioning	Charon – linear solve with preconditioning
4	Ref	Ref	Ref
8	89	87	89
12	73	74	78
16	61	66	66
20	54	49	54
24	45	40	45

References

1. Alvin, K., Barrett, B., Brightwell, R., Dosanjh, S., Geist, A., Hemmert, S., Heroux, M., Kothe, D., Murphy, R., Nichols, J., Oldfield, R., Rodrigues, A., Vetter, J.: On the path to Exascale. International Journal of Distributed Systems and Technologies 1(2), 1–22 (2010)
2. Shalf, J., Dosanjh, S., Morrison, J.: Exascale Computing Technology Challenges. In: Palma, J.M.L.M., Daydé, M., Marques, O., Lopes, J.C. (eds.) VECPAR 2010. LNCS, vol. 6449, pp. 1–25. Springer, Heidelberg (2011)
3. Stevens, R., White, A., et al.: Scientific Grand Challenges: Architectures and Technology for Extreme-scale Computing Report (2011),
   `http://extremecomputing.labworks.org/hardware/reports/`
   `FINAL_Arch&TechExtremeScale1-28-11.pdf`
4. Ang, J., Brightwell, R., Dosanjh, S., et al.: Exascale Computing and the Role of Co-design. In: High Performance Scientific Computing with special emphasis on Current Capabilities and Future Perspectives. Springer, Heidelberg (2011)
5. Rodrigues, S.D., Hemmert, S.: Co-design for High Performance Computing. In: Proceedings of the International Conference on Numerical Analysis and Applied Mathematics, Rhodes, Greece (2010)
6. Hu, S., Murphy, R., Dosanjh, S., Olukoton, K., Poole, S.: Hardware/Software Co- Design for High Performance Computing. In: Proceedings of CODES+ISSS 2010 (2010)
7. Geist, Dosanjh, S.: IESP exascale challenge: co-design of architectures and algorithms. International Journal of High Performance Computing 23(4), 401–402 (2009)

Will MPI Remain Relevant?

George Bosilca

While the Message Passing Interface (MPI) is the ubiquitous standard used by parallel applications to satisfy their data movement and process control needs, the over-arching domination of MPI is no longer a certainty. Drastic changes in computer architecture over the last several years, together with a decrease in overall reliability due to an increase in fault probability, will have a lasting effect on how we write portable and efficient parallel applications for future environments. The significant increase in the number of cores, memory hierarchies, relative cost of data transfers, and delocalization of computations to additional hardware, requires programming paradigms that are more dynamic and more portable than what we use today. The MPI 3.0 effort addresses some of these challenges, but other contenders with latent potential exist, quickly flattening the gaps between performance and portability.

Y. Cotronis et al. (Eds.): EuroMPI 2011, LNCS 6960, p. 8, 2011.
© Springer-Verlag Berlin Heidelberg 2011

Exascale Algorithms for Generalized MPI_Comm_split*

Adam Moody, Dong H. Ahn, and Bronis R. de Supinski

Lawrence Livermore National Laboratory,
Livermore, CA 94551, USA
{moody20,ahn1,bronis}@llnl.gov

Abstract. In the quest to build exascale supercomputers, designers are increasing the number of hierarchical levels that exist among system components. Software developed for these systems must account for the various hierarchies to achieve maximum efficiency. The first step in this work is to identify groups of processes that share common resources. We develop, analyze, and test several algorithms that can split millions of processes into groups based on arbitrary, user-defined data. We find that bitonic sort and our new hash-based algorithm best suit the task.

Keywords: MPI, MPI_Comm_split, Sorting algorithms, Hashing algorithms, Distributed group representation.

1 Introduction

Many of today's clusters have a hierarchical design. For instance, typical multi-core cluster systems have many nodes, each of which has multiple sockets, and each socket has multiple compute cores. Memory and cache banks are distributed among the sockets in various ways, and the nodes are interconnected through hierarchical network topologies to transmit messages and file data.

Algorithmic optimizations often reflect the inherent topologies of these hierarchies. For example, many MPI implementations [1] [2] [3] use shared memory to transfer data between processes that run within the same operating system image, which typically corresponds to the set of processes that run on the same compute node. This approach is considerably faster than sending messages through the network, but the implementation must discover which processes coexist on each node. Some collective algorithms consider the topology of the network to optimize performance [4] [5]. As another example, fault-tolerance libraries must consider which processes share components that act as single points of failure, such as the set of processes running on the same compute node, the same network switch, or the same power supply [6] [7].

The first step in these algorithms identifies the processes that share a common resource. A process can often obtain information about the resource on which it

---

* This work was performed under the auspices of the U.S. Department of Energy by Lawrence Livermore National Laboratory under contract DE-AC52-07NA27344. (LLNL-CONF-484653).

Y. Cotronis et al. (Eds.): EuroMPI 2011, LNCS 6960, pp. 9–18, 2011.
© Springer-Verlag Berlin Heidelberg 2011

runs. However, obtaining information about the resources that other processes in the job use is usually more difficult. We could issue a gather operation to collect the resource information from all processes. However, this approach is prohibitively expensive in both time and memory with millions of processes.

Alternatively, we can almost directly offload the problem to MPI_Comm_split since each resource is often assigned a unique name. The challenge lies in mapping the resource name, which may be arbitrary data like a URL string, into a unique integer value that can be used as an MPI_Comm_split color value. We could hash the resource name into an integer and then call MPI_Comm_split specifying the hash value as the color. However, the hash function may produce collisions, in which case, processes using different resources would be assigned to the same group. We would need to refine this group, perhaps by applying a different hash function and calling MPI_Comm_split again in a recursive manner. This process is both cumbersome and inefficient.

We propose a cleaner, faster interface by extending MPI_Comm_split to enable the user to provide arbitrary data for color and key values along with user-defined functions that can be invoked to compare two values. MPI_Comm_split allows one to split *and* to reorder processes. In our generalized interface, the caller may specify special parameter values to disable either the split or re-order functions. When the reorder function is disabled, processes are ordered in their new group according to their rank in the initial group. The reorder function is often unnecessary and, by allowing the caller to disable it, we can split processes in logarithmic time using a fixed amount of memory under certain conditions. Our key contributions in this paper are: a generalized MPI_Comm_split operation; a scalable representation for process groups; implementation of collectives using that representation; implementation of several MPI_Comm_split algorithms; and large-scale experiments of those algorithms. While existing algorithms for MPI_Comm_split require $O(N)$ memory and $O(N \log N)$ time in a job using N processes, we present algorithms that require as little as $O(1)$ memory and $O(\log N)$ time.

The rest of this paper is organized as follows. Section 2 discusses the linked list data structure that we use to represent groups and illustrates how to implement collectives using it. Section 3 presents several algorithms for a generalized MPI_Comm_split, and Section 4 presents experimental results.

2   Groups as Chains

Each process in current MPI implementations typically stores group membership information as an array that contains one entry for each process in the group. This array maps a group rank ID to an MPI process address, such as a network address, so that a message can be sent to the process that corresponds to a given rank. Each process can quickly find the address for any process in the group using this approach. However, it requires memory proportional to the group size, which is significant at large scales.

To represent process groups in a scalable way, we store the group mapping as a doubly-linked list that is distributed across the processes of the group. Each

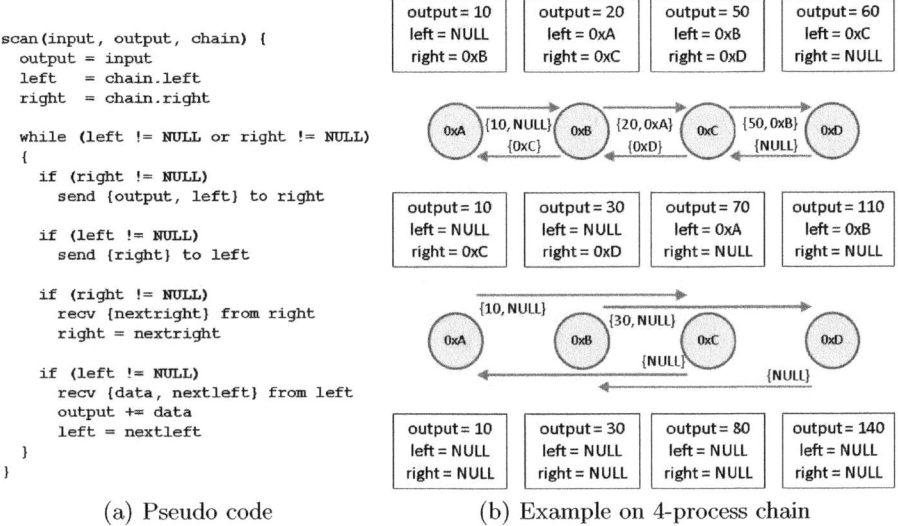

```
scan(input, output, chain) {
    output = input
    left   = chain.left
    right  = chain.right

    while (left != NULL or right != NULL)
    {
        if (right != NULL)
            send {output, left} to right

        if (left != NULL)
            send {right} to left

        if (right != NULL)
            recv {nextright} from right
            right = nextright

        if (left != NULL)
            recv {data, nextleft} from left
            output += data
            left = nextleft
    }
}
```

(a) Pseudo code   (b) Example on 4-process chain

Fig. 1. Inclusive scan on a chain

process represents a node in the list, and each records a small, fixed-size portion of the group mapping consisting of the number of processes in the group, its rank within the group, its process address, and the addresses of the processes with a rank one less and one more than its own. We develop our algorithms on top of MPI, so we simply record MPI rank IDs as process addresses. We refer to this doubly-linked list as a *chain*.

Conceptually, we align the chain horizontally with increasing ranks from left to right. Given a particular process as a reference, the *left neighbor* is the process with rank one less and the *right neighbor* is the process with rank one greater. The first rank of the group stores a NULL value as the address of its left neighbor, and the last rank of the group stores this NULL value for its right neighbor.

Although this chain data structure limits the destinations to which a process can directly address messages, many collectives can be implemented in logarithmic time by forwarding process addresses along with the messages that contain the data for the collective. For example, in Figure 1 we illustrate how to implement a left-to-right inclusive scan operation. With N processes in the chain, this collective executes in $\lceil \log N \rceil$ rounds, where in round $i \in [0, \lceil \log N \rceil)$, each process exchanges messages with left and right partners with ranks 2^i less and 2^i greater than its own. Each process first sets its left and right partners to be its left and right neighbors in the chain, and each process initializes its current scan result to the value of its contribution to the scan. Then a process sends its current scan result to its right partner and appends the address of its left partner to the message. The process also sends the address of its right partner to its left partner.

Each process combines the scan data that it receives from its left partner with its current scan result and sets its next left partner to the address included with that message. Each process also receives the incoming message from its right partner to obtain the address of its next right partner. If the address for the process on either side is NULL, the process does not exchange messages with a process on that side. However, it forwards the NULL address values as appropriate. After $\lceil \log N \rceil$ rounds, the current scan value on each process is its final scan value.

One could similarly implement a right-to-left scan, and we use this technique to implement a *double scan*, which executes a left-to-right scan simultaneously with a right-to-left scan. Our algorithms use double scans to implement inclusive scans, exclusive scans, and associative reduction operations that require $O(1)$ memory and run in $O(\log N)$ time. Further, we can implement tree-based collectives on a chain, including gather, scatter, broadcast, and reduction algorithms. Our group representation does not directly support general point-to-point communication, but it is sufficient for all of the algorithms that we present.

3   Algorithms

3.1   Serial Sort Algorithms

Many existing MPI implementations first gather all color/key/rank tuples into a table at each process to implement `MPI_Comm_split`. Each process extracts the entries in the table with its color value and places those entries into another list. Finally, each process sorts this list by key and then by rank using a serial sort such as `qsort`. If N is the number of MPI processes, each process uses $O(N)$ memory to store the table and $O(N \log N)$ time to execute the sort.

We implement two variants of this algorithm. Our first variant, *Allgather-Group*, executes an allgather using the chain to collect data to each process. Our second variant, *AllgatherMPI*, calls `MPI_Allgather`. AllgatherMPI relies on the optimized MPI library to collect the data to each process to show how much we could optimize the communication in AllgatherGroup. Neither of these algorithms will scale well to millions of processes in terms of memory or time. However, we include them as a baselines since they emulate the algorithms used in existing MPI implementations [1] [2] [3].

3.2   Parallel Sort Algorithms

Our second approach to split processes uses a parallel sort. Given a chain of mixed colors and unordered keys, we can split and sort the chain into ordered groups using a parallel sort, a double scan, and a few point-to-point messages. First, each process constructs a data item that consists of its color value, its key value, its rank within the input group, and its process address. We redistribute these data among the processes using a parallel sort, such that the ith process from the start of the chain has the data item with the ith lowest color/key/rank tuple. Each process next exchanges its sorted data item with its left and right neighbors to determine group boundaries and neighbor processes. We then use a double scan to determine rank IDs and the size of each group. A final point-to-point message sends this information back to the process that originally contributed the data item. Mellor-Crummey et al. proposed this approach for a similar operation in Co-Array Fortran [8].

For this approach, we implement four different parallel sort algorithms. In our first algorithm, *GatherScatter*, we use a tree communication pattern to gather all data items to a root process. We sort items during each merge step of this gather operation so that the items are sorted after the final merge at the root. We then scatter the items from the root to the processes. Similar to our allgather algorithms, GatherScatter uses $O(N)$ memory at the root, but it executes the sort in $O(N)$ time instead of $O(N \log N)$.

In our second parallel sort, we implement an algorithm that is similar to the one that Sack and Gropp describe [9]. In this scheme, we gather the color/key/rank tuples to a subset of processes that then execute Cheng's algorithm to sort them [10]. We then scatter the data items back to the full process set. In our tests, we set the maximum number of data items that a process may hold to a fixed value, M. We use several different values ranging from 128 to 8192, and we label each algorithm as *ChengM*. The number of processes, P, that perform the sort is an important parameter this algorithm. These algorithms use $O(\frac{N}{P})$ memory and $O(P \log N + \log^2 N + \frac{N}{P} \log P)$ time.

Third, we implement Batcher's bitonic sort [11], which we label *Bitonc*. This algorithm uses $O(\log N)$ memory and $O(\log^2 N)$ time.

Finally, for standard `MPI_Comm_split`, in which the color and key values are integers, we implement a divide-and-conquer form of radix sort. This algorithm, *Radix*, splits chains into subchains based on the most significant bits of the key values. We then recursively sort each subchain, and rejoin the sorted subchains into a single, sorted chain. Radix uses $O(1)$ memory and $O(\log N)$ time.

3.3   Hash-Based Algorithm

Our third method to split processes employs hashing. This method avoids sorting processes when only a split is required. We first hash the color value of the calling process to one of a small number of bins, ensuring that we assign the same color value to the same bin on each process. Then, we execute a double exclusive scan on the chain. For each direction, the scan operates on a table that includes an entry for each bin. Each entry contains two values. The first value encodes the address of the process that is assigned to that bin and is next in line along a certain direction (left or right) from the calling process. The second value counts the number of processes along a certain direction from the calling process that belong to that bin.

Each process initializes all table entries to a NULL address and a zero count, except for the entry corresponding to its bin, in which case it sets the address field to its own address and sets the count field to one. We then perform the double exclusive scan operation, after which the result of the left-to-right scan lists the address of the next process to the left and the number of processes to

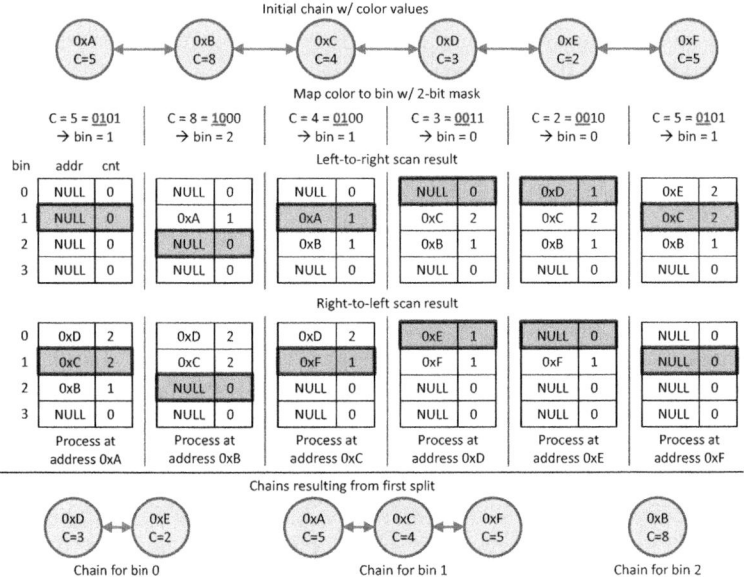

Fig. 2. Splitting a 6-process chain using 4 bins

the left of the calling process for each bin. Similarly, the result of the right-to-left scan lists the address of the next process to the right and the number of processes to the right of the calling process for each bin.

We then create chains that consist only of the processes mapped to a given bin. Each process uses the address and count fields from the table entry corresponding to its bin and assigns the process address from the left-to-right scan to be its left neighbor and the process address from the right-to-left scan to be its right neighbor. It sets its rank to be the value of the count field from the the left-to-right scan and it adds one to the sum of the count fields from the left-to-right and right-to-left scans to compute the total number of processes in its chain. This operation splits the input chain into a set of disjoint chains, potentially creating a new chain for each bin. This new chain may contain processes with different colors. However, the hash function guarantees that all processes with the same color value are in the same chain. An example split operation is illustrated in Figure 2 in which the hash function uses the first two bits of a 4-bit color value to select one of four bins.

We then iteratively apply this split operation to the chains produced in the prior step. Each iteration uses a new hash function so that processes that have different colors eventually end up in separate chains. For this work, we pack the color value into a contiguous buffer and then apply Jenkin's one-at-a-time hash [12] [13]. For different iterations, we apply the same hash function but

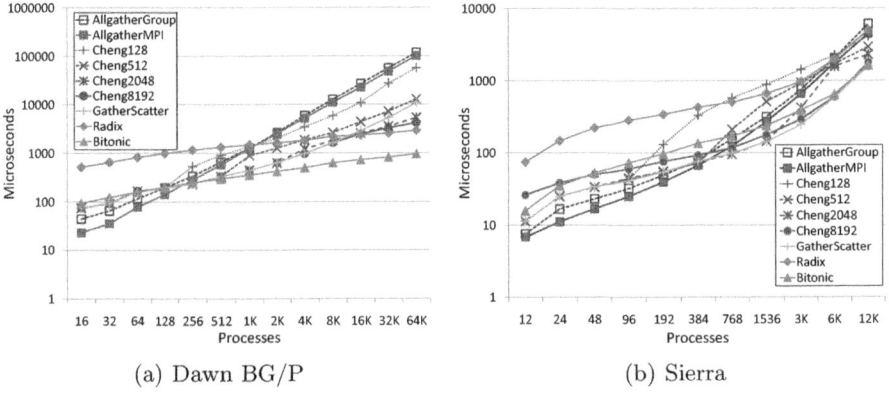

(a) Dawn BG/P   (b) Sierra

Fig. 3. Time for reorder without split

rotate the bytes of the packed color value and mask different regions of the hash value to obtain new bin numbers. If needed, we invoke a sort algorithm to finish splitting and reordering the chains.

We implement two variants of this algorithm, *Hash* and *Hash64*. Hash repeatedly applies the hash operation until the initial chain is completely split. Hash64 iterates until the chain is completely split or its length falls below a threshold of 64 processes, at which point, we invoke AllgatherGroup to finish the split. Each version stops iterating if a single color value is detected throughout the chain. We check for this condition using an allreduce whenever a split iteration does not reduce the length of the chain. If we need to reorder the chain after completing the hash iterations, we use Bitonic sort. When reordering is required, these algorithms have the same time and memory complexity as Bitonic. When only a split is required, these algorithms use $O(1)$ memory. Due to the nature of hash functions, one may only determine probabilistic upper time bounds. However, one can show strict lower time bounds of $\Omega(\log^2 N)$ when the number of groups equals the number of processes and $\Omega(\log N)$ when the number of groups is small and independent of the number of processes.

4   Results

We test each algorithm on two clusters at Lawrence Livermore National Laboratory. We use Dawn, an IBM BlueGene/P system that has 128K cores on 32K nodes. The second system, Sierra, has over 1,800 compute nodes, each with two Intel Xeon 5660 hex-core chips for a total over 21,600 cores. The Sierra nodes are connected with QLogic QDR Infiniband.

We first investigate the performance of the various sorting algorithms. Disabling the split, and using an integer value as the key, we show the time required to complete a reorder operation on each platform in Figure 3. The two platforms produce significantly different results. The plots all follow clear, distinct trends

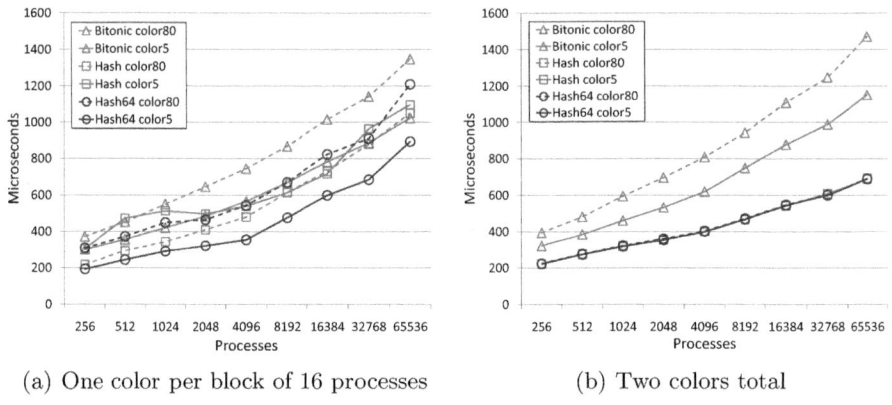

(a) One color per block of 16 processes   (b) Two colors total

Fig. 4. Time for split without reorder

on Dawn. However, on Sierra, the plots generally converge at higher process counts, at which we conject that network contention limits performance. On both machines, the serial sort algorithms are best at small scale, but more scalable algorithms soon outperform them. On Dawn, Radix and Bitonic sort show the best scaling trends. As expected, Radix sort, with its $O(\log N)$ complexity, scales the best. However for all scales tested, Bitonic always has better performance. At 16 processes, Bitonic is 5.5 times faster than Radix. The difference is reduced to 3.0 times at 64K processes, but the hidden constants associated with the big-O notation are too high for Radix to surpass Bitonic. At 64K processes on Dawn, Bitonic sort is 100 times faster than the serial sort algorithms and 4.4 times faster than the fastest Cheng sort. Bitonic is still fast on Sierra, although the apparent contention limits its performance. Regardless, on both machines, the best approach is to use a serial sort algorithm for small scale and to switch to Bitonic sort at large scale.

We next focus on the task of just splitting processes. We compare Bitonic, the fastest parallel sort algorithm, to Hash and Hash64. To see how different color datatypes impact the algorithms, we use character strings of length 5 and of length 80 for color values. Figure 4 shows the results for Dawn. When the number of process groups (the number of distinct colors) is on the order of the number of processes, we find that the hash-based algorithms perform on par with Bitonic sort. However, when the number of groups is small, the hash-based algorithms outperform Bitonic with speedups between 1.7 and 2.1 at 64K processes, depending on the length of the color value. The size of the color value affects the performance of Bitonic, but it has little impact on Hash and only impacts Hash64 in cases where it must call AllgatherGroup. Since the hash operation always maps the color to an integer, its communication costs are not affected by the size of the color value. However, the sort algorithms send the color value in each message, so the cost of these algorithms increases with the size of the color value. Hash or Hash64 perform the best in all cases shown.

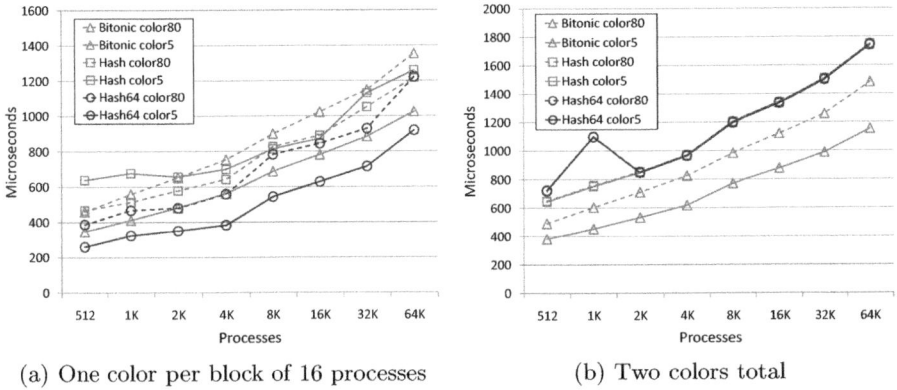

(a) One color per block of 16 processes   (b) Two colors total

Fig. 5. Time for split with reorder

We also tested Bitonic, Hash, and Hash64 for splitting and reordering processes in the same operation. The results from Dawn are shown in Figure 5. We used an integer value for the key and character strings of different lengths for color values. As shown in Figure 5(a), with many groups, and when the groups are roughly equal in size, the timing results look very similar to Figure 4(a). However, with only a small number of groups, as shown in Figure 5(b), then both hash algorithms require more time than Bitonic. In this case, we incur overhead to execute the hash algorithm to split the initial chain, but since the resulting subchains are relatively long, the split does not significantly reduce the cost of sorting. Since we cannot know the size of the resulting groups *a priori*, Bitonic is the best option whenever reordering is required. The peak for Hash64 at 1K processes in Figure 5(b) is an artifact from a bug that invoked AllgatherGroup instead of Bitonic even though the chain was longer than 64 processes after the split. This bug only affected the data points for 512 and 1K processes in Figure 5(b).

5   Conclusions

Developers will soon need scalable algorithms to split millions of processes into groups based on arbitrary, user-defined data. In this work, we developed several algorithms that represent groups as a doubly-linked list, and we investigated their performance through large-scale experiments. We found that bitonic sort and a new hash-based algorithm offer the best results. We find that the hash-based algorithm is up to twice as fast as bitonic sort when only splitting processes. Compared to algorithms used in current MPI implementations, these new algorithms reduce memory complexity from $O(N)$ to as little as $O(1)$, and they reduce run time complexity from $O(N \log N)$ to as little as $O(\log N)$.

Although we focus on algorithms for a generalized MPI_Comm_split interface, our findings also apply to the simpler, standard MPI_Comm_split function. Thus, we expect MPI implementations to benefit from our results. Further, we can implement these algorithms and group representations directly in applications that need fast methods to identify sets of processes. With this approach, applications can create lightweight groups without the overhead of creating full MPI communicators.

References

1. Argonne National Laboratory, MPICH2, http://www.mcs.anl.gov/mpi/mpich2
2. Network-Based Computing Laboratory, MVAPICH: MPI over Infiniband and iWARP, http://mvapich.cse.ohio-state.edu
3. Gabriel, E., Fagg, G.E., Bosilca, G., Angskun, T., Dongarra, J., Squyres, J.M., Sahay, V., Kambadur, P., Barrett, B.W., Lumsdaine, A., Castain, R.H., Daniel, D.J., Graham, R.L., Woodall, T.S.: Open MPI: Goals, concept, and design of a next generation MPI implementation. In: Kranzlmüller, D., Kacsuk, P., Dongarra, J. (eds.) EuroPVM/MPI 2004. LNCS, vol. 3241, pp. 97–104. Springer, Heidelberg (2004)
4. Kandalla, K., Subramoni, H., Vishnu, A., Panda, D.K.: Designing Topology-Aware Collective Communication Algorithms for Large Scale Infiniband Clusters: Case Studies with Scatter and Gather. In: The 10th Workshop on Communication Architechture for Clusters, CAC 2010 (2010)
5. Faraj, A., Kumar, S., Smith, B., Mamidala, A., Gunnels, J.: MPI Collective Communications on The Blue Gene/P Supercomputer: Algorithms and Optimizations. In: 17th IEEE Symposium on High Performance Interconnects, HOTI 2009, pp. 63–72 (August 2009)
6. Moody, A., Bronevetsky, G., Mohror, K., d. Supinski, B.R.: Design, Modeling, and Evaluation of a Scalable Multi-level Checkpointing System. In: Proceedings of the 2010 ACM/IEEE International Conference for High Performance Computing, Networking, Storage and Analysis, SC 2010, pp. 1–11. IEEE Computer Society, Washington, DC (2010)
7. Gomez, L.A.B., Maruyama, N., Cappello, F., Matsuoka, S.: Distributed Diskless Checkpoint for Large Scale Systems. In: CCGRID 2010, pp. 63–72 (2010)
8. Mellor-Crummey, J., Adhianto, L., Scherer III, W.N., Jin, G.: A New Vision for Coarray Fortran. In: Proceedings of the Third Conference on Partitioned Global Address Space Programing Models, PGAS 2009, pp. 5:1–5:9. ACM, New York (2009)
9. Sack, P., Gropp, W.: A Scalable MPI_Comm_split Algorithm for Exascale Computing. In: Keller, R., Gabriel, E., Resch, M., Dongarra, J. (eds.) EuroMPI 2010. LNCS, vol. 6305, pp. 1–10. Springer, Heidelberg (2010)
10. Cheng, D.R., Edelman, A., Gilbert, J.R., Shah, V.: A Novel Parallel Sorting Algorithm for Contemporary Architectures. In: Submitted to ALENEX 2006 (2006)
11. Batcher, K.E.: Sorting Networks and their Applications. In: AFIPS Spring Joint Computer Conference, vol. 32, pp. 307–314 (1968)
12. Jenkins, B.: Algorithm Alley: Hash Functions. Dr. Dobb's Journal of Software Tools 22(9), 107–109, 115–116 (1997)
13. Jenkins, B.: Hash Functions for Hash Table Lookup (2006), http://burtleburtle.net/bob/hash/doobs.html

Order Preserving Event Aggregation in TBONs

Tobias Hilbrich[1], Matthias S. Müller[1],
Martin Schulz[2], and Bronis R. de Supinski[2]

[1] Technische Universität Dresden, ZIH, D-01062 Dresden, Germany
{tobias.hilbrich,matthias.mueller}@tu-dresden.de
[2] Lawrence Livermore National Laboratory, Livermore, CA 94551
{bronis,schulzm}@llnl.gov

Abstract. Runtime tools for MPI applications must gather information from all processes to a tool front-end for presentation. Scalability requies that tools aggregate and reduce this information so tool developers often use a Tree Based Overlay Network (TBON). TBONs aggregate multiple associated events through a hierarchical communication structure. We present a novel algorithm to execute multiple aggregations while, at the same time, preserving relevant event orders. We implement this algorithm in our tool infrastructure that provides TBON functionality as one of its services. We demonstrate that our approach provides scalability with experiments for up to 2048 tasks.

1 Introduction

We need scalable tools to develop correct, high performance MPI applications. Event based tools intercept application events, such as the issuing of an MPI call, and analyze the intercepted data. Since MPI applications can use many processes, scalability is critical for these tools. Tool developers often use Tree Based Overlay Networks (TBONs), which provide scalable tree-based communication from the MPI processes to the tool front end, to overcome this challenge through extra tool processes that offload analyses from the application tasks.

We target improved scalability for MPI correctness tools [5][12] with the Generic Tool Infrastructure (GTI), a new modular, scalable infrastructure. GTI provides a TBON as one of its components and uses event aggregation to achieve scalability. Our approach supports (mostly) transparent aggregation that must preserve the orders in which some events occur. For example, if a correctness check analyzes an MPI_Send event that uses a user defined datatype, we must first process the events that create and commit the datatype. Otherwise, a correctness tool would report that using the datatype in the MPI_Send call was erroneous. Thus, GTI must aggregate multiple event types while preserving relevant intraprocess orders. Existing TBON infrastructures do not provide this mechanism, so we present a novel scalable algorithm that preserves relevant event orders across multiple TBON aggregations. Our contributions include:

– A basic algorithm for order preserving aggregation;

Y. Cotronis et al. (Eds.): EuroMPI 2011, LNCS 6960, pp. 19–28, 2011.

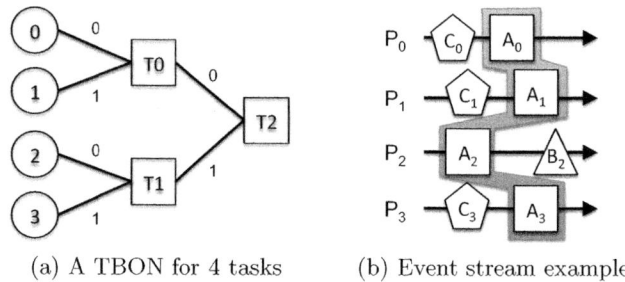

(a) A TBON for 4 tasks   (b) Event stream example

Fig. 1. TBON aggregtion example

- The concept of "channel IDs" as a scalable means to store which tasks could have provided information to an event;
- A channel ID based queuing algorithm for order preserving aggregation.

Existing tools encounter scaling issues that are already visible at a scale below 1024 tasks. We demonstrate the scalability of our approach and show measurements with up to 2048 tasks. The remainder of the paper is organized as follows. Section 2 introduces TBON event aggregation and the difficulties in preserving event ordering. In Section 3, we present a preliminary algorithm to overcome those difficulties. Section 4 introduces our channel ID concept. We show this concept allows us to overcome the scalability limitations of the preliminary algorithm in Section 5.

2   TBON Aggregation

TBONs improve tool scalability by distributing complex analyses across a hierarchy of processes. Figure 1(a) shows a TBON with four application processes and three tool processes. Usually, the leaves generate events that TBON nodes propagate towards the root. However, to improve scalability TBON tree nodes must aggregate multiple events when propagating upwards in the tree.

Aggregations support hierarchical event processing in which an analysis on a tree node forwards the intermediate result instead of the original events. Figure 1(b) shows an event aggregation example in which the four tasks in (a) create different event types over time, with each shape and letter representing a different type. In our example, we can aggregate the four events of type A, which could for example be matching MPI_Barrier calls.

Events arrive in some order on each TBON node and we must wait until all events that participate in a given aggregation arrive, as Figure 2 illustrates for the example node T_1. Figure 2(a) shows an order in which the events could arrive. The event A_3 (the event of type A from process 3) arrives first, while A_2 only arrives after events B_2 and C_3 arrived. In order to maintain intraprocess event order, we must not process and forward C_3 until the aggregation finishes. Thus,

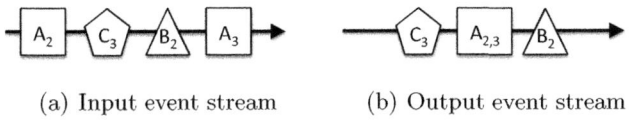

(a) Input event stream   (b) Output event stream

Fig. 2. Event processing on T_1

if the event processing on T_1 preserves order, only the stream in Figure 2(b) can result as output, in which $A_{2,3}$ is the aggregated result of A_2 and A_3.

3   Order Preserving Aggregation

Our basic algorithm aggregates events and preserves relevant event orders. It first determines if an event should be processed based on the state of ongoing aggregations. If so, an aggregation that processes the event reports which tasks participated in a successful aggregation. Thus, when we process an event, we return these three values:

wasSuspended   Further input is required to complete the aggregation so we suspend processing of events from this process;
finishedAggregation Aggregation completed successfully;
reopenedProcesses   List all processes for which we suspended event process-ing of a successfully completed aggregation.

These values allow us to determine which events to aggregate without requir-ing an *a priori* distinction. For example, we can aggregate events of an MPI collective while distinguishing communicators. An aggregation can evaluate the communicator argument to distinguish events from different communicators.

The algorithm in Figure 3 sketches the order preserving aggregation that forms the main loop on each TBON node. We denote an event of some type that was created on task p as e_p. When we aggregate events, we replace a set of events $e_{p_1}, e_{p_2}, \ldots, e_{p_n}$ by a new event that we denote as $e_{(p_1,p_2,\ldots,p_n)}$. Our algo-rithm receives new events and determines whether it can process them without violating event order. If not, it queues the event.

Each iteration of the main loop, which runs until the tool shuts down, first determines if a queued event can be processed (lines 6-14). Our algorithm must determine whether an event is part of an ongoing aggregation (*suspended*). We also must determine if an event with information from a particular task is in the list of queued events (*blocked*) and whether we can process events of each task (*open*). The *states* vector tracks tasks that are currently *suspended*. The *tempStates* vector tracks which tasks in *states* must be set to the *blocked* state.

Lines 15-17 copy the information on whether a task is in the *blocked* or *open* state from the *tempStates* vector to the *states* vector. If the algorithm cannot

Input: P Number of tasks

1 $queue = newQueue()$
2 $states[P] = (open, open, \ldots)$; // Process suspension states
3 **while** $running$ **do**
4      $eventToProcess = NULL$
5      $tempStates[P] = (open, open, \ldots)$; // Used to update states
    // Can a queued event be processed ?
6   **for** $each\ event\ e_{(p_1,\ldots,p_n)}\ in\ queue$ **do**
7          $canBeProcessed = true$
8   **for** $p \in \{p_1, \ldots, p_n\}$ **do**
9   **if** $states[p] == suspended\ ——\ tempStates[p] == blocked$ **then**
10                 $tempStates[p] = blocked$
11                 $canBeProcessed = false$

12   **if** $canBeProcessed == true\ \&\&\ toReceive == true$ **then**
13             $queue.erase(e_{(p_1,\ldots,p_n)})$
14             $eventToProcess = e_{(p_1,\ldots,p_n)}$

    // Update states
15   **for** $i \in \{0, \ldots, P-1\}$ **do**
16   **if** $states[i] \neq suspended$ **then**
17             $states[i] = tempStates[i]$

    // Receive an event (if necessary)
18   **if** $eventToProcess == NULL$ **then**
19         $e_{(p_1,\ldots,p_n)} = receiveEvent()$
20         $eventToProcess = e_{(p_1,\ldots,p_n)}$
21   **for** $p \in \{p_1, \ldots, p_n\}$ **do**
22   **if** $states[p] \neq open$ **then**
23                 $eventToProcess = NULL$
24                 $queue.pushBack(e_{(p_1,\ldots,p_n)})$
25   **break**

    // Process the selected event
26   **if** $eventToProcess \neq NULL$ **then**
27         $e_{(p_1,\ldots,p_n)} = eventToProcess$
28         $(wasSuspended, finishedAggregation, reopenedProcesses) = processEvent(e_{(p_1,\ldots,p_n)})$
29   **if** $wasSuspended == true$ **then**
30   **for** $p \in \{p_1, \ldots, p_n\}$ **do**
31                 $states[p] = suspended$

32   **if** $finishedAggregation == true$ **then**
33             $\{q_1, \ldots, q_m\} = reopenedProcesses$
34   **for** $p \in \{q_1, \ldots, q_m\}$ **do**
35                 $states[p] = open$

Fig. 3. Order preserving event suspension algorithm for multiple aggregations

process a queued event, it receives a new one. Lines 18-25 check whether all tasks for which the newly received event contains information are in the *open* state. If so, the algorithm can process the event, otherwise it queues the event.

Lines 26-35 process the selected event, if any, and update the task states. As described above, each aggregation returns *wasSuspended*, *finishedAggregation*, and *reopenedProcesses* after processing an event. We change the states of tasks in the *states* vector to or from *suspended* based on these values.

4   Scalability

Order preserving execution of aggregations must not impose excessive overhead. The algorithm in Figure 3 must determine which tasks provided information to a newly received event. At scale, storing this information is too expensive, as the number of processes that an event can involve increases as the TBON propagates events towards the root.

GTI extends the algorithm in Figure 3 with a coarse grained locking mechanism. A *channel ID* stores which processes provided information for an event. This ID represents the path that was taken by an event through the TBON, starting at the node that created the event and ending at the node that is currently processing it. The ID stores this path in terms of the channels through which the current node of the TBON received the event. The example TBON in Figure 1(a) shows the index for the individual input channels of each node (0 or 1). To illustrate channel IDs, consider the output event stream $B_2, A_{2,3}, C_3$ of T_1 from Figure 2(b). When these events arrive at node T_2, the root of the TBON, B_2 has channel ID 1.0, which means that T_2 received it from channel 1 and T_1 received it from channel 0. Whereas $A_{2,3}$ has the channel id 1, as T_2 received it from channel 1, which created the record.

We store the channel ID for each event instead of a list of tasks that provided information to it. The size of the ID depends on the branching factor of the TBON and its depth, which are both typically logarithmic in the number of MPI processes. To cover also extreme cases, GTI stores the ID in a set of 64 bit values. However, a single value is normally sufficient. Consider a balanced binary TBON, each ID requires one bit, where we need one additional state to store whether or not a given entry in the ID is used. So we need two bits for each level of the TBON to store from which channel the event was received. Thus, a 64 bit value provides storage for 32 levels; such a TBON would span 2^{32} tasks, which exceeds the size of any existing system.

Our scalable algorithm uses the event channel ID to compute whether it can process the event without violating the order property. We use a *channel tree* to store the suspension state for sets of tasks. We also store a queue of events for each node that the algorithm cannot currently process. Each node in the channel tree represents a certain channel ID. The algorithm can process an event if none of its nodes ancestors or successors are suspended. This represents a worst-case evaluation of the tasks that can provide information to an event. We sketch the scalable algorithm as follows:

- If we receive a new event, we search the channel tree for the node that represents the channel ID of the event:
    - We add the searched for node and any of its ancestors if it is not present in the channel tree;
    - If any node on the path from the root to the found node has a non-empty queue, we queue the new event at the first such node on the path;
    - If the found node, any of its descendants, or any of its ancestors is suspended, we queue the new event at the found node;
    - If any descendant of the found node has a queued event, we queue the new event at the found node;
- When we process an event:
    - If an aggregation determines that it needs further input, we suspend the node to which the event belongs in the channel tree;
    - If an aggregation finishes, we unsuspend all nodes related to events of the successful aggregation.
- Before receiving a new event, we determine if we can process a queued one:
    - We process events in the order in each queue;
    - We can process a queued event of a node if its ancestors and descendants are not suspended and its suspension tree descendants do not have queued events;
    - We rebalance a queue when we remove an event from it: we push each event into the node corresponding to its channel ID until the first event in the queue belongs to the current node in the channel tree or the queue is empty; when a node with a non-empty queue is on the path from a node that we rebalance to the node to which an events channel ID belongs, we add the event to the node with non-empty queue instead.

We illustrate the scalable channel ID based algorithm with the example from Figure 1. Assume that the output event stream of T_1 arrives at T_2, which runs the extended algorithm. A possible output event stream of T_0 is $A_{0,1}, C_1, C_0$. Figure 4 shows a possible input event stream for T_2 along with the channel IDs for each event. When T_2 processes event B_2, it adds the node 1.0 to the channel tree. As this event is not part of any aggregation, the algorithm does not suspend any nodes, which leads to the state in Figure 5(a). When T_2 processes $A_{0,1}$, an aggregation starts and we suspend node 0 in the channel tree ((b), suspended node in red). When the algorithm processes C_1 and C_0, we queue both events in their respective nodes ((c) and (d), indirectly suspended nodes in yellow). In its next step, the algorithm processes $A_{2,3}$, which completes the ongoing aggregation and unsuspends all nodes (e). Before the algorithm receives a new event, it must process all queued events, which results in the state that (f) shows. Finally, when the algorithm processes C_3, the channel tree reaches the state in (g).

The algorithm initiates a queue rebalancing if events cannot be queued in the nodes to which they belong. For example, starting from the configuration in (b), if a further event X with channel id 0 arrives before C_1 and C_0, the algorithm will queue X in node 0. Thus, it will also queue C_0 and C_1 on that node. In that case the algorithm rebalances the queue of node 0 after it processes X, as

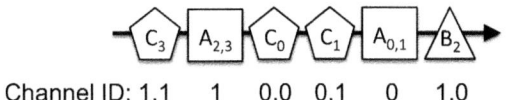

Fig. 4. Input event stream for T_2

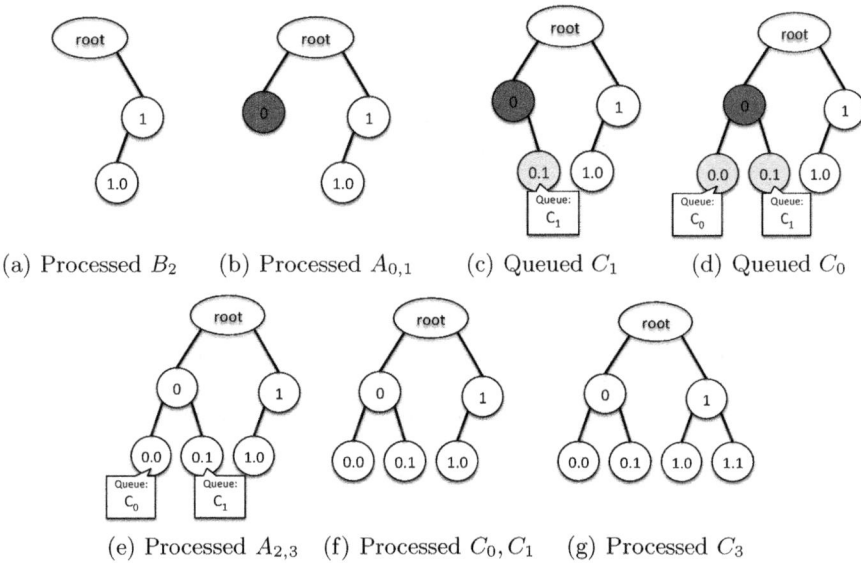

(a) Processed B_2 (b) Processed $A_{0,1}$ (c) Queued C_1 (d) Queued C_0

(e) Processed $A_{2,3}$ (f) Processed C_0, C_1 (g) Processed C_3

Fig. 5. Channel ID based order preservation for the event stream from Figure 4

neither C_0 nor C_1 belong to that node. This action moves the events from the queue in node 0 to the nodes to which these events belong.

This extension to the algorithm in Figure 3 removes its scalability limitations, as each event only carries a channel ID instead of a list of processes. However, it could produce cyclic wait-for situations in which no aggregation can complete when aggregating multiple event types due to its coarser suspension granularity. We have not encountered this issue. However, the GTI implementation of channel ID based suspension includes a timeout mechanism that would abort sets of aggregations to resolve these deadlocks if they ever do occur.

5   Performance Results

We demonstrate the scalability of our algorithm through results for two synthetic test cases on a 16 cores per node Opteron Linux cluster with 864 nodes and a QDR InfiniBand network. Figure 6(a) shows slowdowns for a benchmark in which we aggregate a single type of event. The event is a no-op that does not perform interprocess communication. We normalize the slowdown to the fastest

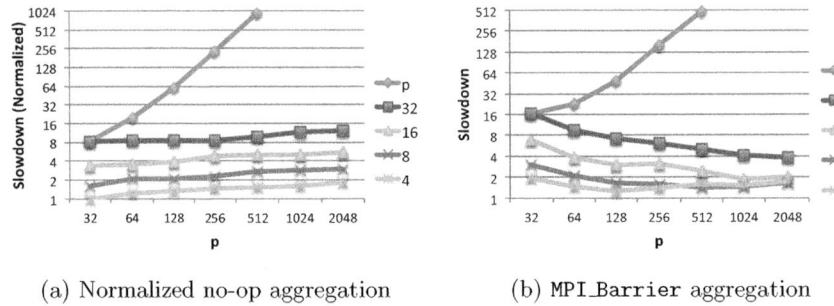

(a) Normalized no-op aggregation   (b) `MPI_Barrier` aggregation

Fig. 6. Slowdowns for order preserving event aggregation

measured run for 32 tasks and use five different tree configurations. We use fan-ins of $p, 32, 16, 8$ and 4 for the TBON nodes, where p is a single tool layer with one tool process that receives events from all application tasks. Except for p, the slowdown increases slowly with increasing process counts; it increases by a factor of two from 32 to 2048 processes with a fan-in of 4. For 2048 tasks the GTI based tool aggregates over 10,000 events per task per second. Figure 6(b) shows slowdowns for a test case that aggregates `MPI_Barrier` events. The slowdown is computed based on the uninstrumented version of the benchmark. As the communication cost of the barrier events increases with scale, the tool overhead actually decreases for all fan-ins other than p.

6   Related Work

GTI uses the TBON concept of infrastructures like MR-Net [8] or Lilith [3] and extends it to provide order preserving aggregation. System monitoring tools such as ganglia [7], performance monitoring and observation systems such as EventSpace [2], and debuggers such as Ladebug [1] also use TBON aggregation. Aggregation is also important in database design such as aggregation schemes [9] and sensor networks such as directed diffusion [4]. None of these approaches considers the preservation of order when aggregation introduces new events. However, Teo et al. study the memory requirements for enforcing different types of event ordering in parallel simulations [11].

Our algorithm operates on parallel event streams on each TBON node. The input is a serialization of these streams, with some parallel events being aggregated into new ones. The algorithm determines how to reorganize the input stream without violating event order. As opposed to the ordering defined by Lamport [6], we only maintain intraprocess order; aggregations require detailed knowledge of the event semantics to maintain interprocess order. Stephens summarizes the many approaches to process parallel streams and related theory [10].

7   Conclusions

We present a novel algorithm for order preserving execution of aggregations in TBONs. Our basic algorithm has limited scalability since it requires knowledge of which processes contributed to each specific event. We use channel IDs to extend the algorithm to a coarse grained locking approach. A channel ID stores which TBON node created an event and what path it took from that node to its current location. We use channel IDs to determine when evaluation of an event would violate the order property. As the channel ID is a worst-case approximation of which processes could have contributed to an event, it could cause unnecessary event queuing, which is a topic that we will study more closely in future work. Finally, our performance measurements for up to 2048 processes demonstrate the scalability of our order preserving aggregation algorithm.

Acknowledgments. Part of this work was performed under the auspices of the U.S. Department of Energy by Lawrence Livermore National Laboratory under Contract DE-AC52-07NA27344. (LLNL-CONF-490684).

References

1. Balle, S.M., Brett, B.R., Chen, C.-P., LaFrance-Linden, D.: A New Approach to Parallel Debugger Architecture. In: Fagerholm, J., Haataja, J., Järvinen, J., Lyly, M., Råback, P., Savolainen, V. (eds.) PARA 2002. LNCS, vol. 2367, pp. 139–758. Springer, Heidelberg (2002)
2. Bongo, L.A., Anshus, O.J., Bjørndalen, J.M.: EventSpace – Exposing and Observing Communication Behavior of Parallel Cluster Applications. In: Kosch, H., Böszörményi, L., Hellwagner, H. (eds.) Euro-Par 2003. LNCS, vol. 2790, pp. 47–56. Springer, Heidelberg (2003)
3. Evensky, D.A., Gentile, A.C., Camp, L.J., Armstrong, R.C.: Lilith: Scalable Execution of User Code for Distributed Computing. In: Proceedings of the 6th IEEE International Symposium on High Performance Distributed Computing, HPDC 1997, p. 305. IEEE Computer Society, Washington, DC (1997)
4. Intanagonwiwat, C., Govindan, R., Estrin, D.: Directed diffusion: a scalable and robust communication paradigm for sensor networks. In: Proceedings of the 6th Annual International Conference on Mobile Computing and Networking, MobiCom 2000, pp. 56–67. ACM, New York (2000)
5. Krammer, B., Müller, M.S.: MPI Application Development with MARMOT. In: Joubert, G.R., Nagel, W.E., Peters, F.J., Plata, O.G., Tirado, P., Zapata, E.L. (eds.) PARCO. John von Neumann Institute for Computing Series, vol. 33, pp. 893–900. Central Institute for Applied Mathematics, Jülich (2005)
6. Lamport, L.: Time clocks, and the ordering of events in a distributed system. Commun. ACM 21, 558–565 (1978)
7. Massie, M.L., Chun, B.N., Culler, D.E.: The Ganglia Distributed Monitoring System: Design, Implementation And Experience. Parallel Computing 30, 2004 (2003)
8. Roth, P.C., Arnold, D.C., Miller, B.P.: MRNet: A Software-Based Multicast/Reduction Network for Scalable Tools. In: Proceedings of the 2003 ACM/IEEE Conference on Supercomputing, SC 2003, p. 21. ACM, New York (2003)

9. Shatdal, A., Naughton, J.F.: Adaptive parallel aggregation algorithms. SIGMOD Rec. 24, 104–114 (1995)
10. Stephens, R.: A survey of stream processing. Acta Informatica 34, 491–541 (1997)
11. Teo, Y.M., Onggo, B.S.S., Tay, S.C.: Effect of Event Orderings on Memory Requirement in Parallel Simulation. In: Proceedings of the Ninth International Symposium in Modeling, Analysis and Simulation of Computer and Telecommunication Systems, MASCOTS 2001, pp. 41–48. IEEE Computer Society, Washington, DC (2001)
12. Vetter, J., de Supinski, B.: Dynamic Software Testing of MPI Applications with Umpire. In: ACM/IEEE Conference on Supercomputing, November 4-10, p. 51 (2000)

Using MPI Derived Datatypes in Numerical Libraries

Enes Bajrović and Jesper Larsson Träff

Faculty of Computer Science, University of Vienna
Nordbergstrasse 15/3C, A-1090 Vienna, Austria
{bajrovic,traff}@par.univie.ac.at

Abstract. By way of example this paper examines the potential of MPI user-defined datatypes for distributed datastructure manipulation in numerical libraries. The three examples, namely gather/scatter of column-wise distributed two dimensional matrices, matrix transposition, and redistribution of doubly cyclically distributed matrices as used in the Elemental dense matrix library, show that distributed data structures can be conveniently expressed with the derived datatype mechanisms of MPI, yielding at the same time worthwhile performance advantages over straight-forward, handwritten implementations. Experiments have been performed with on different systems with mpich2 and OpenMPI library implementations. We report results for a SunFire X4100 system with the mvapich2 library. We point out cases where the current MPI collective interfaces do not provide sufficient functionality.

1 Introduction

The derived (or user-defined) datatype mechanism of MPI is a powerful and concise mechanism for describing arbitrary, MPI process local layouts of data in memory in a way that such (possibly) noncontiguous layouts can be used in all MPI communication operations (point-to-point, one-sided and collective) the same way that unstructured, contiguous data of primitive datatypes can be communicated [5, Chapter 4]. The advantage of derived datatypes is to free the application programmer from explicit packing and unpacking of noncontiguous application datastructures before and after communication operations. By describing such datastructures as MPI datatypes the MPI library can in a nontrivial way take care of the necessary packing (if needed at all), exploit special hardware support, and interact with the underlying (collective) communication algorithms, relieving the user from tedious detail work and providing a potential (and sometimes real and considerable) performance benefit, as well as saving memory in the application.

Much research has been devoted in the MPI community to make the datatype mechanism perform well and fulfill some of its promise [1,2,9,12,13]. Despite much progress, the mechanism still seems not to be used to its full advantage, partly because skepticism as to its efficiency still lingers, partly perhaps because of ignorance of its power and potential.

Y. Cotronis et al. (Eds.): EuroMPI 2011, LNCS 6960, pp. 29–38, 2011.
© Springer-Verlag Berlin Heidelberg 2011

In this paper we consider the use of derived datatypes in (dense) numerical computations, where basic datastructures are vectors and two-dimensional matrices. Extending on the intended usage of MPI derived datatypes to describe and improve the handling of local datastructures, we use the mechanism for describing virtual, distributed datastructures *and* for accomplishing collective transformations on such structures. Particular examples of this are matrix transposition (of column distributed matrices) and transformations between sequentially and doubly cyclically block distributed matrices, as used in the Elemental dense matrix library [6]. We show that somewhat complex transformations that would have to be done locally can be accomplished as part of the communication by using the datatype mechanism to determine in what order elements are sent and received, thus shifting the actual restructuring work to the datatype mechanism. This usage is somewhat analogous to the random permutation algorithm in [10], where permutation is accomplished by communication with randomly selected destinations. Other work on using derived datatypes, partly in the same direction as described here include [3,4]. In [11] the intimate connection between collective communication interfaces and datatypes was used to generalize the expressivity of the collective interfaces to situations that currently cannot be handled well by MPI. Such examples will also be seen in the following.

The MPI derived datatype mechanism makes it possible to describe arbitrarily complex layouts of data in memory, often in a very concise way. However, the mechanism can sometimes be tedious to use and require a deep understanding of the functionality that is possibly not possessed by the application programmer. Unfortunately, there is currently no way in MPI to utilize compiler support for automatically generating datatype descriptions, and also no commonly accepted tools for assisting in setting up the type constructor calls [7,8]. Having thus a sometimes high conceptual overhead, and a high construction overhead, derived datatypes are mostly useful for statically defined structures, where the overhead can be amortized over many MPI communication operations with the same layout. The mechanism is not well suited to sparse, dynamic structures, where changes are frequent and has high sequential overhead e.g. involving traversal of a linked structure. But for many dense numerical computations this should not be an impediment to using derived datatypes.

In the rest of the paper we explain three increasingly complex situations where derived datatypes provide both a conceptual as well as a potential performance advantage along the lines explained above. We benchmark these example applications, and discuss the advantages. We take the stance of the application programmer and use the datatypes at face value; this paper is not about the datatype mechanism *per se* or any potential optimizations thereto.

2   Distributed Matrix Operations

We consider simple, but nevertheless relevant operations involving two-dimensional $n \times n$ matrices that are distributed over p MPI processes in various ways. Matrices are stored C-like in row-major order. We provide solutions for both

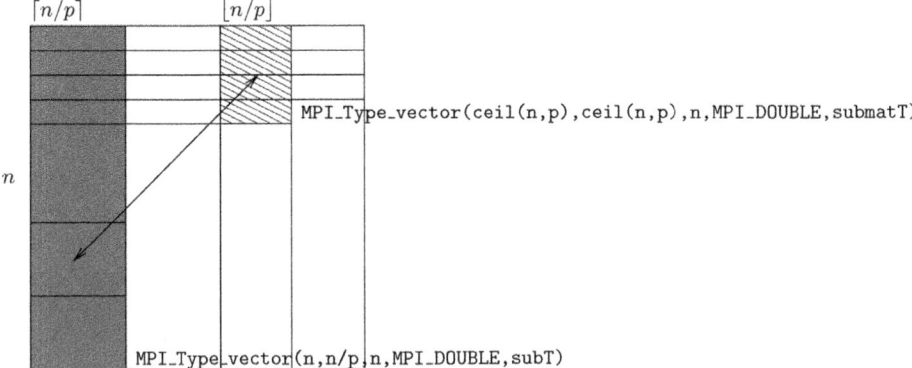

Fig. 1. An $n \times n$ matrix distributed as $n \times \lceil n/p \rceil$ and $n \times \lfloor n/p \rfloor$ submatrices over p processes. Each submatrix is described by an MPI vector datatype like subT. Also illustrated is the submatrix (block) movement for transposition of such matrices as captured by the submatT datatype.

the easier case where p divides n, as well as the more involved case where p does not divide n. This is particularly illustrative of certain shortcomings in the MPI collective interfaces as discussed in more detail in [11]. In Section 3 we benchmark the implementations and compare the performance to straightforward, hand-coded implementations not using derived datatypes.

2.1 Gathering/Scattering Column-Wise Matrices

In the first example, we assume that the matrix is distributed roughly evenly, column-wise over the p processes, such that each process stores either $\lfloor n/p \rfloor$ or $\lceil n/p \rceil$ consecutive columns, that is either an $n \times \lfloor n/p \rfloor$ or an $n \times \lceil n/p \rceil$ matrix. Assume that a root process has to collect the full matrix from the distributed columns or, conversely, initially has the full matrix and has to distribute the columns to the other processes. A natural approach is to describe the columns corresponding to the submatrices by a derived datatype, and then use a gather operation to collect the submatrices, automatically putting them in their correct position in the full matrix (scattering the matrix is completely analogous).

We first assume that $p|n$ (p divides n). An $n \times n/p$ submatrix can be described as an MPI vector type with n blocks (corresponding to the n rows), a blocksize of n/p elements (corresponding to each row of n/p elements), and a stride of n units (corresponding to the columns of the full matrix). This is illustrated in Figure 1. In order to get the ith submatrix into its right position by the MPI_Gather call, the extent of the vector datatype must be such that the ith submatrix starts at displacement in/p (times the size of the element type). The MPI_Type_vector constructor per definition assigns the vectortype an extent of $(n-1)n + n/p$ (times the extent of the element type). This must be changed by a resize operation. The full code for gathering the submatrices is thus

```
MPI_Type_vector(n,n/p,n,MPI_DOUBLE,&colT); // all columns of submatrix
MPI_Type_create_resized(colT,0,n/p*sizeof(double),&subT);
MPI_Type_commit(&subT);
```

```
MPI_Gather(columns,n*n/p,MPI_DOUBLE,matrix,1,subT,root,comm);
```

Note that the submatrices that are sent to the root can be treated as contiguous buffers. When p does not divide n this does not work. In this case, two receive datatypes would be needed (one for the $\lceil n/p \rceil$ columns, one for the $\lfloor n/p \rfloor$ columns), but there is no version of MPI_Gather or MPI_Gatherv that allow for different datatypes. The analogon of MPI_Alltoallw is currently missing from MPI (and might not be desirable, either). A different solution is possible, though. Define instead a datatype describing a single column of either of the matrices, and send and receive single columns using the MPI_Gatherv collective to set the displacements right.

```
MPI_Type_vector(n,1,n,MPI_DOUBLE,&colT); // one column of full matrix
MPI_Type_create_resized(colT,0,sizeof(double),&subT);
MPI_Type_commit(&subT);
```

```
c = (rank<n%size) : n/p+1 : n/p; // number of columns for this rank
MPI_Type_vector(n,1,c,MPI_DOUBLE,&col1T); // one column of local matrix
MPI_Type_create_resized(col1T,0,sizeof(double),&sub1T);
MPI_Type_commit(&sub1T);
```

```
for (x=0,i=0; i<size; i++) {
   displs[i] = i*n/p+x; if (i<n%size) x++; counts[i] = n/p+x;
}
MPI_Gatherv(columns,c,sub1T,matrix,counts,displs,subT,root,comm);
```

A further disadvantage of this solution compared to the $p|n$ case is that both the receiving root process as well as all sending processes now access data in strides, which may be inefficient. Thus, there may be an unnecessary performance difference from the regular case where $p|n$.

2.2   Matrix Transpose

The next example shows how datatypes and collective operations can jointly accomplish a matrix transpose without having to do any explicit, local reordering of submatrices (Examples 4.14 and 4.15 in [5] use datatypes for process local matrix transposition). Again, the matrix is distributed across the processes as columns, and each process has either $\lceil n/p \rceil$ or $\lfloor n/p \rfloor$ columns, as shown in Figure 1. The transposed matrix is stored in the same way. We first consider the case where $p|n$.

The transpose can accomplished by an MPI_Alltoall operation. Each process has to send n/p rows to each of the other processes, and these have to be transposed. This transposition is accomplished directly if the $n/p \times n/p$ submatrix is sent as n/p columns instead of as n/p rows, receiving these columns as rows. As in the previous example each column can easily be described by an MPI vector type and put together to a datatype describing the whole submatrix in column major oder. This is shown in the code below:

```
MPI_Type_vector(n/p,1,n/p,MPI_DOUBLE,&colT); // single column of submatrix
MPI_Type_create_resized(colT,0,sizeof(double),&colresT); // single element
MPI_Type_contiguous(n/p,colresT,&submT); // tile together
MPI_Type_create_resized(submT,0,n/p*n/p*sizeof(double),
                        &submatT); // resize again to full submatrix
MPI_Type_commit(&submatT);

MPI_Alltoall(localmatrix,1,submatT,
             transposed,n/p*n/p,MPI_DOUBLE,MPI_COMM_WORLD);
```

When p does not divide n the irregular MPI_Alltoallw function is called for, since there are now four different sizes and types of the submatrices. Since the submatrices differ in both number of rows and number of columns, the trick of specifying only a single column of the submatrices to be sent (or received) and using the possibly more efficient (and in any case less tedious) MPI_Alltoallv operation will not work. Other matrix operations where submatrices have different shapes and where other collective operations are required (e.g. allgather) may thus be difficult to implement currently in MPI since only the all-to-all operations have the fully general, type parameterized MPI_Alltoallw variant.

2.3   Elemental Cyclically Distributed Matrices

Our last example is concerned with the matrix distribution employed in the Elemental dense matrix library [6]. Elemental matrices are stored in a doubly cyclic fashion inside blocks of size $n/r \times n/c$, where the number of MPI processes is factored into $p = rc$. Each process stores a submatrix block $(\alpha_{i+k'r,j+k''c})$ for $0 \le k' < n/r, 0 \le k'' < n/c$ as shown in Figure 2. To make the points, it suffices here to assume that $p|n$, implying that also both $r|n$ and $c|n$. We will in fact assume that both $c^2|n$ and $r^2|n$, but these are not fundamental restrictions.

Process$(0,0)$	Process$(0,1)$	$\cdots$
$(\alpha_{0+k'r,0+k''c})_{0 \le k' < n/r, 0 \le k'' < n/c}$	$(\alpha_{0+k'r,1+k''c})_{0 \le k' < n/r, 0 \le k'' < n/c}$	$\cdots$
Process$(1,0)$	Process$(1,1)$	$\cdots$
$(\alpha_{1+k'r,0+k''c})_{0 \le k' < n/r, 0 \le k'' < n/c}$	$(\alpha_{1+k'r,1+k''c})_{0 \le k' < n/r, 0 \le k'' < n/c}$	$\cdots$
Process$(2,0)$	Process$(2,1)$	$\cdots$
$(\alpha_{2+k'r,0+k''c})_{0 \le k' < n/r, 0 \le k'' < n/c}$	$(\alpha_{2+k'r,1+k''c})_{0 \le k' < n/r, 0 \le k'' < n/c}$	$\cdots$
Process$(3,0)$	Process$(3,1)$	$\cdots$
$(\alpha_{3+k'r,0+k''c})_{0 \le k' < n/r, 0 \le k'' < n/c}$	$(\alpha_{3+k'r,1+k''c})_{0 \le k' < n/r, 0 \le k'' < n/c}$	$\cdots$
$\cdots$	$\cdots$	$\cdots$

Fig. 2. The doubly cyclical Elemental matrix distribution with $p = rc$

We here show how to convert between consecutively numbered, block distributed matrices and the doubly cyclical blocks of Elemental using MPI derived datatypes. First the processes are identified by the coordinates in the Cartesian grid as shown in Figure 2, for which the MPI Cartesian topology functionality can be used [5, Chapter 7]. All processes store a block of the same size and each process gets a contribution from each other process in the course of the transformation, so a regular MPI_Alltoall call could ideally be used. However, the start positions of both sent and received blocks are not uniformly strided, and therefore an irregular MPI_Alltoallv must be used instead. This problem with lack of MPI expressivity was discussed in more detail in [11]. The blocks to be sent and received are described by datatypes. Each process sends a block of $n/r^2 \times n/c^2$ elements to each other process, consisting of the elements $(\alpha_{i+k'r,j+k''c})$ for $0 \leq k' < n/r^2, 0 \leq k'' < n/c^2$ for process (i,j). The elements from process (i,j) are received as a submatrix that can again be described by a vector datatype and stored at row jn/r^2 in column in/c^2. In other words the c and r strided elements are picked via the send datatype, and received as contiguous submatrices via the receive type.

```
// shape of Elemental submatrix
MPI_Type_vector(n/(c*c),1,c,MPI_DOUBLE,&ErowTvec);
MPI_Type_create_resized(ErowTvec,0,n/c*sizeof(double),&ErowT);
MPI_Type_vector(n/(r*r),1,r,ErowT,&EblockTfull);
MPI_Type_create_resize(EblockTfull,0,1*sizeof(double),&EblockT);
MPI_Type_commit(&EblockT);

// shape of sequentially numbered block
MPI_Type_vector(n/(r*r),n/(c*c),n/c,MPI_DOUBLE,&BlockTfull);
MPI_Type_create_resize(BlockTfull,0,n/(c*c)*sizeof(double),&BlockT);
MPI_Type_commit(&BlockT);

for (i=0; i<p; i++) {
   sdipls[i] = (i/c)*(n/c)+i%c;  rdispls[i] = (i/c)*(c*n)/(r*r)+i%c;
}
MPI_Alltoallv(matrixblock,1,sdispls,EblockT,
              elementalblock,1,rdispls,BlockT,comm);
```

3   Experiments

The three examples have been implemented as outlined above. We now present benchmark results of these implementations and compare to straight-forward, manual solutions that do not use datatypes, but do all communication on contiguous buffers *and* with the same collective operations. The basetype is MPI_-DOUBLE. Benchmarks have been run on a small shared-memory Sun system with OpenMPI, a larger Sun SunFire X4100 cluster with mvapich2-1.4, and a small AMD/InfiniBand cluster with mpich2. The qualitative results are similar (under OpenMPI there were some problems with the larger matrix sizes), and we only discuss the results from the SunFire cluster.

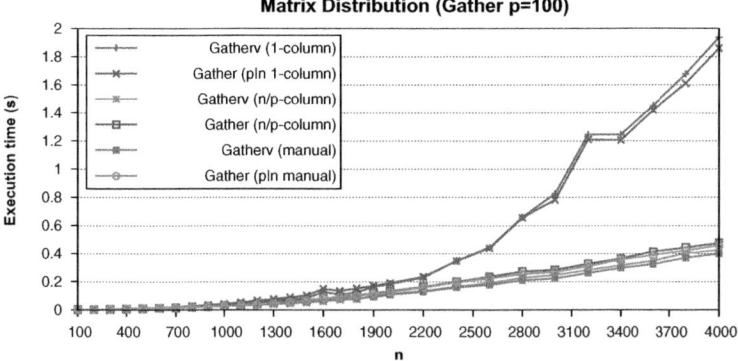

Fig. 3. Performance of column-wise matrix gathering on a SunFire 4100 cluster under `mvapich2-1.4` with $p = 100$ as a function of matrix order n

3.1 Column Matrices

We implemented both gathering and scattering of the column distributed matrices for any p and n. In the hand-written solution without datatypes, the $n/p \times n$ column matrices are reordered into n consecutive chunks that are gathered/scattered as consecutive blocks. This is compared with two versions using datatypes. The first describes the column matrices by a single vector of n/p-element blocks, and works only when p divides n, whereas the other version communicates single columns and works for any p and n. We can thus compare the manual version to single column vectors for any p and n, and for the case where $p|n$ to the single column vector and the n/p-column vector. For the $p|n$ case regular MPI_Gather or MPI_Scatter operations can be used, otherwise MPI_Gatherv and MPI_Scatterv are needed.

Results for $p = 100$ are shown in Figure 3 which makes it possible to compare the six implementation variants. For this instance there is no visible performance difference between the variants with regular and irregular collective operations. The single column implementation variant where both receiving and sending processes have to handle a vector datatype describing a single matrix column is more than a factor 4 slower than the hand-coded variant, where all communication is from contiguous buffers of MPI_DOUBLE. The datatype implementation with a vector describing all n/p columns is slightly better than the hand-coded version, but works only when $p|n$. The results for the matrix scatter example are similar, but are left out here.

3.2 Matrix Transposition

For the matrix transposition example we present results for six versions. A manual implementation where a local transposition into a contiguous buffer is performed is contrasted to the datatype version where the submatrices are described by nested datatypes as described in Section 2.2, both for the $p|n$ case that can

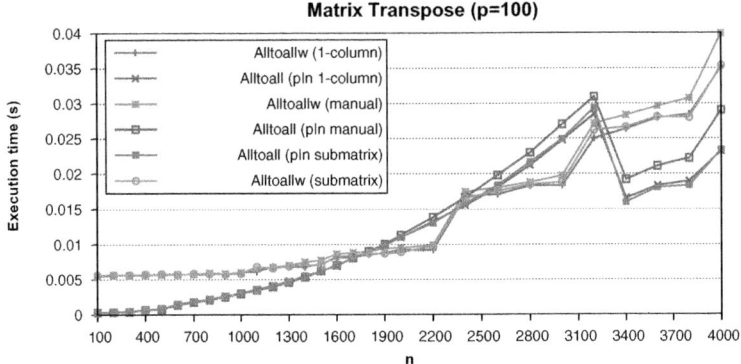

Fig. 4. Performance of column-wise matrix transposition on a SunFire 4100 cluster under `mvapich2-1.4` with $p = 100$ as a function of matrix order n

be implemented with MPI_Alltoall and the general case where MPI_Alltoallw has to be used. In addition a version where datatypes describe only a single column has been implemented; this can be made to work, if one is willing to use $n \times n$ space per process for the full matrix instead of just $n \times n/p$.

The results for $p = 100$ are shown in Figure 4 for increasing matrix orders n. The regular, $p|n$ variant with submatrix datatypes is significantly, about 20%, faster than the hand-coded variant. Likewise, the general version with MPI_Alltoallw and datatypes is similarly faster than the hand-coded, general version. For matrix orders between 1900 and 3200 the MPI_Alltoall implementations are worse than the implementations with MPI_Alltoallw, which is probably due to a suboptimal switch point from an improved to a direct all-to-all algorithm. The hand-coded variants are straight-forward and certainly not optimal; the point with these experiments is that better performance can be achieved by simply describing the problem with the appropriate datatypes. Getting similar, or better performance by hand might easily require considerably more work, and could be less performance portable.

3.3   Cyclically Distributed Matrices

The redistribution from sequentially stored, blocked matrices to the Elemental format has been implemented for the $p|n$ case. Matrix order n is chosen as kp^2 for $k = 1, \ldots, 8$. Results are shown in Figure 5, and indicate that for this instance the version employing datatypes is about a factor of 1.2 faster than the hand-written version. The hand-written implementation performs a tedious (multiple nested loops) restructuring of the submatrices into contiguous buffers at both sending and receiving sides, and employs MPI_Alltoall for the communication. For this transformation derived datatypes provide both conceptual and performance benefits.

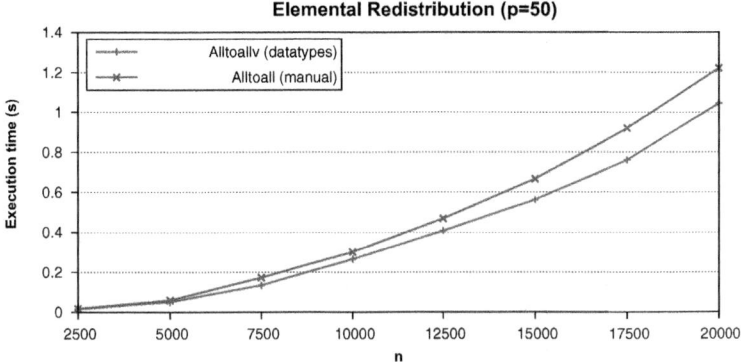

Fig. 5. Performance of matrix redistribution into Elemental format on a SunFire 4100 cluster under `mvapich2-1.4` with $p = 50$ as a function of matrix order $n = kp^2, k = 1, \ldots, 8$

4 Concluding Remarks

This study exemplified use of MPI derived datatypes in connection with collective operations to describe and effect redistribution operations on distributed two-dimensional data structures as sometimes found in numerical applications. Some such data structures, submatrices in particular, can conveniently be described with the MPI datatype constructors (including the convenience functions for arrays that were not used here), and can thus in principle alleviate the user from tedious, detailed hand-coding work. In all three examples we showed that performance could, with a standard MPI library and no extra effort, be improved over straight-forward hand-coded implementations, often considerably.

Acknowledgment. The second author thanks George Bosilca whose remarks on an earlier version of [11] have also influenced the examples described in the present paper.

References

1. Byna, S., Gropp, W.D., Sun, X.-H., Thakur, R.: Improving the performance of MPI derived datatypes by optimizing memory-access cost. In: IEEE International Conference on Cluster Computing (CLUSTER 2003), pp. 412–419 (2003)
2. Byna, S., Sun, X.-H., Thakur, R., Gropp, W.D.: Automatic memory optimizations for improving MPI derived datatype performance. In: Mohr, B., Träff, J.L., Worringen, J., Dongarra, J. (eds.) PVM/MPI 2006. LNCS, vol. 4192, pp. 238–246. Springer, Heidelberg (2006)
3. Hoefler, T., Gottlieb, S.: Parallel zero-copy algorithms for fast fourier transform and conjugate gradient using MPI datatypes. In: Keller, R., Gabriel, E., Resch, M., Dongarra, J. (eds.) EuroMPI 2010. LNCS, vol. 6305, pp. 132–141. Springer, Heidelberg (2010)

4. Lu, Q., Wu, J., Panda, D.K., Sadayappan, P.: Applying MPI derived datatypes to the NAS benchmarks: A case study. In: 33rd International Conference on Parallel Processing Workshops (ICPP 2004 Workshops), pp. 538–545 (2004)
5. MPI Forum. MPI: A Message-Passing Interface Standard. Version 2.2, September 4 (2009), http://www.mpi-forum.org
6. Poulson, J., Marker, B., Hammond, J.R., Romero, N.A., van de Geijn, R.: Elemental: A new framework for distributed memory dense matrix computations. ACM Transactions on Mathematical Software (2011) (conditionally accepted)
7. Renault, É.: Extended MPICC to generate MPI derived datatypes from C datatypes automatically. In: Cappello, F., Herault, T., Dongarra, J. (eds.) PVM/MPI 2007. LNCS, vol. 4757, pp. 307–314. Springer, Heidelberg (2007)
8. Renault, E., Parrot, C.: MPI pre-processor: Generating MPI derived datatypes from C datatypes automatically. In: International Conference on Parallel Processing Workshops (ICPP), pp. 248–256 (2006)
9. Ross, R.J., Miller, N., Gropp, W.D.: Implementing fast and reusable datatype processing. In: Dongarra, J., Laforenza, D., Orlando, S. (eds.) EuroPVM/MPI 2003. LNCS, vol. 2840, pp. 404–413. Springer, Heidelberg (2003)
10. Sanders, P.: Random permutations on distributed, external and hierarchical memory. Information Processing Letters 67(6), 305–309 (1998)
11. Träff, J.L.: A (radical) proposal addressing the non-scalability of the irregular MPI collective interfaces. In: 16th International Workshop on High-level Parallel Programming Models and Supportive Environments (HIPS 2011), International Parallel and Distributed Processing Symposium (IPDPS), page 42 (2011)
12. Träff, J.L., Hempel, R., Ritzdorf, H., Zimmermann, F.: Flattening on the fly: Efficient handling of MPI derived datatypes. In: Margalef, T., Dongarra, J., Luque, E. (eds.) PVM/MPI 1999. LNCS, vol. 1697, pp. 109–116. Springer, Heidelberg (1999)
13. Wu, J., Wyckoff, P., Panda, D.K.: High performance implementation of MPI derived datatype communication over InfiniBand. In: 18th International Parallel and Distributed Processing Symposium (IPDPS 2004), page 14 (2004)

Improving MPI Applications Performance on Multicore Clusters with Rank Reordering

Guillaume Mercier and Emmanuel Jeannot

Université de Bordeaux, INRIA, LaBRI
351, cours de la Libération F-33405 Talence, France
{guillaume.mercier,emmanuel.jeannot}@labri.fr

Abstract. Modern hardware architectures featuring multicores and a complex memory hierarchy raise challenges that need to be addressed by parallel applications programmers. It is therefore tempting to adapt an application communication pattern to the characteristics of the underlying hardware. The MPI standard features several functions that allow the ranks of MPI processes to be reordered according to a graph attached to a newly created communicator. In this paper, we explain how the MPICH2 implementation of the MPI_Dist_graph_create function was modified to reorder the MPI process ranks to create a match between the application communication pattern and the hardware topology. The experimental results on a multicore cluster show that improvements can be achieved as long as the application communication pattern is expressed by a relevant metric.

Keywords: Message-Passing, multicore architectures, process placement, rank reordering, communication pattern.

1 Introduction

Parallel programming is the prevalent paradigm for scientific applications. It is widely considered as the sole mean to achieve the computing power sought after by applications. Programming standards and their implementations play here a pivotal role because their efficiency conditions the overall performance. Among the parallel programming standards, the *Message Programming Interface* (MPI) is very popular because of its rich interface. Also, the implementations available manage to bridge the performance gap between the hardware and the applications. As for the hardware, the most widespread architecture to build parallel computers is based on the cluster paradigm. This trend has gained a huge momentum since its inception more than a decade ago and is still very strong. The machines used to build clusters have however changed from SMP-based nodes to more complex multicore ones, altering the way applications should be programmed. To harness such architectures is a difficult undertaking. NUMA effects, memory hierarchies and cores/cpus physical location within a node force the programmer to finely apprehend the hardware. The side effect is a decrease of performance portability: whilst any MPI code will run on such machines, only

Y. Cotronis et al. (Eds.): EuroMPI 2011, LNCS 6960, pp. 39–49, 2011.
© Springer-Verlag Berlin Heidelberg 2011

those specifically tailored to fit the hardware will benefit from its full performance. MPI being hardware-agnostic, no function in the interface can help the
programmer to retrieve information about the hardware and convey it up to
the application. Some workarounds do exist at best: process managers can enforce the binding of MPI processes onto specific cores and the logical topology
mechanism can be used to communicate application-specific information (such
as a communication pattern for instance) to the implementation. Concerning
the latter point, implementations that go beyond a trivial work are not aplenty.
Only some vendors MPI implementations (such as the ones provided by HP [1]
or NEC [2]) are tailored for specific classes of hardware and propose topology routines implementations taking advantage of the underlying specific fabric.
Generic MPI implementations addressing a wider spectrum of hardware however
manage to feature optimizations taking advantage of multicore nodes. Indeed,
collective communication operations are usually designed and implemented in a
hierarchical, two-levels, fashion so as to yield better performance [3]. But nothing is done in the topology mechanisms department to allow the programmer
to map an application communication pattern onto the underlying hardware.
In this paper, we propose an enhanced implementation of one MPI function:
MPI_Dist_graph_create. In our expanded version, the ranks of the MPI processes calling this function are reordered to allow an application communication
pattern to match as best as possible the underlying physical topology of a multi-
core cluster. This paper is organized as follows: Section 2 will expose the issue of
mapping a communication pattern onto a hardware architecture and compares
different existing techniques. Technical details are discussed in Section 3 while
results are analyzed in Section 4. Section 5 will describe previous existing works
while Section 6 concludes this paper and discusses future directions.

2   Matching a Communication Pattern to the Hardware Architecture: Issues and Techniques

2.1   General Overview of the Problem

During an MPI application, data are exchanged among the various participating
processes. The MPI programming paradigm is flat: each process may communicate with any other in the application. However, depending on pairs of processes,
the amount of data sent and received (in either terms of bytes/volume or number
of messages) may be irregular. Hence, each MPI application possesses a so-called
communication pattern which can be considered as an intrisic characteristic [4]
of the *affinity* between processes (here, we assume that this pattern is deterministic and does not change between executions). On the other hand, the communication channels in a multicore, NUMA nodes-based cluster are heterogeneous.
Internode communication using a network is slower than intranode communication using shared memory. The novelty with multicore NUMA nodes is that
communication performance is also heterogeneous within the node itself. The
various levels of cache memory and the NUMA effects when accessing the main

memory induce this. It is therefore rather *intuitive* to seek to adapt a potentially irregular communication pattern to the also heterogeneous (performance-wise) underlying hardware architecture.

2.2   Core Binding vs. Rank Reordering

There are two different methods to achieve this goal. The first one is called the *core binding* technique [5]. A binding algorithm determines on which physical core a specific MPI process should be located and pinned, so as to improve the overall communication performance (*e.g.* MPI processes are mapped according to the communication pattern and the hardware topology so as to minimize communication cost). An MPI application does not need to be modified: this binding information is provided by the user to the process manager which in turn enforces this user-defined binding policy at runtime. Legacy MPI applications can thus take advantage of this technique, if sufficient information is provided to the binding algorithm (which might imply an instrumentation of the application code, for example to build the communication pattern). However, this approach lacks transparency since the user has to use MPI implementation-specific command line options. Also, modifying, in a standard fashion, the binding during the course of an application is difficult. With the second method, called *rank reordering*, a new communicator is created with application-specific information attached to it. Ranks of the MPI processes belonging to this communicator can be *reordered*, meaning that they can be changed to fit some application constraints. Thus, the ranks of the MPI processes belonging to this newly created communicator could be determined to match the communication pattern to the underlying physical architecture. A reordering algorithm is necessary, playing a similar role as the binding algorithm of the first method. Legacy MPI applications would have to be slightly modified to issue a call to the ranks reordering MPI routine and use the newly produced communicator. Such reordering should be performed before application data is loaded into the MPI processes, otherwise data movements would be necessary. However, relying on a standard MPI call ensures portability, transparency and dynamicity since it can be issued multiple times during an application execution. These aspects aside, both methods yield the same performance improvements.

3   A Non-trivial Implementation of MPI_Dist_graph_create

This paper focuses on the *rank reordering* technique. Several MPI functions can reorder processes ranks. It is the case of MPI_Dist_graph_create, part of the standard since MPI 2.2 [6]. This function is meant to replace the non-scalable MPI_Graph_map function. MPI_Dist_graph_create takes as arguments a set of pointers (sources, destinations, degrees and weights) that characterize a graph. Hence, random application communication patterns can be passed to the implementation using these pointers. We modified the current

MPI_Dist_graph_create implementation available in MPICH2 [7] in order to allow the given input graph to be mapped onto another graph we build and that describes the underlying architecture. Such a problem is known as a *graph embedding problem*. In our case, the optimization criterion is the minimization of communication costs. Our approach is three-fold: first we gather information about the hardware topology, then we access the application communication pattern and at last we solve our graph embedding problem with a tailored algorithm.

3.1   Gathering the Hardware Information

To gather hardware information raises portability issues because we need to address the largest possible spectrum of architectures. No standard tool currently exist to perform this task. Our version of MPI_Dist_graph_create uses to the Hwloc library (version 1.1.1) [8] that offers a generic and portable interface to retrieve hardware information. Thanks to Hwloc, we manage to gain insights of a NUMA node structure (e.g cache hierarchies, number of processors, location of processing units within sockets, etc.). On each multicore node, one process extracts the hardware information, then a global root process gathers all these data. That is, our current implementation is *centralized*, which might impact scalability. Hwloc being currently unable to provide us information about the network topology, we consider it as flat, as in the MPI model. Now, this information has to be represented in a convenient way. Since multicore nodes are organized hierarchically, a relevant data structure is a *tree*, where leaves represent processing units. To create the data structure that represents a cluster of multicore nodes is a straightforward process: we add a new level encompassing all the subtrees representing the various NUMA nodes at the top level of the structure. This corresponds to our vision of a flat network topology.

3.2   Communication Pattern Information and Metrics

There are two cases to consider for an application communication pattern, First, newly developed MPI applications can directly use MPI_Dist_graph_create. In this case, the programmer has to provide the pattern information thanks to the function arguments. Indeed, the programmer is supposed to possess some knowledge of the organization of communication. But it is not always the case, especially when using collective communication, because the pattern will depend on algorithms known only by the designers of the MPI implementation. Hence, switching from one MPI implementation to another is likely to influence the application pattern. In the case of applications for which the pattern is unknown to the user, some information can be gathered by the means of instrumentation. Therefore, we introduced a lightweigt trace system in MPICH2 to retrieve the pattern information. We trace the data exchanged at the MPI application level to obtain the most implementation-independent data. Of course, a prior execution of the application is mandatory to generate a pattern data file. It contains information for each pair of processes. We actually use two different *metrics* to assign weights to the edges of the pattern graph. The first metric is the global amount of data (a.k.a *Data Size*) while the second one is the number of exchanged messages (a.k.a *Number of Messages*). We also implemented a helper routine that directly reads the pattern file output by our trace system in order to fill the arguments of MPI_Dist_graph_create according to the chosen metric.

3.3 The TREEMATCH Matching Algorithm

In order to solve our *graph embedding problem*, we implemented a new algorithm called TREEMATCH [9]. The TREEMATCH algorithm is a graph algorithm which takes into account the affinity of the processes expressed as a communication matrix to bind these processes to the topology. It works recursively on each level of the memory hierarchy (following a bottom-up approach) and groups processes in such a way that the cost of remaining communications is minimized. TREEMATCH extracts a tree from the communication matrix representing a communication pattern and matches this tree to the hardware topology tree. Finally, the algorithm outputs a permutation of the processes σ such that process rank i (in the original communicator) is mapped on core σ_i. This algorithm is called by the global root process that possesses both the hardware information and the pattern information: indeed, the implementation is currently fully centralized.

4 Performance Improvements Evaluation

We carried out a series of experiments to assess the performance improvements induced by the use of our enhanced MPI_Dist_graph_create function. All tests are executed on a cluster composed of 68 nodes linked with an Infiniband interconnect (HCA: Mellanox Technologies, MT26428 ConnectX IB QDR). Each node is composed of two INTEL XEON NEHALEM X5550 cpus featuring 4 2.66 GHz cores each. The 8 Mbytes of L3 cache are shared between the four cores of a CPU. There are 24 GB of DDR3 RAM at 1.33 GHz on each node. As for the software, the operating system is SLES 11 and the MPI implementation is MVAPICH2 1.7 (alpha 1) [10]. All of the benchmarks involve 64 processes (8 nodes connected to the same Infiniband switch are used) and each process is bound to its dedicated core. The baseline chosen to compare the process placement policies is the *Serial Ranking* policy where the process rank number n (in MPI_COMM_WORLD) is placed on the node number m with $m = n/8$ ($n \in [0, 63]$ in our case). Such a policy is enforced by default by most resource schedulers when providing a machine file to the user after reserving nodes (e.g PBS/Torque). Also, the execution times do not take into account the time spent in the MPI_Dist_graph_create function called at the begining of each benchmark (less than 140 milliseconds in our experiments with 64 processes). The tests are run several times in a row: the *Serial Ranking* case (without reordering) is followed by the reordered cases, using the two metrics listed in Sec. 3.2.

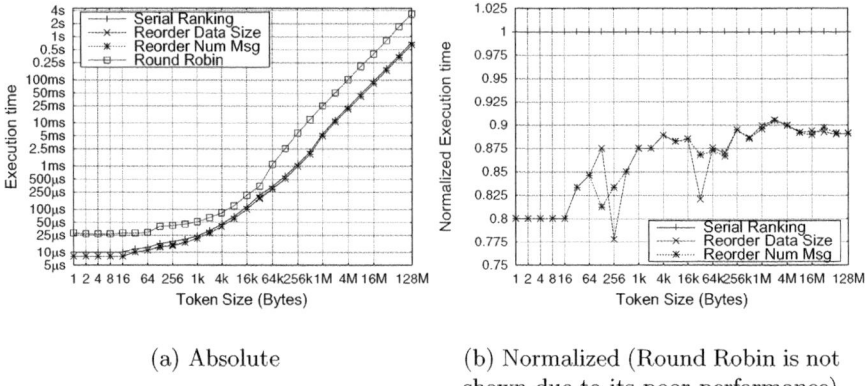

(a) Absolute   (b) Normalized (Round Robin is not
                                shown due to its poor performance)

Fig. 1. Ring pattern execution times

This first execution actually creates the communication pattern file we need to initialize the pointer arrays of the MPI_Dist_graph_create function called in the following runs.

4.1   The Ring Pattern Benchmark

The first benchmark is purposely designed to showcase the benefits of the re-ordering technique. The communication pattern features several rings of processes of equal sizes. A token is exchanged in each ring, that stops circulating when received back by the process that initially sent it (a.k.a the *ring leader*). Then, all ring leaders exchange the token with a call to MPI_Allgather. The test is run with 8 rings composed of 8 processes each. For this test, we make experiments with two non-reordered cases: the first one is when the *Serial Ranking* policy (as described above) is used and the second one is when a so-called *Round Robin* policy is used. With this policy, core number i of node number j executes process rank n (in MPI_COMM_WORLD), where: $n = (8 * i) + j$ and $(i, j) \in [0, 7]^2$ (cores and nodes are numbered linearly). The process-to-core binding policy within nodes is the same for both policies. We use the Round Robin case in order to show that a suboptimal process mapping/binding policy can be effectively corrected by reordering. In our example, since 8 nodes of the cluster are used and given the communication pattern, most of the traffic goes through the network in the Round Robin case. In the reordered cases, most of communication uses shared memory, much faster than the network. Indeed, Figure 1 shows that our algorithm is able to compute a relevant reordering and that we are able to exploit the nodes internal structure more finely since the execution times achieved by either the *Data Size* or *Number of Messages* cases are 10% to 20% faster than in the *Serial Ranking* case. Since the amount of data and the number of messages grow proportionaly, both metrics yield the same results.

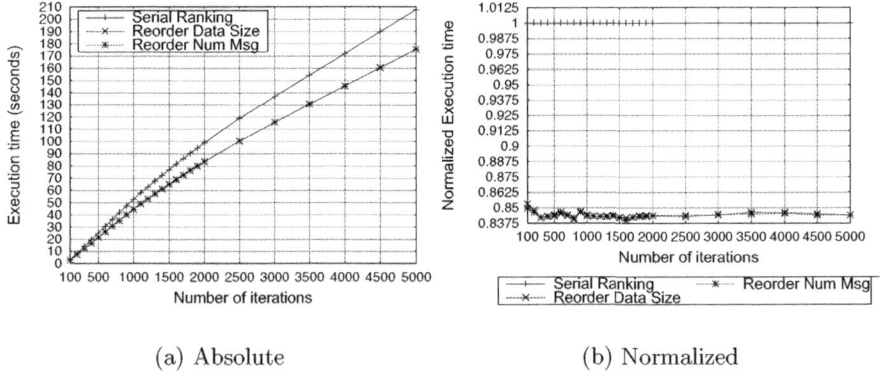

(a) Absolute   (b) Normalized

Fig. 2. ZEUS-MP Execution times

4.2   ZEUS-MP

The second set of experiments involve a real application called ZEUS-MP. It is a computational fluid dynamics code for the simulation of astrophysical phenomena that solves magnetohydrodynamics equations. The three-dimensional computational domain is organized in tiles where the boundary data is exchanged with MPI messages between neighbours. We used the 2.1.2 [11] version of ZEUS-MP. Originally, this application uses the MPI cartesian topology mechanism, but without reordering. Also, ZEUS-MP is not able to take into account the underlying physical architecture thanks to options or arguments passed to the program for instance. We ran ZEUS-MP for various iteration counts and measured the execution times. Figure 2 shows the benefits of using reordering. Since our modified version of MPI_Dist_graph_create allow the application to better exploit the underlying multicore architecture, the cost of communication is reduced. The overall execution times are decreased by more than 15%. Both metrics yield the same results. This results shows that our approach is relevant and allows the user to better exploit the nodes internal structure, without possessing a prior knowledge of their physical topology. To manage to get equivalent results, the programmer would have to: 1– understand the application behaviour, leading to the use the *Serial Ranking* policy and 2– provide an adequate process-to-core binding when running it.

4.3   RSA-768 – The Block Wiedemann Algorithm

The 768-bits, 232-digits number RSA-768 is factorized since December 12, 2009 [12]. Several algorithms and applications were used to achieve this result and one particular step consists to find dependencies between the rows of a sparse matrix using a *Block Wiedemann* algorithm. We benchmarked a simplified version of this Block Wiedemann step provided by one of the authors of [12]. It is

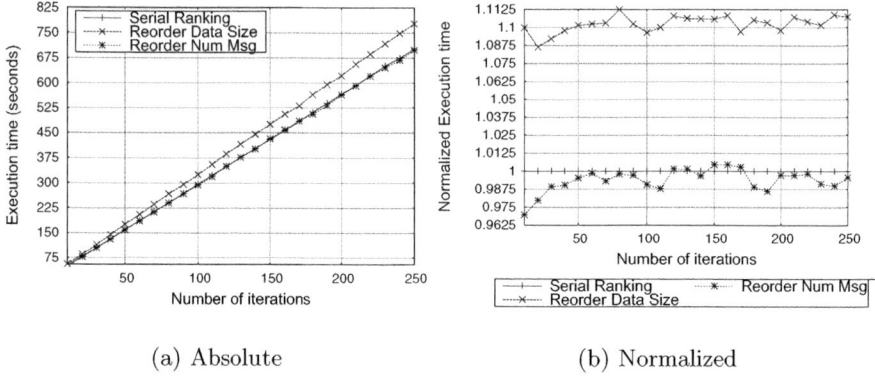

(a) Absolute   (b) Normalized

Fig. 3. Block Wiedemann step – RSA-768 Execution times

a relevant target for our work because it is a *communication bound* application. This application takes into account the underlying physical architecture thanks to parameters passed to the program (i.e the number and layout of the cores). This is a difference with our previous application case (ZEUS-MP). This version of *Block Wiedemann* is designed to be used with the *Serial Ranking* placement policy. Figure 3 shows the results obtained. Out of the two metrics, only *Number of Messages* manages to slightly improve the results obtained with *Serial Ranking*. The execution times are decreased by less than 2%. This results demonstrate that some room for improvement exists even for applications that are tuned to exploit the underlying architecture. All these results therefore advocate for reordering, and demonstrate how this technique can alleviate some of the burden of tuning an MPI application to an underlying multicore system. Thanks to MPI_Dist_graph_create, this tuning is performed transparently, automatically and in a portable fashion.

5   Related Works

The placement of MPI processes on processors in order to match the communication pattern to the underlying hardware architecture has been previously examined. The problem is introduced in [1] and an algorithm, based on the Kernighan-Lin heuristic [13], is described as well as results for some benchmarks. However, this work is tailored for a specific vendor hardware and is thus not suitable for a generic case. Also, the author optimizes some of the routines creating *cartesian topologies* but left unaddressed the more generic *graph topology* case. The experiments show dramatic improvements, but are restricted to benchmarks only communicating but not doing any computation. In particular, the *Jacobi method for a Poisson problem solver test* results are consistent with of our own ring test results. Topology mechanisms implementation issues are also discussed in [2]. Both cartesian and graph topologies are addressed by this work, and the algorithm is based on the same Kernighan-Lin heuristic. The optimization criterion considered is either the total communication cost (as in the TREEMATCH algorithm) or the optimal load balance. But here again, it is a work designed for a specific vendor hardware. The approach is thus less generic than our. Besides the Kernighan-Lin and TREEMATCH, there are other algorithms that can solve the *graph embedding problem*. SCOTCH [14] is a graph coupling framework but not optimized for our case where we work only on *trees* (see Sec. 3.3). A previous version of our work, which concerned the core binding technique, did in fact use SCOTCH [5]. TREEMATCH, however, outperforms SCOTCH in terms of execution times and is therefore better suited for runtime optimizations. MPIPP [15] is another framework aiming at optimizing an application execution on the underlying hardware. MPIPP relies on an external tool to generate the hardware information while we manage to perform this at runtime. A comparison between TREEMATCH and MPIPP can be found in [9]. Also, the MPIPP framework uses the core binding technique, as well as a couple of vendors such as Cray [16], HP [17] and probably IBM [18].

6   Conclusion and Future Works

In this paper, we showed that using rank reordering can allow MPI applications to transparently exploit clusters of multicore nodes. The application communication pattern is matched to the underlying hardware, thus reducing the cost of application communication. The communication pattern is usually expressed as the overall amount of bytes exchanged among processes but we experienced that other metrics are more relevant in our particular environment, that is using MVAPICH2 as the MPI implementation and TREEMATCH as the algorithm to solve the *graph embedding problem*. Indeed, using the number of messages to characterize the communication pattern yields better results than the amount of bytes. We plan to understand why the *Number of Messages* metric outperforms in some cases the *Data Size* one. Also, our MPI_Dist_graph_create implementation is currently centralized, which is not scalable. We would like to implement a distributed version, but this might imply to distribute the TREEMATCH algorithm itself. This algorithm could also integrate new optimization criteria such as the ones listed in [6]. Currently, we do not take into account the physical topology of the network. We only exploit the nodes internal structure. There are plans to expand HWLOC in order to provide such information, we are hence looking forward to take advantage of it. Also, we currently lack some quantitative information about NUMA effects. Indeed, a more recent release of HWLOC (1.2) features latency matrices in order to assess performance between cores. It is another piece of information that we want to exploit. At last, we plan to work on the extraction of the communication pattern information without relying on a previous run of the complete application. A static analysis of the MPI code could be performed at compile time to generate the needed information.

References

1. Hatazaki, T.: Rank reordering strategy for MPI topology creation functions. In: Alexandrov, V.N., Dongarra, J. (eds.) PVM/MPI 1998. LNCS, vol. 1497, pp. 188–195. Springer, Heidelberg (1998)
2. Träff, J.L.: Implementing the MPI process topology mechanism. In: Supercomputing 2002: Proceedings of the 2002 ACM/IEEE Conference on Supercomputing, pp. 1–14. IEEE Computer Society Press, Los Alamitos (2002)
3. Zhu, H., Goodell, D., Gropp, W., Thakur, R.: Hierarchical Collectives in MPICH2. In: Ropo, M., Westerholm, J., Dongarra, J. (eds.) PVM/MPI. LNCS, vol. 5759, pp. 325–326. Springer, Heidelberg (2009)
4. Ma, C., Teo, Y.M., March, V., Xiong, N., Pop, I.R., He, Y.X., See, S.: An Approach for Matching Communication Patterns in Parallels Applications. In: Proceedings of 23rd IEEE International Parallel and Distributed Processing Symposium (IPDPS 2009). IEEE Computer Society Press, Rome (2009)
5. Mercier, G., Clet-Ortega, J.: Towards an efficient process placement policy for MPI applications in multicore environments. In: Ropo, M., Westerholm, J., Dongarra, J. (eds.) PVM/MPI. LNCS, vol. 5759, pp. 104–115. Springer, Heidelberg (2009)
6. Hoefler, T., Rabenseifner, R., Ritzdorf, H., de Supinski, B.R., Thakur, R., Träff, J.L.: The scalable process topology interface of mpi 2.2. Concurrency and Computation: Practice and Experience 23, 293–310 (2011)
7. Argonne National Laboratory: MPICH2 (2004), http://www.mcs.anl.gov/mpi/
8. Broquedis, F., Clet-Ortega, J., Moreaud, S., Furmento, N., Goglin, B., Mercier, G., Thibault, S., Namyst, R.: hwloc: a Generic Framework for Managing Hardware Affinities in HPC Applications. In: Proceedings of the 18th Euromicro International Conference on Parallel, Distributed and Network-Based Processing (PDP 2010), Pisa, Italia. IEEE Computer Society Press, Los Alamitos (2010)
9. Jeannot, E., Mercier, G.: Near-optimal placement of MPI processes on hierarchical NUMA architectures. In: D'Ambra, P., Guarracino, M., Talia, D. (eds.) Euro-Par 2010. LNCS, vol. 6272, pp. 199–210. Springer, Heidelberg (2010)
10. Design of High Performance MVAPICH2: MPI2 over InfiniBand. In: Proceedings of the 6th IEEE International Symposium on Cluster Computing and the Grid. IEEE Computer Society, Los Alamitos (2006)
11. Hayes, J.C., Norman, M.L., Fiedler, R.A., Bordner, J.O., Li, P.S., Clark, S.E., Ud-Doula, A., Mac Low, M.-M.: Simulating Radiating and Magnetized Flows in Multiple Dimensions with ZEUS-MP. The Astrophysical Journal Supplement 165, 188–228 (2006)
12. Kleinjung, T., Aoki, K., Franke, J., Lenstra, A.K., Thomé, E., Bos, J.W., Gaudry, P., Kruppa, A., Montgomery, P.L., Osvik, D.A., te Riele, H., Timofeev, A., Zimmermann, P.: Factorization of a 768-bit RSA modulus. In: Rabin, T. (ed.) CRYPTO 2010. LNCS, vol. 6223, pp. 333–350. Springer, Heidelberg (2010), http://www.springerlink.com
13. Kernighan, B.W., Lin, S.: An efficient heuristic procedure for partitioning graphs. Bell System Technical Journal 49, 291–307 (1970)
14. Pellegrini, F.: Static Mapping by Dual Recursive Bipartitioning of Process and Architecture Graphs. In: IEEE Proceedings of SHPCC 1994, Knoxville, pp. 486–493 (1994)

15. Chen, H., Chen, W., Huang, J., Robert, B., Kuhn, H.: Mpipp: an automatic profile-guided parallel process placement toolset for smp clusters and multiclusters. In: Egan, G.K., Muraoka, Y. (eds.) ICS, ACM, pp. 353–360 (2006)
16. National Institute for Computational Sciences: (MPI Tips on Cray XT5), http://www.nics.tennessee.edu/user-support/mpi-tips-for-cray-xt5
17. Solt, D.: A profile based approach for topology aware MPI rank placement (2007), http://www.tlc2.uh.edu/hpcc07/Schedule/speakers/hpcc_hp-mpi_solt.ppt
18. Duesterwald, E., Wisniewski, R.W., Sweeney, P.F., Cascaval, G., Smith, S.E.: Method and System for Optimizing Communication in MPI Programs for an Execution Environment (2008), http://www.faqs.org/patents/app/20080288957

Multi-core and Network Aware MPI Topology Functions

Mohammad Javad Rashti[1], Jonathan Green[1], Pavan Balaji[2],
Ahmad Afsahi[1], and William Gropp[3]

[1] Queen's University, Kingston, ON, Canada
[2] Argonne National Laboratory, Argonne, IL, USA
[3] University of Illinois at Urbana-Champaign, IL, USA

Abstract. MPI standard offers a set of topology-aware interfaces that can be used to construct graph and Cartesian topologies for MPI applications. These interfaces have been mostly used for topology construction and not for performance improvement. To optimize the performance, in this paper we use graph embedding and node/network architecture discovery modules to match the communication topology of the applications to the physical topology of multi-core clusters with multi-level networks. Micro-benchmark results show considerable improvement in communication performance when using weighted and network-aware mapping. We also show that the implementation can improve communication and execution time of the applications.

Keywords: MPI, virtual topology, physical topology, multi-core, network.

1 Introduction

With the emerging many-core architectures and high performance interconnects offering more parallelism and performance, clusters are expected to move towards exascales in the next few years [1]. Such systems are becoming increasingly hierarchical in their node architecture and interconnection network. Communication at various hierarchies demonstrate different performance levels. It is therefore critical for the communication libraries to efficiently handle the communication demands of High Performance Computing (HPC) applications on such hierarchical systems.

Message Passing Interface (MPI) [2] is the predominant messaging standard for HPC applications. MPI provides a set of interfaces that are designed to assist the library to construct a virtual topology out of the application's communication pattern, and to support remapping (reordering) the processes to the available cores in a way that optimizes performance. However, most MPI libraries merely provide a trivial implementation of these functions and lack the support for the remapping feature. In this work, we have designed the MPI non-distributed topology functions (MPI_Graph_create and MPI_Cart_create) for efficient process remapping over hierarchical clusters. We have integrated the node physical topology with network architecture and used graph embedding tools inside MPI library to override the current trivial implementation of the topology functions and efficiently reorder the initial process mapping. We have evaluated our implementation on two different InfiniBand [3] clusters using micro-benchmarks and MPI applications. Micro-benchmark results

Y. Cotronis et al. (Eds.): EuroMPI 2011, LNCS 6960, pp. 50–60, 2011.

show up to 60% communication time improvement for Cartesian topologies with data exchange between the neighbors. The application results show up to 48% communication time improvement, and up to 26% runtime improvement.

The rest of this paper is organized as follows. We briefly describe the MPI topology functions and the motivation behind this work in Section 2. Related work is covered in Section 3. Section 4 discusses our design and implementation. Section 5 presents the experimental results, and Section 6 concludes the paper.

2   Background and Motivation

MPI defines a set of virtual-topology definition functions for graph and Cartesian structures [2]. MPI_Graph_create and MPI_Cart_create are collective calls that accept a virtual topology and return a new MPI communicator enclosing the desired topology. If the user opts for reordering, the function may reorder the process ranks for an efficient process-to-core mapping. The topology accepted by these functions is in a non-distributed form, meaning that all nodes have a full view of the entire structure and pass the whole information to the function. Recently, distributed graph topology functionality has been added to the MPI standard [2] to support large-scale systems. In these functions, each node has a limited neighborhood view of the graph, and all processes collectively construct the virtual topology in a distributed fashion.

Although process topology functionality is not new to the MPI standard, HPC applications that utilize such functionality use them mainly for the constructed virtual topology (e.g., a Cartesian topology). Thus, the ability of this interface to support process reordering for better communication performance has widely remained unutilized, mostly because MPI implementations merely construct the virtual topology, and have no process remapping for performance improvement. In this paper, we focus on the design and implementation of MPI non-distributed topology functions to improve the performance of the applications. We will cover the distributed topology functions in a future work.

3   Related Work

There has been some past research on topology-aware communication in MPI. The authors in [4] look at the algorithms for mapping virtual topology to physical topology in MPI using a hierarchical tree structure to represent the hardware topology. This work defines a cost function that is the sum of the communication costs over the links. This paper presents an implementation for a specific machine (an HP server), with architecture similar to multi-core machines.

The work in [5] presents a set of graph embedding algorithms with hardware topology represented using a hierarchical tree similar to [4], but with a more comprehensive mathematical analysis for different architectures. A major contribution is the optimization of two different cost functions. The first function is the sum of the communication costs over the interconnection links, and the second is a load balancing function that minimizes the number of expensive communications by any one process. This work experiments with a set of NEC SX-6 machines, each with up to eight shared-memory vector processors, connected through a proprietary network.

The author in [6] argues that the MPI topology-aware functionality at the time lacks precision and accuracy. The paper suggests an extension to MPI where weighted communication graphs can be used in order to produce a better solution when using the topology functionality. The paper also suggests extensions to allow dynamic process reordering. No implementation of the suggested functionality is attempted.

The authors in [7] summarize the work in [5] and [6], suggesting some additional changes to MPI as in [6]. The paper also shows that a non-trivial implementation of the MPI topology functions can provide great performance gains on SMP systems.

In [8], functionality is implemented to map weighted communication graphs to weighted node architecture graphs using *Scotch* graph partitioning software [9]. The communication graph weights are chosen based on the total communication volume. Unlike our work in this paper, the work in [8] does not use or implement MPI topology functions. It rather calculates the mapping outside MPI, before the application is started. It also does not consider network hierarchy.

In [10], the authors propose *TreeMatch* algorithm to calculate a near-optimal mapping of processes to resources on a NUMA cluster. Similar to the work in [8], this paper is concentrated on node architecture and does not consider network hierarchy. It does not use or implement MPI topology functions either. It rather calculates the mapping using MPICH2 process manager and hwloc [14]. The paper presents MPI results using simulation and NAS benchmarks on a 4 NUMA-node 96-core machine.

Recently, the authors in [11] have explained the distributed topology functions in MPI 2.2 and discussed possible methods for implementing them in the future. In [12] the authors propose an automated framework to detect regular communication patterns in applications. The framework finds the dimensions of a possibly regular pattern and maps it to mesh/torus processor topologies.

4   Design and Implementation of MPI Topology Functions

4.1   Design of the Graph Topology Function

Our design of both MPI graph and Cartesian topology functions is based on an underlying graph structure. We also use graphs to represent both virtual and physical topologies inside the MPI implementation. Using graphs at the underlying layer, we can use static mapping of virtual to physical topology graphs in order to find the sub-optimal mapping of processes to processor cores.

Virtual topology is constructed as a graph in which vertices represent processes and links represent the existence (or significance) of the inter-process communications. We use the normalized total communication volume between two processes as the metric for communication significance. MPI_Cart_create and MPI_Graph_create functions do not support weighted edges, meaning there is no differentiation among edges of the virtual topology. This is a critical shortcoming, since it is usual that the communication between some pairs is more significant than others. To realize how supporting weighted graphs can increase the effectiveness of process reordering, we use edge replication in MPI_Graph_create input to account for weighted edges. This approach, although not much scalable, can support realistic non-uniform communication patterns.

The graph representation for the cluster's physical topology consists of two distinct but integrated parts: node architecture and network architecture. The node architecture graph includes weighted edges that represent the communication performance between any two cores. We assume higher communication performance between the cores with closer proximity. Our representation of the network comprises network distances between any two cores. The network architecture includes weighted graph edges that represent the communication performance of the network path between the nodes on which the cores reside.

4.2 Implementation of Topology Functions

In this section, we present details about the implementation of the MPI_Graph_create and MPI_Cart_create functions inside MVAPICH2 [13]. We use the Scotch graph processing library [9] to map virtual to physical topology graphs in MPI_Graph_create. The library defines an undirected source graph, which can represent the virtual topology. Each vertex and edge of the source graph is weighted, to account for the computation and communication weight of the corresponding process and link, respectively. Similarly, the target machine architecture is represented by an undirected architecture target graph with weights for vertices and edges representing the processing power of the processor and communication performance of the link, respectively [9].

The *hardware locality* (hwloc) library [14] provides a portable abstraction of the underlying machine architecture. It detects architectural components of the nodes such as processor sockets, cores, caches, memory, SMT and NUMA architecture. The architecture is represented as a tree, with nodes at the top level and logical cores at the leaves. This library can assist MPI to construct the physical topology of the machine.

To have a complete view of the cluster's physical architecture and go beyond a flat network assumption, we add a network discovery part in our physical architecture discovery. This module extracts the network distance between any two nodes in the cluster and merges that information with the node architecture extracted by hwloc. In this paper, our network discovery module uses InfiniBand subnet manager [3] tools (i.e., *ibtracert* utility) to discover the network distance between Infiniband nodes.

4.2.1 Implementation of Graph Topology Function

To supply MPI with the application virtual topology, we extract the exact amounts of data transfer between processes by profiling the applications using probes inside the MPI library. We give the highest weight to the maximum pairwise communication volume. The normalized weights range from 1 to 10. We use edge replication to represent edge weights. MPI_Graph_create calls the Scotch library if the user opts for reordering. Scotch builds a weighted graph out of the user-supplied graph. The topology table of a server node is also created using hwloc at each process. The communication performance between two logical cores on a node is calculated based on the depth of their common ancestor in the node topology tree. For example, if two cores reside on different sockets, their common ancestor will be the node itself, with the lowest depth in the tree, translating into the lowest communication performance.

The process with rank zero (in MPI_COMM_WORLD) will perform a network discovery operation using *ibtracert* to extract the distance between any two InfiniBand nodes. This information is then scattered to other processes to be integrated into the node architecture to have the full physical topology architecture. Topology graph edge loads (input to Scotch) represent the performance of the communication path between the connecting vertices. Network distance, defined as the number of hops between two nodes, is used to calculate the physical topology edge loads. The farthest nodes get the load of 1 on their graph edge. The closest nodes get the maximum network hop count as their edge load. The intra-node graph loads are calculated in a way that are always larger than the closest network distance, indicating the fact that intra-node communication is cheaper than inter-node communication.

Fig. 1 shows an example of how graph loads are assigned based on the system and network architecture. *N1-N4* are multi-core (here, 2-way quad-core) nodes, connected through three switches (*S1-S3*) in a tree-like network. In this architecture, *d1* (the path between two cores of the same socket) will have the highest load value in the graph, while the path between *N2* and *N3* (*d4*) will have the lowest load value, indicating the lowest performance path in the network. Thus we have: $d1 > d2 > d3 > d4 = 1$.

4.2.2 Cartesian Topology Implementation

Since the Scotch library does not support Cartesian topology, we internally convert the MPI_Cart_create topology to a graph and use MPI_Graph_create for reordering. In this conversion, the user view of the Cartesian topology remains intact. Cartesian topologies do not specify weights for graphs. Therefore, the converted topology will be a non-weighted graph.

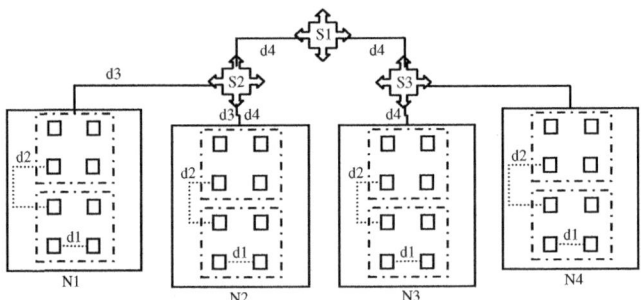

Fig. 1. An example of physical topology distances

5 Experimental Results

We have conducted our experiments on two clusters. The first cluster has four servers, each with two quad-core 2GHz AMD Opteron 2350 processors (a total of 32 cores). The servers have 512KB L2 cache per core, 2MB shared L3 cache per processor and 8GB memory. They are interconnected through Mellanox ConnectX InfiniBand cards [15] via three Mellanox InfiniBand switches, similar to Fig. 1. The nodes run Linux Fedora 12 kernel version 2.6.31. The machines use OFED 1.5.2 to access the InfiniBand network. For the second cluster, we have 16 servers, each with two hexa-core 3GHz Intel Xeon X5670 processors (a total of 192 cores). There is a 12MB multi-level cache per processor, and 24GB memory per machine. The servers use Mellanox ConnectX2 InfiniBand cards. Eight servers are connected to a Mellanox

InfiniBand switch, and the remaining servers are connected to another switch. Each of these switches is connected to two upper layer InfiniBand switches. The machines run Redhat Enterprise Linux 5 with kernel version 2.6.18-194. The nodes use OFED version 1.5.2. On both clusters, we use MVAPICH2 1.5 [13] as our code-base.

5.1 Micro-benchmark Results

We start our evaluation with two micro-benchmarks that put processes in Cartesian arrangements such as Torus and Hypercube. The tests are run on the first (32-core) cluster. The first micro-benchmark (*Cartesian-model exchange*) constructs a 2D/3D torus or a 5D hypercube. Each process runs 1000 iterations, each with a computation followed by exchanging messages with the neighbors in all dimensions. We report the average iteration time. We find the virtual-topology graph of each test by profiling it in an initial run. The graph is then supplied to the same program as input for MPI_Graph_create. To evaluate the effect of graph weights, we carry significantly heavier communication on one of the dimensions.

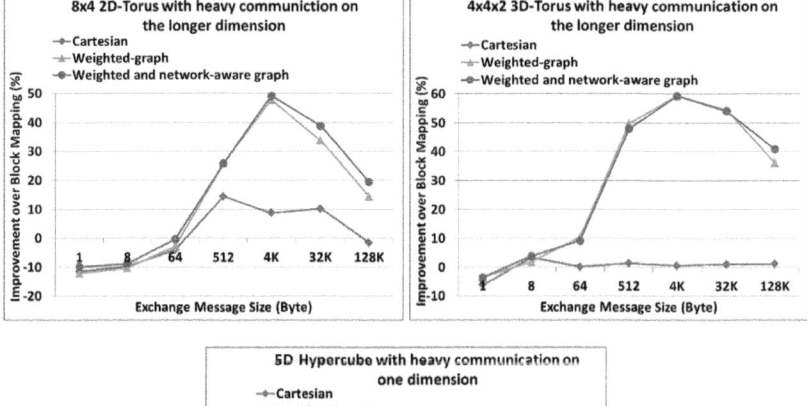

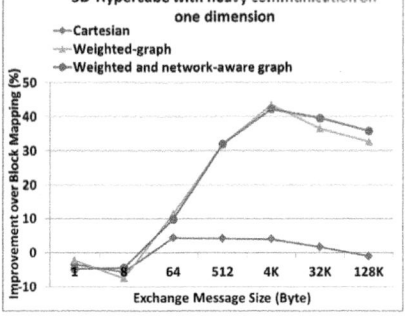

Fig. 2. Runtime improvement of topology-aware mapping over block mapping for 2D-torus, 3D-torus and 5D-hypercube in the Cartesian-model exchange micro-benchmark

Fig. 2 shows the improvement of the topology-aware mapping compared to *block* mapping for an 8×4 2D-torus, a 4×4×2 3D-torus and a 5D-hypercube. The processes communicate more heavily on one dimension (the longer dimension for torus). For the micro-benchmarks, to particularly show the effect of network-aware mapping, we put two subsequent nodes on different switches. As shown in Fig. 2, for all topologies the weighted graph shows significant improvement compared to the block mapping. It is because with a weighted graph the library differentiates between heavier and lighter dimensions and will try to map the processes of the heavier dimension on one node to take advantage of shared-memory performance. On the other hand, the non-weighted Cartesian topology does not make any differentiation between different dimensions. Consequently, it may map the heavier dimension across the network leading to worse performance. The other observation is the improvement when using network-aware mapping, especially for larger message. It shows that even when the difference in network distance is as little as two switches we can observe some improvement.

The second micro-benchmark (*dimensional collectives*) examines the case where processes form a Cartesian arrangement and perform collective communications on one dimension of the topology. We arrange 32 processes in an 8×4 2D-torus and engage them in MPI_Alltoall collective operations on the longer dimension. Fig. 3 clearly shows the improvement when using topology-aware mapping. The reason for on par performance between weighted graph and non-weighted graph is that the communication is done only on one dimension; therefore both graphs result in the same mapping. The network-aware mapping also shows similar improvement, because all eight processes of the collective dimension are mapped to the same node.

Fig. 3 also shows the results for a 16×2 2D-torus. These results are reported since the longer dimension (on which collectives are performed) has the length of 16, which does not fit into one node, therefore network communication is inevitable even after topology-aware remapping. As the results suggest, network-aware mapping shows improvement compared to other graph mappings, especially for larger message sizes.

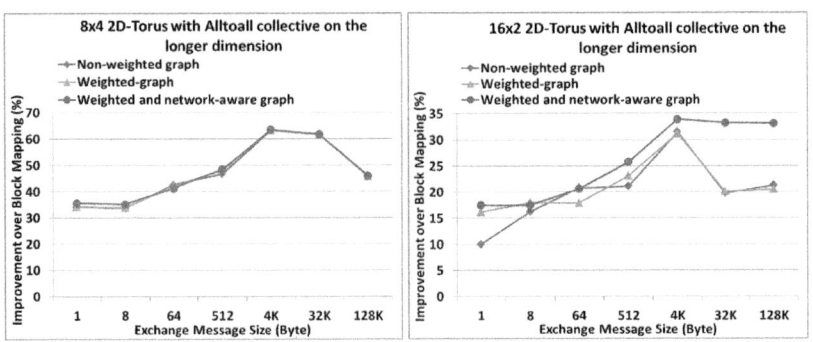

Fig. 3. Runtime improvement of topology-aware mapping over block mapping for 2D-torus in the dimensional collective (Alltoal) micro-benchmark

5.2   Evaluation Results for MPI Applications

To see the effect of our implementation on MPI applications, we have adapted some MPI application benchmarks (NAS and LAMMPS) to use MPI graph topology function. We profiled the original applications to discover their virtual topology graph in order to supply them to MPI_Graph_create function. In LAMMPS, processes communicate in a 2D-torus. Processes in CG.32 also form a logical 8×4 2D-torus, however they do not always follow the torus links for communication. Processes in MG.32 communicate in the form of a reordered 5D-hypercube. To be consistent, regardless of application's logical structure, we always use a graph topology function.

Fig. 4 shows the improvement of the topology-aware mapping for these applications compared to block and cyclic mappings on the 32-core cluster. This excludes the time in MPI_Graph_create to create the communicator. For most applications, the benefit over cyclic mapping is considerable while there is less benefit over block mapping. This is because processes mostly communicate with their adjacent ranks, thus the ideal mapping is close to block mapping. In LAMMPS (especially the *friction* workload), where the communication volume is not equal on different links, we can see more improvement with weighted and network-aware graphs, compared to non-weighted graph where sometimes it is even worse than the block mapping. We are currently investigating the reasons behind the poor performance of the weighted graph results for LAMMPS-Couple.

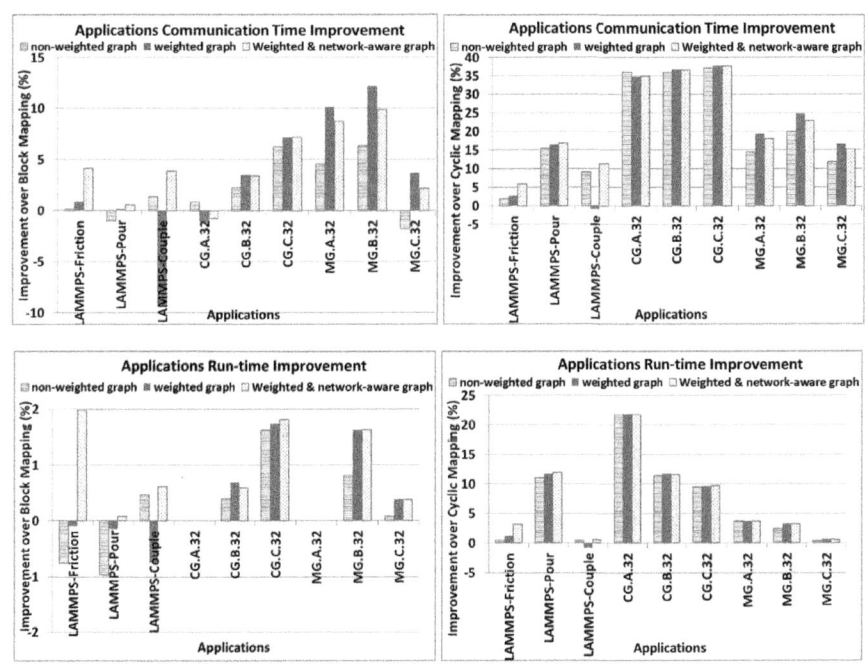

Fig. 4. Application communication time and runtime improvement of topology-aware mapping over block and cyclic mappings on the 32-core cluster

To show the benefits of our design on a larger test bed, we have also evaluated our work on a second cluster with 16 nodes (using 8 cores per node, for a total of 128-cores) using some of the application workloads. The results are presented in Fig. 5.

In Fig. 5 we observe more improvement for most of the workloads, which indicates the performance scalability with the machine size. Higher improvement in the 128-core case for some workloads such as LAMMPS-*friction* is partially because some of the neighbors that would fall on the same node in block mapping in 32-core case fall on different nodes in 128-core case. Therefore reordering is more effective for the latter case. LAMMPS-*couple* shows a considerable difference in communication pattern between 32-core and 128-core cases. While for 32 cores, processes communicate to their neighbors almost symmetrically, the pattern becomes asymmetric in 128-core case (a process mostly communicates with two partners). Thus we see higher difference between non-weighted and weighted/network-aware results for 128 cores. Such behavior is also observed for *pour* workload to some extent, leading to more improvement compared to block mapping.

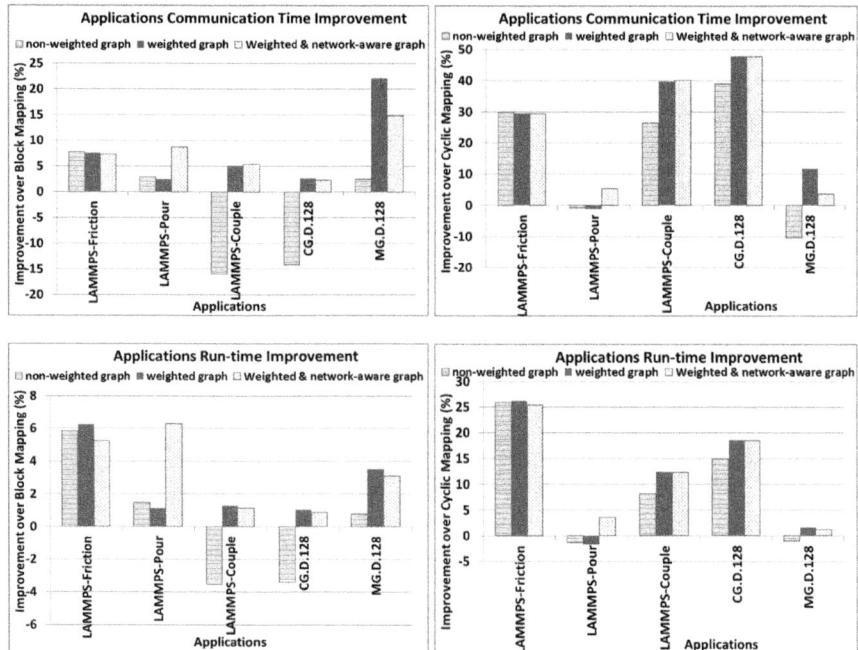

Fig. 5. Application communication time and runtime improvement of topology-aware mapping over block and cyclic mappings on the 128-core cluster

5.3   Implementation Overhead

Our implementation of MPI_Graph_create imposes an overhead compared to the trivial implementation. Table 1 shows the approximate overhead and its scalability in LAMMPS application. This one-time overhead is amortized in application runtime.

Table 1. Time to create the communicator in MPI_Graph_create for LAMMPS

System	Job size (#processes)	Trivial (ms)	Non-weighted Graph (ms)	Weighted Graph (ms)	Network-aware Graph (ms)
	8	0.3	7.3	7.3	7.9
Cluster 1	16	0.3	7.6	7.7	8.1
	32	0.5	8.6	8.7	9
Cluster 2	128	5.1	31.3	31.7	31.7

6 Conclusions and Future Work

In this paper, we presented design and implementation of MPI non-distributed graph and Cartesian functions in MVAPICH2 for multi-core nodes connected through multi-level InfiniBand networks. The Cartesian-model micro-benchmarks show that the effect of reordering process ranks can be significant, and when the communication is heavier on one dimension the benefits of using weighted and network-aware graphs (instead of non-weighted graph / Cartesian functions) are considerable. We also modified some MPI applications with MPI_Graph_create. The evaluation results show that MPI applications can benefit from topology-aware MPI_Graph_create.

As for the future work, we intend to evaluate the effect of topology awareness on other MPI applications, design a more general communication cost/weight model for graph mapping, and design and implement MPI distributed topology functions for more scalability.

Acknowledgments. This work is supported in part by Natural Sciences and Engineering Research Council of Canada, Canada Foundation for Innovation, Ontario innovation Trust, U.S. Department of Energy and National Science Foundation. We thank Mellanox Technologies and HPC Advisory Council for the resources.

References

1. IESP: International Exascale Software Project, http://www.exascale.org/
2. MPI Forum: MPI: A Message-Passing Interface Standard, version 2.2 (September 2009)
3. IBTA: InfiniBand Architecture Specification (2007),
    http://www.infinibandta.org/
4. Hatazaki, T.: Rank reordering strategy for MPI topology creation functions. In: Alexandrov, V.N., Dongarra, J. (eds.) PVM/MPI 1998. LNCS, vol. 1497, pp. 188–195. Springer, Heidelberg (1998)
5. Träff, J.L.: Implementing the MPI Process Topology Mechanism. In: ACM/IEEE Conference on Supercomputing, pp. 1–14. IEEE CS, Los Alamitos (2002)
6. Träff, J.L.: SMP-aware Message Passing Programming. In: 17th IEEE Parallel and Distributed Processing Symposium (IPDPS), p. 56. IEEE CS, Washington (2003)
7. Berti, G., Träff, J.L.: What MPI could (and cannot) do for mesh-partitioning on non-homogeneous networks. In: Mohr, B., Träff, J.L., Worringen, J., Dongarra, J. (eds.) PVM/MPI 2006. LNCS, vol. 4192, pp. 293–302. Springer, Heidelberg (2006)
8. Mercier, G., Clet-Ortega, J.: Towards an Efficient Process Placement Policy for MPI Applications in Multicore Environments. In: Ropo, M., Westerholm, J., Dongarra, J. (eds.) PVM/MPI. LNCS, vol. 5759, pp. 104–115. Springer, Heidelberg (2009)
9. Pellegrini, F.: Scotch and libScotch 5.1 User's Guide. Bacchus team, INRIA Bordeaux Sud-Ouest (2010),
    https://gforge.inria.fr/docman/view.php/248/5709/scotch_user5.1.pdf
10. Jeannot, E., Mercier, G.: Near-Optimal Placement of MPI processes on Hierarchical NUMA Architectures. In: Proceedings of EuroPar 2010 Conference, Italy (2010)

11. Hoefler, T., Rabenseifner, R., Ritzdorf, H., de Supinski, B.R., Thakur, R., Träff, J.L.: The Scalable Process Topology Interface of MPI 2.2. J. Concurr. Comp.-Pract. E., vol. 23(4), pp. 293–310. John Wiley & Sons, Ltd., Chichester (2010)
12. Bhatele, A., Gupta, G., Kale, L.V., Chun, I.H.: Automated Mapping of Regular Communication Graphs on Mesh Interconnects. In: 17th International Conference on High Performance Computing (HiPC). IEEE CS, Washington (2010)
13. MVAPICH: MPI Over InfiniBand, 10GigE/iWARP and RoCE, http://mvapich.cse.ohio-state.edu/
14. Broquedis, F., Clet-Ortega, J., Moreaud, S., Furmento, N., Goglin, B., Mercier, G., Thibault, S., Namyst, R.: hwloc: A Generic Framework for Managing Hardware Affinities in HPC Applications. In: Proceedings of the 18th Euromicro International Conference on Parallel, Distributed and Network-Based Processing (PDP 2010), Italy (2010)
15. Mellanox Technologies, http://www.mellanox.com/

Scalable Node Allocation for Improved Performance in Regular and Anisotropic 3D Torus Supercomputers*

Carl Albing[1,2], Norm Troullier[2], Stephen Whalen[2,3], Ryan Olson[2], Joe Glenski[2], Howard Pritchard[2], and Hugo Mills[1]

[1] University of Reading, Reading, Berkshire, UK
[2] Cray Inc., Saint Paul, MN, USA**
[3] University of Minnesota, Minneapolis, MN, USA

Abstract. MPI application performance can vary based on the scheduler's placing of ranks, whether between nodes or on cores in the same multi-core chip. MPI applications, by default, are at the mercy of the application placement software decision that assigns nodes to a job. We describe herein the general approach of node ordering for allocation in a 3D torus, how it improved MPI application performance, even in the face of an anisotropic interconnect. We demonstrate, quantitatively, that our topologically-based ordering results in improved performance for several MPI applications running on a Top10 supercomputer.

Keywords: resource management, application placement, scheduling, topology placement, software performance.

1 Introduction

"Mapping tasks of a parallel application onto physical processors of a parallel system is one of the very essential issues in parallel computing. It is critical for today's supercomputing system to deliver sustainable and scalable performance." - Hao Yu, et al. [14]

Large scientific parallel applications may use thousands or tens of thousands of cores when they run, but even so such jobs are only a fraction of the total machine size on today's largest HPC systems. Maintaining high utilization is important for these very large (i.e., costly) machines. While allocations of contiguous blocks of nodes seems intuitive, utilization can suffer. This leads to

* This material is based upon work supported by the Defense Advanced Research Projects Agency under its Agreement No. HR0011-07-9-0001. Any opinions, findings and conclusions or recommendations expressed in this material are those of the author(s) and do not necessarily reflect the views of the Defense Advanced Research Projects Agency.
** We are very grateful for the help of many people in the benchmarking, MPI, ALPS and support groups within Cray Inc.

Y. Cotronis et al. (Eds.): EuroMPI 2011, LNCS 6960, pp. 61–70, 2011.
© Springer-Verlag Berlin Heidelberg 2011

assignment of individual, non- contiguous nodes. This, in turn, introduces performance variabilities as a job may get a very different set of nodes and have different neighboring applications each time it is run.

Improved performance of MPI applications can be achieved through application placement in massively parallel 3D-torus supercomputers using allocation strategies based on an ordered, one-dimensional sequence of nodes. We are able to compute these new orderings with minimal overhead at system startup and thus the system software that is responsible for placement incurs no run-time penalty in its placement decisions and no user action is required to see this benefit.

The effects were significant enough [4] that they have been incorporated into product features for Cray's Application Level Placement Scheduler (ALPS). The effects may vary enough by machine size and typical job types that the feature was made configurable on a per system basis.

We had devised what we hoped would be a general solution benefiting most applications. We began from the simple approach of assigning nodes in order from a list of available, unassigned nodes (a "free list"). We ordered that list according to dimensions of the torus. Our work has shown that application performance was improved over random placement or other orderings. The amount of that effect depended on both the application and the "size" (dimensions) of the underlying torus. We devised an improved placement approach which doubled the bisection bandwidth of collections of nodes, described below. We used this solution successfully on several Cray installations.

We encountered a new system which performed well with our node ordering at small scale. When we enlarged the system to a petaFLOPS scale system, the previous mix of applications performed poorly. This was investigated and we came to understand that the problem grew from system scale in combination with our placement approach. The situation was further complicated by an anisotropic interconnect. Anisotropic interconnects exhibit transfer rates which differ depending on the direction traversed. In a 3D torus this means network traffic will travel at different speeds in each of the dimensions: x, y, and z. After analysis, discussion and prototyping, we devised a new ordering for the nodes on the free list to address the scaling issues. With our new ordering we re-ran the benchmarks to find restored or improved performance.

This paper makes the following contributions: we describe the general approach of node ordering for allocation and placement, our attempt at an ordering to improve bisection bandwidth, how this succeeds at small scale but how it fails at larger scale; we describe the design and implementation of a new ordering for dealing with a 3D torus at large scale; we describe how we balance the trade-offs between large and small jobs, considering intra-job communication while seeking to minimize job-to-job interaction to provide a broadly useful if sub-optimal solution; we describe the adjustment we made for the anisotropic interconnect; we demonstrate that this topologically-based ordering results in improved scaling for a variety of parallel applications running on a Top10 supercomputer.

The rest of the paper is organized as follows: Section 2 gives brief explanation of various techniques that have been used for the placement of applications within a torus with a discussion of their limitations; Section 3 presents our approach to balance the various trade-offs for optimal placement; Section 4 describes issues related to the anisotropisms for Cray XE systems; Section 5 presents a comparison of run times on real MPI applications to demonstrate the effectiveness of our new topology node ordering in placement scheduling; finally, Section 6 concludes the paper with a summary of our results and future directions of our investigations.

2   Related Work

There are some similarities between application placement and memory allocation schemes. The classic First Fit and Best Fit algorithms [12,6] are reasonable approaches for both [5]. However, node allocation differs significantly in the topologies from which it allocates; memory is a linear resource whereas nodes in parallel systems are typically multi-dimensional (e.g. 3D torus, hypercube).

Laxmikant Kalé and others make the argument that earlier work done for clusters of tens or hundreds of processors is insufficient to cover peta–scale and exa–scale systems. In 2009 Bhatele and Kalé presented "... a study showing that with the emergence of very large supercomputers, typically connected as a 3D torus or mesh, topology effects have become important again." [8]

Some researchers have approached the problem of node selection, or placement, from the application's point of view, attempting to place each particular application in a space best suited to the topology of the application's algorithms [3,7]. Such an approach may involve exposing application topology (by the developer or discovered by the compiler or other tools) to a user, who may not want to know or specify such arcane details, or it may involve prior runs of the application to detect communication patterns. It also involves attempting to fit the application as an arbitrary, fixed 3D shape onto the available resources.

Focusing on the BlueGene/L system, Krevat, et al. [9], have defined the problem away by assuming that the only way to assign nodes for a toroidal system "must be rectangular and contiguous". Systems that implement such approaches then either suffer potentially low utilization or pay the high cost of process migration between nodes in order to rearrange work in contiguous blocks. Lo et al. [11] demonstrate that better throughput can be achieved with non–contiguous allocation. They make their placements, though, in $2n$ size blocks, trying to allocate the largest block possible. For them the allocation of single nodes is the degenerate case of 2^0 size blocks.

Leung et al. [10] theorized that one-dimensional allocation strategies could be effective for massively parallel supercomputers. They proposed Hilbert curves in two dimensions for assigning node ordering. (Hilbert curves are continuous space-filling curves [1] described by German mathematician David Hilbert, and have properties of locality that make them useful in this and other spacial applications.) We have seen limitations to the Hilbert curve ordering both because of the shape of the curve itself and because it is defined for squares (in 2D) and cubes (in 3D) whereas we need to apply it in systems of unequal dimensions.

That an improved ordering was needed for Cray XT torus systems was demonstrated in the work of Weisser et al. [13]. They, too, used a linear ordering, specific to the shape of their particular system.

Yu et al., [14] use embedding techniques like folding to map from one topology (the application's virtual topology) onto the other (the underlying hardware topology). They do this so that at the application level a mapping (using MPI calls) can be made onto the allocated hardware nodes. Our approach could be described as attempting to do the mapping without knowing the application's (virtual) topology. Such knowledge, after all, may not be available to the application user. Moreover, our approach requires no changes to the application's source code and even provides a certain amount of hardware independence.

3   Our Approach

Whereas "first fit" algorithms require a block of contiguous resource at least as large as the requested resource, we do not require contiguity of nodes in an allocation. We simply assign nodes from a free list in order, skipping nodes allocated to other jobs, until the request is filled. The "in order" initially meant in numerical order by node id (nid). The nid numbers were assigned by lower level software based on physical position within the machine. This did not necessarily correspond to the network topology.

Cray XT and XE networks use a "folded torus" where cabinets are interleaved in the x dimension. Thus cabinets that were physically adjacent would have nid numbers in adjacent ranges, but their locations in the torus would be antipodal. The extra hops required to reach the nodes in these cabinets lead to increased latencies. Moreover, since an allocation of sequential nodes could involve "gaps" in the torus locations, there was an increase in jobs whose communications traveled between and among other jobs. This job-job interaction meant that one job could affect the performance of other jobs and any job's performance could vary from run to run depending on what other jobs were running at the same time.

3.1   First Attempt - MDF

Our first attempt at improving job placement was targeted at reducing the "jumps" in the network locations due to the nid numbering. We changed the order of the free list, building it not in numerical order but based on (x, y, z) location. We ordered the nodes along the shortest axis first, calling this the Minimum Dimension First (MDF) ordering. This increases the number of jobs that could use "wraparound" in the torus, reducing their latencies. MDF showed improvement over a few other orderings as well as the original numerical ordering [4].

On larger systems where no dimension was much smaller than any other, much of this benefit was lost. Consider a job that uses 8192 cores. On a quad–core XT4 system such a job needs 2048 nodes. Even on a very large system (for its time) these 8192 cores would be spread across 22 cabinets or four yz planes in the torus. By contrast, on the newer XE6 system with 24-core nodes that same job uses only 342 nodes, less than four cabinets, all in a single plane.

Increasing the number of cores per node results in a decreasing number of nodes per job. The newer, higher density nodes result in a placement with fewer planes in the torus for a given job. The previous example shows how the three-dimensional allocation (4 full planes of nodes) is reduced to a single 2D plane. Smaller jobs, less than 3000 cores, can be reduced to a one dimensional allocation.

3.2   Increasing Bisection Bandwidth

With such "flattening" of the placement, the bisection bandwidth of the job is reduced. We tried a different ordering of the nodes on the free list to try to "thicken" our allocations. To get even the smallest jobs into the third dimension we ordered the nodes so that allocations would occur along 2x2x2 blocks of nodes. We did not allocate block–size units of nodes, but rather the ordering followed a sequence that traced out a 2x2x2 block then another and so on. Within the 2x2x2 shape we follow a Gray code ordering which allows a contiguous enumeration from one block to the next along the first filled dimension. See Fig. 1.

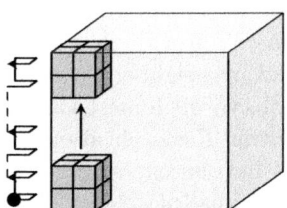

Fig. 1. Node ordering in 2x2x2 blocks

We used this "thicker" ordering on a new Cray XE6 system installation. We ran the benchmarks described below in Sect. 5 and found our baseline performance. Then the installation was expanded three fold, to over 6000 nodes or more than 150,000 cores. The same benchmarks, run on this new larger system, showed degraded performance. See Sect. 5.

This new system was using the new XE6 ("Gemini") interconnect. A few words about its features are needed at this point. The XE6 supports two nodes per router chip. The x and z dimensions have double links, one corresponding to each node; the y dimension has a single link to the neighboring router chip since the two nodes within the router can be considered connected across the y axis. See Fig. 2. Therefore in our node ordering for XE6 systems we always ordered y last, even if it was the smallest dimension, since it would have the least bandwidth to a neighbor. Later, below, we will discuss the anisotropic considerations of the XE6.

What we found with the additional cabinets and the new ordering was that the x dimension had grown so large as to eliminate the torus wrap around for

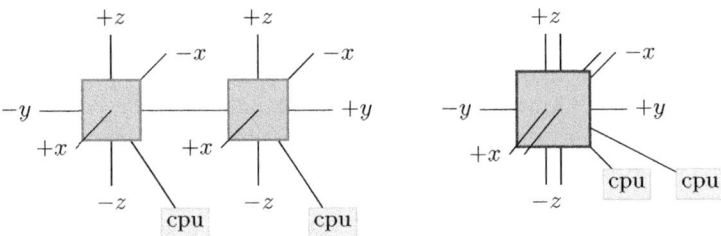

Fig. 2. Comparison of XT SeaStar(left) and XE Gemini(right) interconnects - notice the effective doubling of connections in X,Z dimensions for Gemini

our applications. We could have adjusted our block sizes or choice of axis to increase the chances of wrap around, but that would only be a fix until the next size increase left us with a more planar or non-toroidal placement. We decided to rethink our ordering.

3.3   Towards Optimal Placement

How does one define optimal placement? The objective measure for optimal placement is best performance, measured as least runtime. But what arrangement of the nodes of the job will provide that performance? That will vary from job to job. It is also affected by neighboring jobs. We are looking for a generic solution that will fit most jobs through most job mixes most of the time. Further refinements may be done within the application through mechanisms like MPI's rank reordering, but that is beyond our scope.

We submit that a truly optimal placement algorithm will consist of competing trade-offs: a) minimize job-job interactions, b) minimize intra-job latency, c) maximize bisection bandwidth for both large and small jobs, d) provide consistent job run times, all while e) providing high utilization.

To see why satisfying all these may be impossible, consider the trade-offs. We could maximize bisection bandwidth by spreading the nodes of the job across the machine. Place them far enough apart so they can make use of the torus' wrap-around effect, thereby using additional routes for inter-node traffic and increasing bisection bandwidth. However, with this arrangement latency is made worse by the larger distance to travel and job-job interaction is maximized.

We could minimize job-job interaction by allocating nodes in planes (like MDF ordering). All intra-job communication would stay within the plane, avoiding job-job interaction except on the edges of an allocation. Yet such a planar allocation gives poor bisection bandwidth.

We can minimize intra-job latency with tight placement — the shortest distance between every node in a job. In a 3D torus, this is equivalent to the shortest distance between a set of points in 3-space. The optimal (minimal) theoretical placing would be a ball. The shape of the "ball" will depend on the metric used, in this case, a [weighted] 1-metric, known as Manhattan distance. This means that the containing ball is geometrically octahedral. While that may be fine for minimal distance (and its implication for all-to-all communication), that may not be optimal for applications that do nearest neighbor computation. The nodes farthest out on the octahedron can be further from neighbors. Such a shape may provide the optimal theoretical bisection bandwidth for a single application, but with multiple applications on a system, it is worth noting that octahedrons do not pack well.

Moreover, a 3D torus is neither ball nor octahedral. The nearest approximation would be a cube. Within the cube, a Hilbert curve provides good space-filling density — but only for perfect cubes. Even then, the curve sometimes produces multiple hops in the same dimension which we wish to avoid as that represents reduced bisection bandwidth between those nodes in the sequence. We attempt to illustrate this in the diagrams, below. (Our work was with a 3D torus but the figures show the easier–to–draw 2D Hilbert curve and an alternative ordering. The intent here is to motivate the ideas behind our ordering while not specifically enumerating the ordering we chose.) In Fig. 3, note the run of: (1,2), (1,3), (1,4) on the left hand side of the diagram. Placing an application along such a run would result in reduced bisection bandwidth.

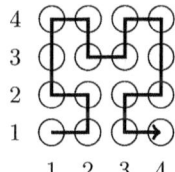

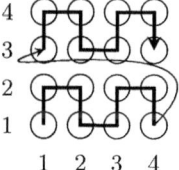

Fig. 3. Hilbert curve on 4x4 grid

Fig. 4. Alternate Weave with toroidal wrap

These orderings certainly reach all the nodes but may result in job-job interaction as nodes "talk across" each other to get to nodes within their respective jobs. For example, in Fig. 3, the nodes at (2,2) and (4,2) might be part of the same job, but if (3,2) is the start of another job, then there will be job-job interaction.

Our compromise on these trade-offs was to emphasize latency improvements by ordering the nodes to keep them close together in the torus, but also to consider bisection bandwidth on how those nodes are ordered to avoid planar layouts. We arrived at our overall solution by dividing the ordering into two sub-problems. First we attempted to define a compact ordering for a subset of nodes. We then used this subset as a template for ordering the entire machine.

Our experience has shown us that most user applications typically use 12000 cores or less. That represents 512 nodes on a 24-core/node system. We took that size, which is a cube of 8x8x8 nodes, as a starting point. By ordering the nodes within that cube for closeness and bisection bandwidth we could make placement along that ordering result in good performance locally, that is, for individual jobs. Our approach was to turn in the path whenever possible. This increases connectivity and bisection bandwidth and lowers hop counts.

We chose our first and last nodes in that cube to be directly above and below each other in the cube. We could use that cube then as a basic building block (BBB), connecting one to the next with no large discontinuities (but see Sect. 4, below).The order of the entire torus could then be planned in terms of the layout of those building blocks. Since our XE6 routers support two nodes per location in the torus, this would be 8x4x8 in terms of (x, y, z) coordinates. For any size system we fill it with BBB units ordering them along the z then y then x axis.

4   Some Issues

But now the anisotropic considerations need to be considered. The y dimension is best at only 4 nodes across (staying on board). Each XE6 Gemini supports two nodes per network interface, only two (x, y, z) locations across. In torus coordinates, this leaves us an 8x2x8 building block. With higher speed in the z dimension there is less distance we can travel in the x dimension in the same time, so to keep it balanced or "cubical", we used 4x2x8 as our BBB size. Since x is the only odd dimension in Cray systems, we may use a BBB with 3 or 5 as the x dimension. This avoids being left with a single 1x2x8 "slice", a planar arrangement with lower bisection bandwidth.

The building blocks allow us to connect from one to another in a single hop except at the point where we have filled a dimension. These locations involve a move to the start of the next block in an adjacent column, adding an additional hop in network flow for jobs that span these discontinuities. A small application placed astride such a spot will experience larger latencies.

5   Placing Real Applications

To measure the effectiveness of our topology node ordering in placement scheduling the following applications from a variety of scientific disciplines were used: CAM, GAMESS, GTC, IMPACT-T, MAESTRO, MILC, and PARATEC. Except for GAMESS, all the others are pure MPI applications. Gamess is a hybrid MPI application. It uses MPI for collectives, but uses a onesided messaging library called DDI for PUT/GET/ACC/FADD operation. More details about all of these codes can be found online [2].

We ran these benchmarks on a 28-cabinet system using our 2x2x2 ordering. After expanding the system to 68 cabinets we ran the benchmarks again. The third run is with a slightly modified ordering of 3x2x2 to avoid a planar ordering on the odd dimension. Finally we ran with the newer BBB ordering. That ordering was a mix of BBB sizes of 4x4x8 and 5x4x8. (We use the node count for each dimension here rather than the xyz dimension to match the 2x2x2 terminology.) The results are shown in Table 1.

The initial run times on the smaller 28 cabinet system are the baseline to which we compared the larger runs. The initial run on the 68 cabinet system showed fractionally better runtime on CAM and GTC, slightly worse runtime on ImpactT and PARATEC, but significantly worse runtime for GAMESS and

Table 1. Wall clock run times, in seconds

Benchmark	28 cab 2x2x2	68 cab 2x2x2	68 cab 3x2x2	68 cab 4|5x4x8
CAM	346	338	340	338
GAMESS	1386	1858	1572	1478
GTC	1234	1226	1226	1226
ImpacT	604	629	600	597
MAESTRO	1969	2233	2105	1982
MILC	909	1355	1237	919
PARATEC	386	413	395	383

MAESTRO and MILC. By enlarging the initial thicker ordering to a 3x2x2 block those latter codes only improved slightly. Not until we use the larger BBB with its internal weave ordering do we see performance return to the baseline for most codes.

The GAMESS code is still slightly worse (6.6%) though much better with the newer ordering than without at the larger scale. This is the quantum chemistry package used for atomic and molecular electronic structure calculations. For the benchmark in question (MP2 gradient), the dominant communication motif is independent one-to-all one-sided accumulate operations. Each process accumulates approximately the same amount of data to each and every other rank. All of the accumulate operations are independent, i.e. there are no synchronizations. The performance is directly related to the total aggregate bandwidth of each rank to every other rank. We suspect that the lower performance in the larger torus regardless of ordering may be due to the effect of additional bisection bandwidth in the smaller torus because of toroidal wrap around.

6   Summary of Results and Future Directions

Our general approach of node ordering for placement and working from a free list provides us with a zero-cost-at-run-time approach that keeps utilization high. We needed an ordering to improve bisection bandwidth and our first attempts succeed on smaller scale systems but failed at larger scale. The expansion of the system to more cabinets was the immediate cause but we could see that increasing core counts per node will have a similar effect — the average job size as a percentage of the torus dimension will decrease, with a corresponding loss of bisection bandwidth. We needed to revise our ordering further.

By balancing the trade-offs between large and small jobs, considering intra-job communication while seeking to minimize job-to-job interaction we came up with an ordering that was broadly useful if less than perfect. We used this two-part compromise ordering and saw our performance on the large scale return to the same speed as the original system.

We continue to investigate ordering trade-offs; we would like to know which is the better size block to use for odd-length dimensions. We have a simulation underway to see how long the effects of the ordering can be sustained. We also plan to combine these effects with rank reordering to measure relative contributions of each. We hope to extend this work to other interconnect topologies.

References

1. Hilbert curve – from wolfram MathWorld (March 2010),
   http://mathworld.wolfram.com/HilbertCurve.html
2. NERSC6 benchmarks (March 2011),
   http://www.nersc.gov/projects/SDSA/software/?benchmark=NERSC6
3. Agarwal, T., Sharma, A., Kal, L.V.: Topology-aware task mapping for reducing communication contention on large parallel machines. In: Proceedings of IEEE International Parallel and Distributed Processing Symposium, Rhodes Island, Greece, p. 110 (2006)
4. Albing, C., Baker, M.: ALPS, topology, and performance: A comparison of linear orderings for application placement in a 3D torus. Cray User Group, Edinburgh, Scotland, UK (May 2010)
5. Bani-Mohammad, S., Ould-Khaoua, M., Ababneh, I.: An efficient non-contiguous processor allocation strategy for 2D mesh connected multicomputers. Information Sciences 177(14), 2867–2883 (2007)
6. Bays, C.: A comparison of next-fit, first-fit, and best-fit. Communications of the ACM 20(3), 191–192 (1977)
7. Bhatele, A., Kale, L.V.: Application-specific topology-aware mapping for three dimensional topologies. In: 2008 IEEE International Symposium on Parallel and Distributed Processing, pp. 1–8. IEEE, Miami (2008)
8. Bhatele, A., Kal, L.V.: An evaluative study on the effect of contention on message latencies in large supercomputers. In: 2009 IEEE International Symposium on Parallel & Distributed Processing, Rome, Italy, pp. 1–8 (May 2009)
9. Krevat, E., Castaos, J., Moreira, J.: Job scheduling for the BlueGene/L system. LNCS, pp. 38–54. Springer, Edinburgh (2002)
10. Leung, V.J., Arkin, E.M., Bender, M.A., Bunde, D., Johnston, J., Lal, A., Mitchell, J.S., Phillips, C., Seiden, S.S.: Processor allocation on Cplant: achieving general processor locality using one-dimensional allocation strategies. In: Proc. 4th IEEE International Conference on Cluster Computing, pp. 296–304 (2002)
11. Lo, V., Windisch, K., Liu, W., Nitzberg, B.: Noncontiguous processor allocation algorithms for mesh-connected multicomputers. IEEE Transactions on Parallel and Distributed Systems 8(7), 712–726 (1997)
12. Russell, J.J.: A simulation of first and best fit allocation algorithms in a modern simulation environment. In: Proc. of 6th Annual CCEC Symposium (2008)
13. Weisser, D., Nystrom, N., Brown, S., Gardner, J., O'Neal, D., Urbanic, J., Lim, J., Reddy, R., Raymond, R., Wang, Y., Welling, J.: Optimizing job placement on the Cray XT3. Lugano, Switzerland (May 2006)
14. Yu, H., Chung, I., Moreira, J.: Topology mapping for blue Gene/L supercomputer. In: Proceedings of the 2006 ACM/IEEE Conference on Supercomputing, SC 2006, Tampa, Florida, p. 116 (2006)

Improving the Average Response Time in Collective I/O

Chen Jin[1], Saba Sehrish[1], Wei-keng Liao[1],
Alok Choudhary[1], and Karen Schuchardt[2]

[1] Northwestern University, Evanston, IL 60202, USA
[2] Pacific Northwest National Laboratory, Richland, WA 99352, USA
{chen.jin,ssehrish,wkliao,choudhar}@eecs.northwestern.edu,
Karen.Schuchardt@pnnl.gov

Abstract. In collective I/O, MPI processes exchange requests so that the rearranged requests can result in the shortest file system access time. Scheduling the exchange sequence determines the response time of participating processes. Existing implementations that simply follow the increasing order of file offsets do not necessary produce the best performance. To minimize the average response time, we propose three scheduling algorithms that consider the number of processes per file stripe and the number of accesses per process. Our experimental results demonstrate improvements of up to 50% in the average response time using two synthetic benchmarks and a high-resolution climate application.

1 Introduction

Parallel I/O systems have always faced challenges to efficiently store and retrieve the ever-growing amount of data. Over the past two decades, researchers have proposed different solutions, such as MPI I/O, to improve the performance. MPI collective I/O requires the participation of all processes that open a shared file. This requirement provides a collective I/O implementation an opportunity to exchange access information and reorganize I/O requests among the processes. Several process-collaboration strategies have been proposed, such as two-phase I/O [1], disk directed I/O [2], and server-directed I/O [3].

Two-phase I/O is a representative collaborative I/O technique that runs in the user space. Its idea is to reorganize the requests among processes, so that the rearranged requests incur the minimal overhead from the underlying file system. The request reorganization is referred to as the communication phase while the read/write system calls constitute the I/O phase. ROMIO, a popular MPI-IO implementation [4], adopts the two-phase I/O strategy [5], which first identifies the aggregate access region and picks a subset of MPI processes as the I/O aggregators, the only processes making I/O calls to the file system. The aggregate access region is a minimum contiguous file region covering all the I/O requests. The region is partitioned into disjointed sub-regions denoted as **file domains**, and each is assigned to a unique I/O aggregator.

Y. Cotronis et al. (Eds.): EuroMPI 2011, LNCS 6960, pp. 71–80, 2011.

MPI collective operations only require the participation of all processes, which should not be confused with the process synchronization. Essentially, once a process participates in a collective operation, it can return from the call without waiting for the completion of other processes. Hence, an optimal request scheduling method for a collective operation should minimize the average response time of all processes. The I/O request scheduling is a key component of the communication phase in the collective I/O implementation. Let us take ROMIO's implementation for the Lustre file system as an example to examine the importance of a scheduling strategy. At the file open, the Lustre driver chooses an equal number of I/O aggregators as the number of the file servers, referred to as Object Storage Targets (OSTs) in Lustre. All stripes of a file are assigned to the aggregators' file domains in a round-robin fashion in order to produce a one-to-one mapping between the aggregators and file servers [6]. Because Lustre adopts an extent-based locking protocol, such assignment can optimally minimize the lock request from each aggregator to the file system [7,8].

In the current ROMIO implementation, each aggregator handles the requests exchange for all the stripes in its file domain. That is, it schedules one stripe at a time in the increasing file offset order of the stripes. We argue that such service scheduling strategy does not necessarily result in the best response time for the non-aggregators. The example in Figure 1(a) shows a collective write from 4 MPI processes to 3 aggregators. The service scheduling using the increasing file offset order is shown in Figure 1(b), where the requests from P_3 are served last as they have the highest offsets. In this case, the average response time is $(4t \times 3 + 5t)/4 = 4.25t$. In Figure 1(c) where P_3's requests are served first, the average response time is reduced to $(t + 5t \times 3)/4 = 4t$. The

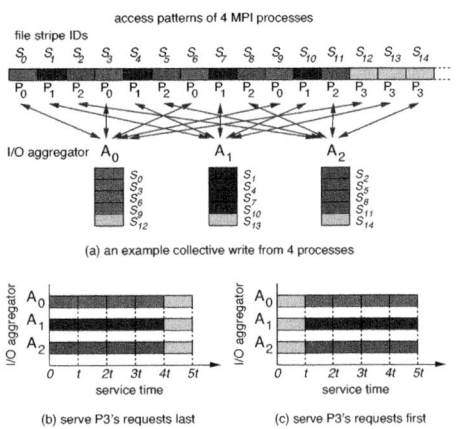

Fig. 1. (a) Collective I/O example with 4 MPI processes and 3 OSTs. (b) Average response time is 4.25t when P_3's requests are served last. (c) Average response time is 4t when P_3's requests are served first.

faster response time means the processes can return earlier from the call and proceed to the successive tasks. This example shows that a different request scheduling order can result in a different average response time and serves as the motivation of our work.

We propose three alternative algorithms for request scheduling: Most Degree First (MDF), Locally Weighted MDF (LW-MDF), and Globally Weighted MDF (GW-MDF). These algorithms prioritize the file stripes based on their access degree, the number of accessing processes. The MDF schedules the stripe with the highest degree to be served first. The LW-MDF assigns a weight to each process using its total number of requests to individual aggregators. The GW-MDF assigns the weight based on all the local weights of a process across the aggregators. The weighted schemes are used to calculate the priority scores for the stripes. Our experiments on the Cray XT4 parallel machine and Lustre file system at the National Energy Research Scientific Computing Center show that the average response time is reduced by 30% for a fixed uneven workload, 50% for a random workload, and 20% for a large-scale climate simulation application.

2   Design and Implementation

Our objective for developing the three alternative scheduling algorithms is to minimize the average service response time, $\overline{T_c}$, of all the MPI processes in a collective I/O. We take into consideration the accessing **degree** of a file stripe, which is defined as the number of processes accessing it. The proposed algorithms do not change the I/O amount on the aggregators and if the cost of I/O phase dominates the collective I/O, then the time on the aggregators will not change significantly. What the proposed algorithms intend to improve is the response time mainly on the non-aggregators. We assume the same cost for the I/O phase irrespective of the stripe permutations carried out to the file system. The request size per aggregator is same as the stripe size, which is between 1MB and 4MB. The stripe size is a multiple of the disk sector size (512 bytes), hence, it will not affect the disk seek time. Based on our experiments on Lustre, the I/O for different stripe permutations costs approximately the same, as long as each aggregator only accesses the same server.

Most Degree First (MDF). Among all the stripes in an aggregator's file domain, the MDF method schedules the stripe with the highest degree first. Intuitively, if the stripes with larger degree stripes are serviced first, then more non-aggregator processes will complete their collective I/O earlier. In ROMIO, at the beginning of a collective I/O, the request information of all processes is made available to all the aggregators. Hence, with MDF each aggregator can calculate and sort the stripes in its file domain independently from other aggregators. Once the scheduling is determined, the two phases are carried out alternatively, one stripe at a time.

Solely utilizing the access degree may not always give the minimal $\overline{T_c}$. For example, the first three stripes of a file are written by P_0 and each of the successive 12 stripes is written by P_1, P_2 and P_3, in an interleaving manner. If there are three aggregators, then each has a file domain consisting of 5 stripes in which the access degree is 1 for the first stripe, and 3 for the remaining four stripes, as illustrated in Figure 2(a). In the MDF algorithm,

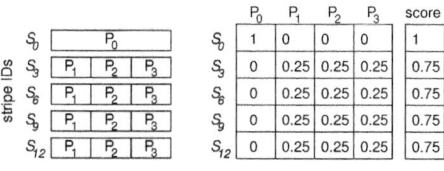

(a) access pattern on A_0 (b) stripe priority score matrix

Fig. 2. An example access pattern and the weight score assignment by the LW-MDF method

P_0's requests are served last and $\overline{T_c} = (5t + 3 \times 4t)/4 = 4.25t$. However, if P_0's requests are served first then $\overline{T_c} = (t + 3 \times 5t)/4 = 4t$. Hence, solely depending on the stripe's degree is not sufficient to achieve the best response time. We propose two additional algorithms with weighted schemes.

Locally Weighted Most Degree First (LW-MDF). As the assumption in the MDF method that each process has equal contribution to the stripe scheduling priority may not produce the best result, the LW-MDF method is designed to assign a weight to each process based on its number of requests in an aggregator's file domain. In each aggregator, the weight of a process is set to the inverse of the number of stripes accessed on that aggregator. For example, in Figure 2, the weight is 1 for P_0, and 0.25 for the others. The priority score of a stripe is then calculated as the sum of all process weights on that stripe. Note that the scores only depend on the local data access pattern on the aggregators. As a result, the LW-MDF method assigns the higher priority to stripe S_0 than other stripes.

In the LW-MDF method, the weights are calculated using only the local information on each aggregator. Consider a case that process P_0 accesses only two aggregators, for instance two stripes on aggregator A_0 and six stripes on aggregator A_1. The weights assigned to P_0 on both aggregators will be $\frac{1}{2}$ and $\frac{1}{6}$, respectively. If A_0 schedules P_0's stripes first but A_1 schedules P_0's stripes later, then P_0's collective I/O will not complete until the six stripe requests on A_1 are processed. In order to deal with this problem, we propose another variant of MDF algorithm that considers the weights of a process across all aggregators.

Globally Weighted Most Degree First (GW-MDF). When a process has a higher number of accesses to an aggregator, it makes little sense to schedule its requests first on other aggregators, as the response time of a process is determined by the slowest aggregator that serves its requests. The GW-MDF method selects the minimum of local weights of a process across all the aggregators to calculate the priority scores for stripes.

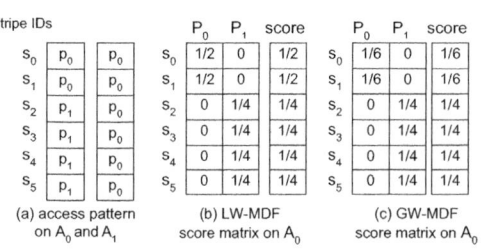

Fig. 3. An example access pattern and the weight score assignment by the GW-MDF method

Consider the example case presented in Figure 3. Process P_0 accesses two stripes on aggregator A_0, and six on aggregator A_1, the weight of P_0 on A_0 is $\frac{1}{2}$ according to LW-MDF and $\frac{1}{6}$ according to GW-MDF. There is another process P_1 that only accesses 4 stripes on aggregator A_0, the weight of P_1 on A_0 is $\frac{1}{4}$. In GW-MDF, the weight of P_1 is higher than P_0's weight ($\frac{1}{6}$), thus, P_1's requests are served first. The average response time $\overline{T_c} = (6t + 4t)/2 = 5t$. However, in LW-MDF, P_0 and P_1 will complete at the same time, and $\overline{T_c} = (6t + 6t)/2 = 6t$.

Therefore, GW-MDF achieves the optimal scheduling in this case. One thing to note, if the processor doesn't have any stripe to access, its score is assigned to zero rather than infinity. Our score algorithm only needs to consider the positive values to exclude the processes with zero score. For highly irregular or unbalanced data access patterns, it is anticipated that GW-MDF can outperform the other two MDF methods. However, in order to find the global minimums, there is an extra communication for gathering the local weights of all processes on every aggregator. This global communication among aggregators adds an overhead to the overall performance.

3   Performance Evaluation

All the proposed scheduling algorithms are implemented in the ROMIO library released along with MPICH version 1.3.2p1. By using two artificial benchmarks and a climate simulation application, our evaluation was carried out on Franklin, a Cray XT4 supercomputer at NERSC. Franklin consists of 9572 computer nodes, each of which runs a 2.3 GHz quad-core AMD Opteron processor and 8 GB memory. The parallel file system is Lustre version 2.2.48 with total of 48 OSTs. In our experiments, the stripe size was configured to 1 MB and the numbers of OSTs are set to 8 and 40 for the artificial benchmarks and GCRM evaluation, respectively. The performance results of MPI collective write operations are presented in this work.

Fixed Uneven Workload. We assign the first half of the processes twice the write amount as the other half. The access pattern is illustrated as an example shown in Figure 4(a), where the first half of the 128 processes write 40 MB of data each, and the second half writes 20 MB. The write amount on each process is further partitioned into 160 smaller pieces whose offsets are interleaved among all processes. For each piece in the first 64 processes, the size is $\frac{1}{4}$ MB, and for each piece in the second 64 processes is $\frac{1}{8}$ MB. This setting produces multiple noncontiguous file regions for each process to access and each stripe is accessed by more than one process. In addition, each process accesses the same number of stripes on each aggregator. This pattern implies that all MDF methods have the same weights and should have similar performance. The results presented in Figure 4(b) clearly demonstrate that the proposed scheduling methods outperform the traditional scheduling of the increasing file offset order. With the experiments running on up to 1024 processes, we observe that all four scheduling methods have the same wall time for the slowest processes. The slowest process is one of the aggregators whose file domain has the most stripes. All MDF methods show the similar improvement as the weights of a process, which contribute to the priority are the same for all stripes. In this case, the priorities are determined by the local access degrees. A reduction of up to 30% in average response time is obtained.

Random workload. Given a fixed file size, the random workload partitions the entire file into pieces with arbitrary lengths which are then assigned to processes based on the Gaussian distribution. Figure 5(a) shows the random workload

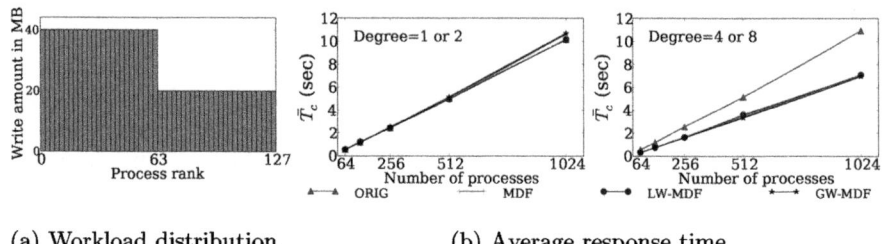

(a) Workload distribution   (b) Average response time

Fig. 4. Access pattern and performance results for the fixed uneven workload

distribution among 128 processes, and Figure 5(b) shows the corresponding histogram. Each process has a random amount of data to write and the processes with ranks close to the median are assigned more workload. Figure 5(c) and (d) illustrate the number of stripes per process and stripe access degree per process, respectively. Each piece of write segment is randomly assigned to a process such that the number of stripes accessed by a process obeys the Gaussian distribution. The stripe degree is also random because the size of each piece is arbitrary. The results in Figure 5(e) indicate that both LW-MDF and GW-MDF perform better than MDF by up to 50%. MDF does not have a noticeable improvement over the original scheduling as the number of stripe accesses in this random pattern varies significantly with the number of processes.

Global Cloud Resolving Model (GCRM). The GCRM is a climate application framework designed to simulate the circulations associated with large convective clouds [9]. Its I/O uses Geodesic Parallel I/O (GIO) library [10], which interfaces parallel netCDF (PnetCDF) [11]. In our experiments, we enable the PnetCDF non-blocking I/O option to aggregate multiple grid data variables into large-sized collective writes. Non-grid variables are excluded from our evaluation, as they are written individually, which does generate uneven file access degrees. There are 11 grid variables and each variable is approximately evenly partitioned among all the processes. We collected results for 3 cases: 640 processes with resolution level 9, 1280 processes with level 10, and 2560 processes with level 11. Resolution levels 9, 10, and 11 correspond to the geodesic grid refinement at about 15.6, 7.8, and 3.9 km, respectively. Figure 6(a) and (b) show the I/O pattern for the 1280-process case and indicates about 80% of the processes access 13 file stripes. The majority of the processes have the same request length, and only a few processes have smaller request lengths. There are two peaks of stripe access degree in Figure 6(c) at stripe ID 0 and 1441 because a few small sized grid variables (40 MB each) are written at these file offsets. Since each variable is evenly partitioned among 1280 processes, there are 32 processes accessing the same stripe, as shown by the right-most bar in the histogram chart, Figure 6(d). The histogram also shows that more than half of the stripes are accessed by 6 distinct processes, while others have even higher degrees.

Both MDF and LW-MDF yield similar performance in Figure 6(e) because more than 80% of the processes have the same request count and hence share

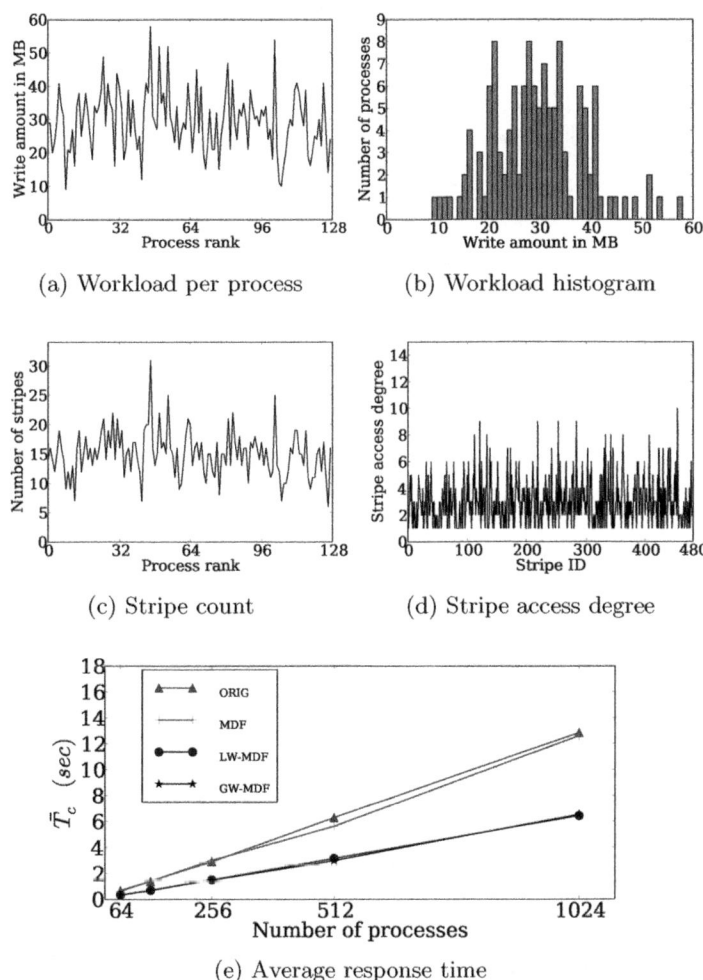

(a) Workload per process

(b) Workload histogram

(c) Stripe count

(d) Stripe access degree

(e) Average response time

Fig. 5. Access pattern and performance results for the random workload

the same weights. Improvement in the average response time is approximately 20%. Compared with the random workload distribution, GCRM's access pattern is relatively regular. As described earlier, GW-MDF requires an additional global communication for finding the minimal weights. However, the benefit of using the global weights is not significant enough to outperform the communication overhead. The similar results between the MDF and the original methods attribute to the balanced I/O workload in the GCRM.

4   Related Work

Scheduling parallel I/O operations has been studied by many researchers to address the speed gap between the CPU and I/O systems. Three off-line heuristics

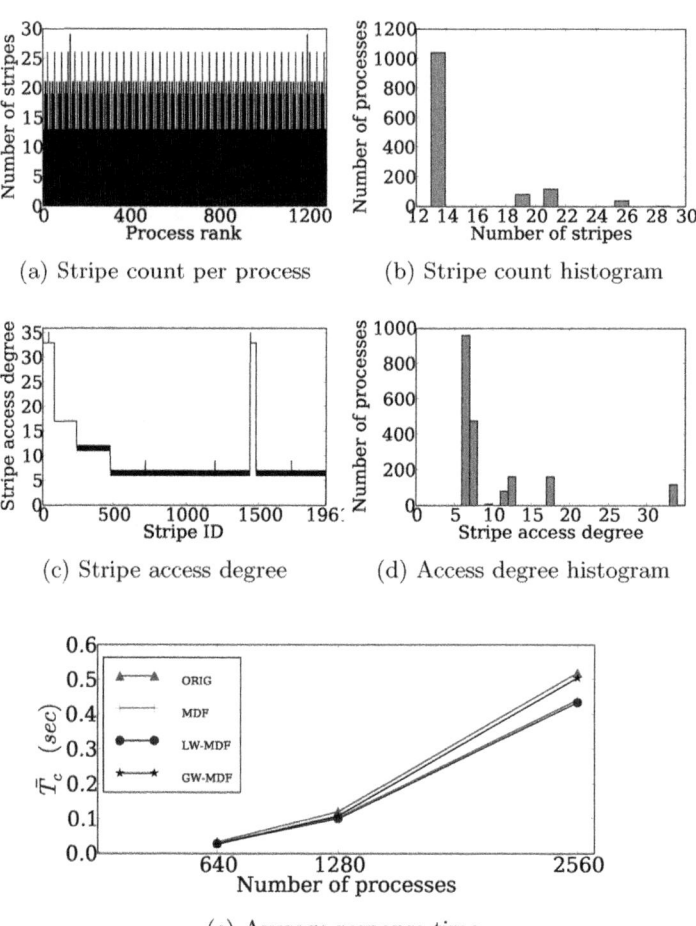

(a) Stripe count per process

(b) Stripe count histogram

(c) Stripe access degree

(d) Access degree histogram

(e) Average response time

Fig. 6. Access pattern and performance results of GCRM

based on graph coloring algorithms are developed to formalize the simultaneous resource scheduling problem [12]. Based on this work, a distributed randomized version using edge coloring method is proposed in [13]. Distributed I/O scheduling in the presence of data replication is presented in [14]. Decentralized I/O scheduling strategies between computer nodes and I/O servers for parallel file systems are developed in [15]. [16] has proposed three different techniques to increase the write bandwidth for collective I/O. The first technique is similar to two-phase collective I/O which aggregate I/O requests from participating processes such that the number of I/O operations provided to the underlying parallel file systems can be minimized. The second technique is to use a designated root process gathering data from all the processes, thus, the communication parallelization can be better utilized in some degree. In the third method, each process writes out data independently. However, there is no I/O scheduling algorithm employed in any of the proposed methods.

5 Conclusion

In collective I/O operations, different I/O request scheduling strategies can give different response time. We use the stripe access degree and request count per process on the I/O aggregators to develop algorithms that improve the average response time of collective I/O operations. Reducing the average response time in collective I/O operations equivalently increases the computational resource utilization in high-performance computing systems. Our performance results show significant improvement in average response time for various data access patterns in collective write operations. In the future, we plan to apply similar approaches for read operations and develop different scheduling methods for different parallel file systems.

Acknowledgment. This work is supported in part by NSF award numbers: OCI-0724599, CNS-0830927, CCF-0621443, CCF-0833131, CCF-0938000, CCF-1029166, and CCF-1043085 and in part by DOE grants DE-FC02-07ER25808, DE-FG02-08ER25848, DE-SC0001283, DE-SC0005309, and DE-SC0005340. A portion of this work was performed under project 57746 funded by DOE's Office of Science under the Scientific Discovery through Advanced Computing program. This research used resources of the National Energy Research Scientific Computing Center, which is supported by the Office of Science of the U.S. Department of Energy under Contract No. DE-AC02-05CH11231.

References

1. del Rosario, J., Brodawekar, R., Choudhary, A.: Improved Parallel I/O via a Two-Phase Run-time Access Strategy. In: The Workshop on I/O in Parallel Computer Systems at IPPS (1993)
2. Kotz, D.: Disk-directed I/O for MIMD Multiprocessors. ACM Transactions on Computer Systems 15(1), 41–74 (1997)
3. Seamons, K., Chen, Y., Jones, P., Jozwiak, J., Winslett, M.: Server-directed Collective I/O in Panda. In: Supercomputing (November 1995)
4. Thakur, R., Gropp, W., Lusk, E.: Users Guide for ROMIO. Technical Report ANL/MCS-TM-234, Argonne National Laboratory (October 1997)
5. Thakur, R., Gropp, W., Lusk, E.: Data Sieving and Collective I/O in ROMIO. In: The Symposium on the Frontiers of Massively Parallel Computation (1999)
6. Ying, L.: Lustre ADIO Collective Write Driver. Lustre Technical White Paper (September 2008)
7. Liao, W., Choudhary, A.: Dynamically Adapting File Domain Partitioning Methods for Collective I/O Based on Underlying Parallel File System Locking Protocols. In: SuperComputing Conference (2008)
8. Liao, W.: Design and Evaluation of MPI File Domain Partitioning Methods under Extent-Based File Locking Protocol. IEEE Transactions on Parallel and Distributed Systems 22(2), 260–272 (2011)

9. Randall, D., Khairoutdinov, M., Arakawa, A., Grabowski, W.: Breaking the Cloud Parameterization Deadlock. Bull. Amer. Meteor. Soc. 84, 1547–1564 (2003)
10. Schuchardt, K., Palmer, B., Daily, J., Elsethagen, T., Koontz, A.: IO Strategies and Data Services for Petascale Data Sets from a Global Cloud Resolving Model. Journal of Physics: Conference Series 78 (2007)
11. Li, J., et al.: Parallel netCDF: A High-Performance Scientific I/O Interface. In: SuperComputing Conference (2003)
12. Jain, R., Somalwar, K., Werth, J., Browne, J.: Scheduling Parallel I/O Operations in Multiple Bus Systems. Journal of Parallel and Distributed Computing 16(4), 352–362 (1992)
13. Durand, D., Jain, A., Tseytlin, D.: Applying Randomized Edge Coloring Algorithms to Distributed Communication: An Experimental Study. In: SPAA (1995)
14. Wu, J., Lin, Y., Liu, P.: Efficient Distributed Algorithms for Parallel I/O Scheduling. In: International Conference on Parallel and Distributed Systems (2005)
15. Isaila, F., Singh, D., Carretero, J., Garcia, F.: On Evaluating Decentralized Parallel I/O Scheduling Strategies for Parallel File Systems. In: VECPAR (2006)
16. Chaarawi, M., Chandok, S., Gabriel, E.: Performance Evaluation of Collective Write Algorithms in MPI I/O. In: Allen, G., Nabrzyski, J., Seidel, E., van Albada, G.D., Dongarra, J., Sloot, P.M.A. (eds.) ICCS 2009. LNCS, vol. 5544, pp. 185–194. Springer, Heidelberg (2009)

OMPIO: A Modular Software Architecture for MPI I/O

Mohamad Chaarawi[1], Edgar Gabriel[1], Rainer Keller[2], Richard L. Graham[3], George Bosilca[4], and Jack J. Dongarra[4]

[1] Department of Computer Science, University of Houston, Houston, TX, USA
{mschaara,gabriel}@cs.uh.edu
[2] High Performance Computing Center Stuttgart (HLRS), Stuttgart, Germany
keller@hlrs.de
[3] Oak Ridge National Laboratory (ORNL), Oak Ridge, TN, USA
rlgraham@ornl.gov
[4] Innovative Computing Laboratory, University of Tennessee, Knoxville, TN, USA
{bosilca,dongarra}@eecs.utk.edu

Abstract. I/O is probably the most limiting factor on high-end machines for large scale parallel applications as of today. This paper introduces OMPIO, a new parallel I/O architecture for Open MPI. OMPIO provides a highly modular approach to parallel I/O by separating I/O functionality into smaller units (frameworks) and an arbitrary number of modules in each framework. Furthermore, each framework has a customized selection criteria that determines which module to use depending on the functionality of the framework as well as external parameters.

1 Introduction and Motivation

Amdahl's law stipulates that the scalability of a parallel application is limited by its least scalable section. For many scientific applications, the scalability limitation comes from the performance of I/O operations. MPI [12], the most popular parallel programming paradigm on clusters today introduced the notion of parallel I/O in version two of the specification. Although its adoption by the end-users has been modest, it has been shown, that in combination with parallel file systems, MPI I/O can significantly improve the performance of I/O operations [7, 14] compared to sequential I/O.

Switching from the sequential Fortran or C I/O routines to MPI I/O potentially requires significant work by application developers, due to the fact that many MPI I/O features do not have counterparts in other I/O specifications. However, application developers are more willing to make drastic investment in rewriting substantial part of the application if the direct effect of the investment is a significant reduction in the application execution time, or a more robust scalability. This is however not always the case. The reasons for the limited performance often observed with MPI I/O is the diversity of existing I/O solutions which make each I/O environment (almost) unique. The performance of parallel I/O operations is influenced by the file system utilized, as well as by the number

Y. Cotronis et al. (Eds.): EuroMPI 2011, LNCS 6960, pp. 81–89, 2011.

of storage servers, the I/O bandwidth of each storage server, the network connectivity in-between the storage servers as well as between the storage servers and compute nodes, and the network interconnect and its OS-level parameters used for the MPI level communication. Additionally, application characteristics such as frequency and volume of I/O operations as well as the algorithm utilized to implement the functionality (e.g., the collective I/O operations), will greatly contribute towards the I/O performance observed by the end-user.

In this paper, we present a new parallel I/O architecture for Open MPI called OMPIO. The goal of OMPIO is to provide the infrastructure that allows to deal with the challenges of parallel I/O in a flexible manner, and consequently allows to optimize the performance of I/O operation for different applications and hardware configurations. At the core of the architecture is the separation of parallel I/O functionality into frameworks. This allows to encapsulate various aspects of parallel I/O into smaller functional units, such as dealing with file system specific operations, individual I/O, collective I/O, or shared file pointer operations. Each framework has typically multiple modules providing the required functionality, each module being designed for different scenarios. We argue, that the selection criteria that determines which module is being used is highly dependent on the functionality provided by a framework and on external parameters such as the file system utilized, hardware configuration, process placement by the batch scheduler or application characteristics.

The remainder of the paper is organized as follows: Section 2 discusses the related work in the area and makes the case why currently existing approaches, provided by most popular MPI I/O libraries, do not offer the required flexibility to deal with the diversity of the available I/O subsystems. Section 3 describes the design of the new OMPIO module and its associated set of frameworks. In section 4 we present a case study where we evaluate two different benchmarks on two different platforms using a PVFS2 and a Lustre file system. The results demonstrates a the available functionality in OMPIO and exposes some of the advantages of the new architecture for collective I/O operations. Finally, section 5 summarizes the paper and presents the ongoing work in this area.

2   Related Work

The most widely used implementation of MPI I/O as of today is ROMIO [16]. ROMIO is part of the MPICH [8] distribution and is the basis for many I/O libraries used in other public domain MPI libraries such as Open MPI and commercial MPI implementations. ROMIO abstracts file systems specific operations using the Abstract-Device Interface for Parallel I/O, called ADIO [15], which reduces the number of routines that have to be implemented in order to support a new file system. ROMIO also has the ability to support multiple file systems simultaneously, e.g., in case an application opens a file on two different file systems. However, the selection criteria which ADIO module shall be used as of today is based on the file system only. Krimpe et al. [9] allowed for non-file system specific selection of some modules by prepending a keyword to the name of the file. The solution presented in this paper has two main advantages compared to ROMIO. First, the usage of different frameworks allows a more fine grained separation of functionality than the approach used in ROMIO. Second, OMPIO introduces the ability to make non-file system specific module selection that do not require any modifications of the end-user application.

In [4], the authors introduced the ability to easily modify parameters of collective I/O operations. However, the work focused entirely on collective I/O, leaving other aspects of parallel I/O unmodified. Furthermore, it is our understanding that the framework described in this paper does not allow for easy deployment of new collective I/O algorithms, but is restricted to modifying parameters of the provided collective read/write operations.

The Adaptable IO System (ADIOS) [11] is an I/O library designed to allow end users to select the best I/O method based on the application's access pattern and the underlying file system and hardware at hand. The access pattern of the application is described in a separate input file, providing some of the functionality that the file view provides in MPI I/O. Thus, the ADIOS library has the ability to utilize POSIX style I/O operations, MPI I/O or any other supported API without having to change the application itself. ADIOS also introduces a file format called BP, which serves as an intermediate format that is easily converted to other standard file formats such as HDF5.

3   The OMPIO Set of Frameworks

The Open MPI Project [5] is an open source implementation of the MPI specification that is developed and maintained by a consortium of academic, research, and industry partners. The internal architecture of Open MPI is built around the Modular Component Architecture (MCA) [1], which allows for compile or run time selection of the components used by the MPI library. A component framework in Open MPI is dedicated to a single task, such as providing parallel job control or performing MPI collective operations. Modules are self-contained software units that can configure, build, and install themselves. Modules adhere to the interface prescribed by the component framework that they belong to, and provide requested services to higher-level tiers and other parts of MPI. This mechanism allows a single Open MPI installation to simultaneously support various network interconnects. The new OMPIO module is a module of the IO framework of Open MPI, and is designed to co-exist with ROMIO, the parallel I/O library used in all released versions of Open MPI. Generally speaking, when a file is being opened, both OMPIO and ROMIO are being queried, and the module returning the higher priority value is used to for the subsequent I/O operations.

The main goals of OMPIO are three fold. First, it increases the modularity of the parallel I/O library by separating functionality into distinct sub-frameworks. Second, it allows frameworks to utilize different run-time decision algorithms to determine which module to use in a particular scenario, enabling non-file system specific decisions. Third, it improves the integration of parallel I/O functions

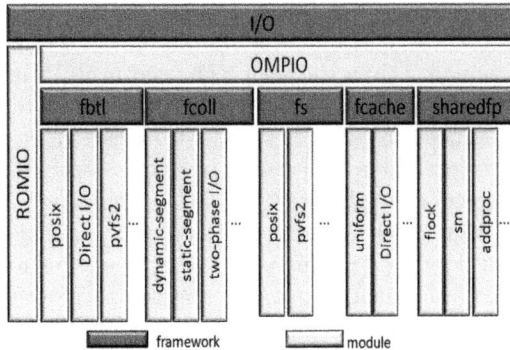

Fig. 1. Overview of the OMPIO component and its frameworks

with other components of Open MPI, most notably the derived data type engine and the progress engine. The integration with the Open MPI progress engine allows for seamless progress of non-blocking I/O operations. The integration with the derived data type engine has multiple advantages, most notably faster decoding of derived data types and the usage of optimized data type to data type copy operations. Furthermore, OMPIO has the ability to use the data conversion functionality of the data type engine, without having to provide the according (fairly complex) functions.

Similarly to the selection logic in other Open MPI frameworks, each sub-framework of the OMPIO component determines in MPI_Init the list of available modules and opens them. Upon opening a file using MPI_File_open, the OM-PIO module initializes each sub-framework for that particular file. A framework will query each available module which in return responds with a priority value indicating its readiness to be used for the given file. As an example, a module providing a POSIX style interface might return a low priority value for most files, indicating that it could be used for the according operations. However, a specific module optimized for the given file system or installation will typically return a higher priority and will be chosen for the subsequent I/O operations. Each sub-framework or module will typically have different rules on when to return a high priority. Conditions include the file systems type, location of participating processes, network parameters or user specified settings. In the following we present briefly each sub-framework and the currently available modules.

3.1 The *file system* Framework (fs)

The *fs* framework abstracts out file manipulation operations such as opening, closing, and deleting a file. The semantics of most of the operations are collective. Furthermore, file system specific info objects have to be interpreted and applied within this module. The *fs* framework has as of today a module providing generic POSIX interface, and separate modules for Lustre and a PVFS2 which allow to modify stripe size and stripe depth when creating a new file.

3.2   The *file byte-transfer layer* **Framework (fbtl)**

The *fbtl* framework provides the abstraction for all individual read and write operations. A module implementing the *fbtl* interfaces has to provide, as of today, blocking and non-blocking read and write operations, as well as a progress function that will be registered with the Open MPI progress engine in order to enforce the progress of pending I/O calls. The interfaces of the read and write operations currently take a list containing tuples of <*memory address, length in bytes, file offset*>. Currently available are *fbtl* modules which provides POSIX semantics, and a module utilizing native PVFS2 read/write operations. Note however, that the current OMPIO implementation only supports blocking operations for both *fbtl*. Support for non-blocking individual operations are expected to be available in the near future.

3.3   The *collective I/O* **Framework (fcoll)**

This framework provides interfaces for collective file I/O operations. In contrary to the other frameworks which are part of the OMPIO set, the *fcoll* framework triggers the selection logic not upon opening a file, but every time the file view is being set.

Collective I/O operations are a very good example for the necessity to have non-file system specific selection logic. As an example, the Lustre file system serving the Jaguar system at Oak Ridge National Laboratory and the Lustre file system at our development cluster at the University of Houston have fundamentally different characteristics, such as number of Object Storage Targets (OSTs), bandwidth of each OST, and network characteristics between compute nodes and OSTs. Despite the fact that both installations utilize the same file system, different algorithms for collective I/O operations have to be used on these two installations in order to maximize the I/O performance of an application, since some optimizations only make sense for certain hardware configurations.

The *fcoll* framework has five different modules to choose from, one module for each of the following algorithms: two-phase I/O, static segmentation, dynamic segmentation, individual algorithm and an algorithm where each I/O node is only receiving requests by a single aggregator process. The first three algorithms have been extended to include a heuristic which automatically determines the number of aggregator processes to be used [2].

The current selection logic is based on an extensive set of tests that has been executed on various platforms and file systems. Among the factors that influence which module is being used is the average contiguous data chunk accessed by each process, gaps size in the file view between processes, and file system characteristics, such as the stripe size and the minimal data required to saturate the read/write bandwidth of one process. We omit here details of this selection logic due to space limitations, more details may be found in [2].

3.4   The *file cache* **Framework (fcache)**

The *fcache* framework provides the ability to set and retrieve information related to the file layout, such as the number of storage servers used, list of storage servers, and stripe depth for each file separately. The main functionality of the *fcache* is to provide a mapping of *<offset into file, length in bytes>* to a list of *<storage server id, local offset on that storage server, local length>*. This allows for various optimizations for example for collective I/O operations. As of today, only a trivial module is available for UFS style file systems which provides only basic information.

3.5   The *shared file pointer* **Framework (sharedfp)**

The *sharedfp* framework provides the functionality required to manage the shared file pointer, allowing for generic and architecture specific optimizations. Although shared file pointer operations have been sparingly used in the community, due to the fact that in the most general case an implementation of shared file pointer operations will be slow, it is well understood that, for particular architectures or settings, efficient implementations do exist. As an example, if the shared file pointer is utilized by processes in a communicator that spans a single physical node, the shared file pointer can be efficiently implemented using a small shared memory segment. Alternatively, some of the strict requirements of a shared file pointer can be relaxed for certain usage scenarios, allowing the utilization of individual files per process. In doing so, the consolidation to a single output file may be delayed to the post-processing step [10].

We have explored a number of shared file pointer algorithms in [10], which are currently being converted into modules in the near future along with the selection logic, which will include process placements as one of the key criteria to determine which module to use.

4   Experimental Results

Two application benchmarks are used for evaluation on two different platforms. The Shark cluster at the University of Houston consists all-in-all of 29 nodes, with a PVFS2 file system consisting of 22 server nodes where each server uses its local disk space as the back-end storage. The stripe size of the file system is 64 kB. The file system uses GE as the network interconnect.

The Deimos PC Farm at TU Dresden has 724 compute nodes with a Lustre file system exported by 11 I/O servers via a separate 4x SDR InfiniBand network. The file system is organized in 48 OSTs with a stripe size of 1 MB.

The first benchmark used is MPI-TILE-IO [13], a test application that implements tile access to a two dimensional dense dataset. This type of workload is seen in tiled displays (for small numbers of tiles) and in some numerical applications. Several parameters that control the file access and 2D distribution of the processes can be modified at runtime. The results shown report two tile sizes of 64 Bytes (2048 x 1600 elements) and 1 MB (20 x 15 elements), which represents a non-contiguous and contiguous access respectively. We report the maximum bandwidth achieved across five executions of every test case.

Table 1. Performance comparison between OMPIO's and ROMIO's default setting using MPI-TILE-IO

Platform/Number of Processes/Tile Size	ROMIO	OMPIO
Shark/81(9x9)/64B	303.8 MB/s	591.1 MB/s
Shark/81(9x9)/1MB	290.1 MB/s	625.4 MB/s
Deimos/256(16x16)/64B	411.6 MB/s	2167.1 MB/s
Deimos/256(16x16)/1MB	517.7 MB/s	2491.2 MB/s

The results shown in table 1 show the results using the MPI-TILE-IO benchmark over Shark with PVFS2 and Deimos with Lustre. ROMIO has been executed with default parameters, i.e. without passing any additional hints or parameters to the library in order to have a base-line number from the performance perspective. In OMPIO we set the optimal cycle buffer size determined for the according file system. OMPIO chooses the two-phase I/O module for the 64 Byte tile size and the dynamic segmentation module for the 1 MB tile size. The heuristic determining the number of aggregators automatically leads to 81 aggregator processes on Shark and 256 aggregators on Deimos in these test cases. Collective I/O operations in ROMIO use the two-phase I/O algorithm with one aggregator per node as the default setting. The results show that OMPIO leads to a performance benefit in these test cases which can be attributed mostly to the different number of aggregators used by OMPIO and the different algorithm used in the first case. However, the main message of this result is not the performance benefit observed due to the different number of aggregators, instead the flexibility to switch seamlessly between different collective I/O module for the same application due to the component architecture of OMPIO.

In the second scenario, we demonstrate the flexibility and modularity of the OMPIO architecture by using the Open Tool for Parameter Optimizations (OTPO) [3] to tune collective I/O operations and parameters for a given test case. OTPO is a tool which can be used to optimize runtime parameters of Open MPI. The tool takes in an input file which contains the names of parameters to be explored along with the according rules on how to modify the parameters, and the name of the benchmark/application to be executed when exploring the parameter space. After the optimization, OTPO reports the set of parameter combination(s) which lead to the lowest execution time. In this particular scenario, we used the Latency-IO micro-benchmark developed as part of the latency test suite [6].

The parameter file that is passed to OTPO contains the different collective I/O algorithms that are available in OMPIO and a some parameters of these modules. Thus, the parameters to be optimized and according values are:

- *fcoll module:* static, dynamic, individual, two-phase
- *number of aggregators:* 5, 10, 20, 40
- *cycle buffer size:* 2 MB, 20 MB, 32 MB, 64 MB, 128 MB

In this case 65 different parameter combinations were generated from the input file, the winning combination was (dynamic, 20, 32 MB). While there were other parameter combinations that provided performance close to the wining combination, only two out of 65 parameter combinations were within 10% of the best performance value. Those combinations were (dynamic, 20, 20 MB) and (static, 20, 32 MB). Overall, 23 out of 65 parameter combinations were within 25% of the best performance.

This type of tuning of collective I/O parameters is possible because of the OMPIO architecture and allows end-users and system administrators to pre-tune a module for a particular application or scenario without having to recompile the MPI library.

5   Conclusion

This paper introduces OMPIO, a newly developed parallel I/O architecture designed for Open MPI. OMPIO introduces a modular architecture for parallel I/O that separates functionality into different sub-frameworks and allows for a highly flexible composition of modules in order to provide MPI I/O functionality, and reduces the barriers to develop new, site-specific modules and configurations. We demonstrate the usability of OMPIO by executing various benchmarks on a PVFS2 and Lustre file system on two different clusters. OMPIO is currently being evaluated by the Open MPI group and should be publicly available by the end of the summer, with the initial intent of serving as a research vehicle into parallel I/O.

The ongoing work includes multiple areas. First and foremost, we are working on implementing the non-blocking I/O operations within the OMPIO framework. This will support most of the operations defined in the MPI-2.2 specification and will open the door for further optimizations for collective I/O operations. Second, we are continuing to improve our collective I/O algorithms, most notably by exploring new grouping strategies for the dynamic and static segmentation algorithms.

References

1. Barrett, B., Squyres, J.M., Lumsdaine, A., Graham, R.L., Bosilca, G.: Analysis of the component architecture overhead in Open MPI. In: Proc. of the 12th European PVM/MPI Users' Group Meeting, Sorrento, Italy, pp. 175–182 (September 2005)
2. Chaarawi, M.: Optimizing Parallel I/O Operations for High Performance Computing. Ph.D. thesis, Department of Computer Science, University of Houston (2011)
3. Chaarawi, M., Squyres, J.M., Gabriel, E., Feki, S.: A tool for optimizing runtime parameters of open MPI. In: Lastovetsky, A., Kechadi, T., Dongarra, J. (eds.) EuroPVM/MPI 2008. LNCS, vol. 5205, pp. 210–217. Springer, Heidelberg (2008)
4. Coloma, K., Ching, A., Choudhary, A., Liao, W., Ross, R., Thakur, R., Ward, L.: A New Flexible MPI Collective I/O Implementation. In: Proceedings of the 2006 IEEE International Conference on Cluster Computing, pp. 1–10 (2006)

5. Gabriel, E., Fagg, G.E., Bosilca, G., Angskun, T., Dongarra, J., Squyres, J.M., Sahay, V., Kambadur, P., Barrett, B.W., Lumsdaine, A., Castain, R.H., Daniel, D.J., Graham, R.L., Woodall, T.S.: Open MPI: Goals, concept, and design of a next generation MPI implementation. In: Kranzlmüller, D., Kacsuk, P., Dongarra, J. (eds.) EuroPVM/MPI 2004. LNCS, vol. 3241, pp. 97–104. Springer, Heidelberg (2004)

6. Gabriel, E., Fagg, G.E., Dongarra, J.J.: Evaluating dynamic communicators and one-sided operations for current MPI libraries. International Journal of High Performance Computing Applications 19(1), 67–79 (2005)

7. Gabriel, E., Venkatesan, V., Shah, S.: Towards high performance cell segmentation in multispectral fine needle aspiration cytology of thyroid lesions. Computational Methods and Programs in Biomedicine 98(3), 231–240 (2009)

8. Gropp, W., Lusk, E., Doss, N., Skjellum, A.: A high-performance, portable implementation of the MPI message passing interface standard. Parallel Computing 22(6), 789–828 (1996)

9. Kimpe, D., Ross, R., Vandewalle, S., Poedts, S.: Transparent log-based data storage in MPI-IO applications. In: Cappello, F., Herault, T., Dongarra, J. (eds.) PVM/MPI 2007. LNCS, vol. 4757, pp. 233–241. Springer, Heidelberg (2007)

10. Kulkarni, K., Gabriel, E.: Evaluating Algorithms for Shared File Pointer Operations in MPI I/O. In: Allen, G., Nabrzyski, J., Seidel, E., van Albada, G.D., Dongarra, J., Sloot, P.M.A. (eds.) ICCS 2009. LNCS, vol. 5544, pp. 280–289. Springer, Heidelberg (2009)

11. Lofstead, J., Zheng, F., Klasky, S., Schwan, K.: Adaptable, metadata rich IO methods for portable high performance IO. In: Proc. of IPDPS 2009, Rome, Italy, May 25-29 (2009)

12. Message Passing Interface Forum: MPI-2.2: Extensions to the Message Passing Interface (September 2009), http://www.mpi-forum.org

13. Ross, R.: Parallel I/O Benchmarking Consortium, http://www.mcs.anl.gov/research/projects/pio-benchmark

14. Ross, R., Nurmi, D., Cheng, A., Zingale, M.: A Case Study in Application I/O on Linux Clusters. In: ACM/IEEE Supercomputing Conference, Denver, CO, USA (2001)

15. Thakur, R., Gropp, W., Lusk, E.: An Abstract-Device Interface for Implementing Portable Parallel-I/O Interfaces. In: Proc. of the 6th Symposium on the Frontiers of Massively Parallel Computation, pp. 180–187. IEEE Computer Society Press, Los Alamitos (1996)

16. Thakur, R., Gropp, W., Lusk, E.: On implementing MPI-IO portably and with high performance. In: Proc. of the 6th Workshop on I/O in Parallel and Distributed Systems, pp. 23–32 (1999)

Design and Evaluation of Nonblocking Collective I/O Operations

Vishwanath Venkatesan[1], Mohamad Chaarawi[1],
Edgar Gabriel[1], and Torsten Hoefler[2]

[1] Department of Computer Science, University of Houston
{venkates,mschaara,gabriel}@cs.uh.edu
[2] Blue Waters Directorate, University of Illinois
htor@illinois.edu

Abstract. Nonblocking operations have successfully been used to hide network latencies in large scale parallel applications. This paper presents the challenges associated with developing nonblocking collective I/O operations, in order to help hiding the costs of I/O operations. We also present an implementation based on the libNBC library, and evaluate the benefits of nonblocking collective I/O over a PVFS2 file system for a micro-benchmark and a parallel image processing application. Our results indicate the potential benefit of our approach, but also highlight the challenges to achieve appropriate overlap between I/O and compute operations.

1 Introduction

Overlapping computation and communication is a standard technique to optimize the performance of parallel applications. This technique allows to hide latencies and improve bandwidth of data transfers to remote processes. This functionality is offered to the user through a special nonblocking interface, which allows to start operations and check for completions later. Benefits of nonblocking operations have been demonstrated for point-to-point [1,2] and nonblocking collective [3,4] operations. The Message Passing Interface (MPI) standard specifies so called "immediate" versions of some operations. MPI-2.2 offers immediate versions of all point-to-point communication calls and MPI-3.0 will add immediate versions of all collective communication functions. Those special functions return with a handle before the operation is completed. The handle can be used to test and wait for completion of the associated operations.

With the advent of data-intensive computing [5], the input/output from/to disk (I/O) of application data can become a significant bottleneck. This does not only include reading the dataset initially and saving it at the end but also periodic application-level checkpoints and out-of-core processing. In addition to this, while the compute and network power of parallel HPC systems is growing steadily, the performance of the I/O subsystem can often not keep up with this growth. Thus, nonblocking I/O interfaces are important to improve application performance.

Y. Cotronis et al. (Eds.): EuroMPI 2011, LNCS 6960, pp. 90–98, 2011.

In this work, we propose a new interface that is similar to the newly introduced nonblocking collective communication operations and show an optimized implementation of this interface. In particular, the contributions of this paper are as follows:

1. We propose a simple extension to the MPI-2.2 standard to enable the user to specify overlap of I/O operations with other computation and communication operations conveniently.
2. We describe a framework to efficiently implement nonblocking collective I/O routines.
3. We demonstrate an implementation of this framework and performance results on a parallel file system.

The outline of the paper is as follows: section 2 presents the technical challenges associated with nonblocking collective I/O operations. Section 3 evaluates the benefits of nonblocking collective I/O operations, followed by a general discussion on nonblocking collective I/O interfaces in section 4. Finally, section 5 summarizes the contributions of the paper and presents the ongoing work in this domain.

2   Challenges of Nonblocking Collective I/O Operations

In the following, we detail the challenges of providing nonblocking collective I/O operations. For this, we describe first the collective I/O algorithm used and then elaborate the extensions introduced in libNBC.

2.1   Collective I/O Algorithm

The collective I/O algorithm used for the prototype implementation of nonblocking collective I/O operations is based on the dynamic segmentation algorithm [6]. This algorithm is an extension of the classical two-phase collective I/O algorithm. Similar to two-phase I/O, the main goal of this algorithm is to combine data from multiple processes in order to minimize the number of discontiguous I/O requests. In contrast to two-phase I/O however, the dynamic segmentation algorithm does not create a globally sorted data array based on the offsets in the file. Instead, each aggregator is assigned a group of processes and performs the data gathering/scattering and sorting only within its group. This allows to execute the shuffle step including the sorting and data gathering/scattering more efficiently, since the Alltoall(v) type communication in the two-phase I/O algorithm is replaced by a number of independent Allgather(v) operations in the dynamic segmentation algorithm.

For very large collective operations, the dynamic segmentation algorithm is split into multiple cycles. This allows to keep the amount of temporary buffer required on the aggregator processes within constant, reasonable limits. Note, that depending on the offsets into the file a process might have to contribute different amounts of data to its aggregator in each cycle.

2.2   A Framework for Nonblocking Collective I/O Operations

A similar problem, the implementation of nonblocking collective operations, has been discussed in [4]. The framework for nonblocking collectives is implemented in the open-source library libNBC. We utilize and extend libNBC in conjunction with Open MPI's OMPIO framework [7] to handle nonblocking collective I/O operations. The central concept in libNBC's design is the collective operation schedule. During initialization of the operation, each process records its part of the collective operation in a local schedule. A schedule contains, among others, send and recv operations and a so called "barrier" which acts as local synchronization object. A barrier in a schedule has the semantics that all operations before the barrier have to be finished before any of the operations after the barrier can be started. The execution of a schedule is nonblocking and the state of the operation is simply kept as a pointer to a position in the schedule. With send, recv, and barrier, one can express many collective communication algorithms; see [4] for further details.

A major difference between collective communication and collective I/O operations stems from the fact, that each process is allowed to provide different volumes of data to a collective read or write operation, without having knowledge on the data volumes provided by other processes. This is not the case for collective communication operations, where either each process provides exactly the same amount of data (e.g. Bcast, Reduce, Allreduce, Gather, Scatter, Allgather, Alltoall etc.) or in case of the vector version of the operations a process knows the communication volumes of all processes communicating with him (Gatherv, Scatterv, Allgatherv, Alltoallv, Alltoallw). This information is, however, essential to determine how much data a process has to contribute within a cycle of the collective I/O operation.

Thus, the first step in most collective I/O algorithms is an Allgather(v) step which determines the overall amount of data each process is contributing along with the according offsets into the file. In the case of the dynamic segmentation algorithm, this communication operation is within each group of an aggregator. This allows every process to determine how much data it has to contribute in every cycle of the algorithm. For nonblocking operations, the challenge is, that upon calling `MPI_File_iwrite_all` the according Allgather(v) operations can not be finished, since this would result in a blocking communication operation when initiating the nonblocking write-all. This is however not possible, since it could lead to a deadlocks.

Thus, the solution developed here consists of a two-step approach: while initiating the nonblocking collective read/write operation, we generate first a schedule which executes the nonblocking Allgather(v) communication step[1] The last step of the Allgather(v) schedule will be executed when the Allgather(v) operation is finished, and creates a new schedule which executes the actual collective

---

[1] Note, that the operation is not exactly an `MPI_Allgatherv`, but consists of multiple Gather(v) operations executed on disjoint groups of processes in the same communicator.

I/O operations. This second schedule contains the data gathering at the aggregator processes, the sorting based on the offsets into the file, and the asynchronous writing to the file.

Associated with that are two further challenges: first, no temporary buffers used within the collective I/O algorithm can be allocated upfront when posting the operation, since the overall amount of data and many of the according buffers are only known at the end of the Allgather(v) step. Therefore, we extended the set of operations supported by the progress engine of libNBC in addition to nonblocking read and write by dynamic memory management functions, which allow to allocate and free buffers as part of the libNBC schedule. Second, due to the fact that the asynchronous I/O operation are implemented using `aio_read` and `aio_write` operations which have their own data structure to identify pending operations, the libNBC progress engine has been extended with the ability to progress multiple, different handles simultaneously, e.g. `MPI_Requests` for communication operations and the internal aio-handles for asynchronous I/O operations.

2.3   Schedule Caching

One of the distinct features of libNBC is its ability to cache a schedule of a collective operation. This allows to speed up execution of operations which are posted repeatedly by an application. I/O operations generally fit the repetitive pattern required for caching a schedule, e.g. in case an application writes periodic checkpoint files. In this scenario an application has two options. The first option is to append the most recent data that has to be written to the end of an existing file. The second option would use a different file for every checkpoint. Both approaches post unique challenges for caching a schedule.

For the first option, the challenge comes from the fact that every collective I/O operation which appends data to an already existing file will lead to new offset values into the file. Moreover, the MPI standard also allows for a process to mix individual and collective I/O calls, which makes predicting the current position of the file pointer of a process impossible. Since the order in which data has to be written to the file depends on the file view and the current position of the individual file pointer, the actual amount of data that a process has to contribute to a particular cycle of the collective I/O is not necessarily repetitive, even if the arguments passed to the MPI function are identical to the previous instance. Thus, caching the schedule would not help in this scenario.

The second scenario where a separate file is used for every checkpoint is equally challenging, due to the fact the schedules would be cached on a per file handle basis. This is in equivalence to the collective communication operations, where the caching is being done on a per communicator basis, although the MPI specification does not providing attribute caching functions on files as of today. Transferring a schedule from one file handle to another file handle can theoretically be done, the challenge being however how to keep a file handle around once a file has been closed, without creating an unnecessary memory overhead.

3   Performance Evaluation

In the following section we evaluate the impact of the nonblocking collective I/O operations. We first describe the execution environment followed by the results obtained with a micro-benchmark and a parallel image processing application.

3.1   Experimental Setting

The system used in these tests is the *crill-cluster* at the University of Houston, which consists of 16 nodes with four 12-core AMD Opteron (Magny Cours) processor cores each (48 cores per node, 768 cores total), 64 GB of main memory and two dual-port InfiniBand HCAs per node. The parallel file system used is PVFS2 with 16 I/O servers and a stripe size of 64 KB. The file system is mounted onto the compute nodes over the Gigabit Ethernet network interconnect of the cluster. The current implementation of nonblocking I/O collective operations is tied to Open MPI and its new parallel I/O framework (OMPIO), mostly for retrieving and maintaining file handle related aspects and for decoding derived data types and the file view. The version of Open MPI executed is equivalent to the Open MPI trunk revision 24640. In the following analysis we focus, for the sake of simplicity, on write operations.

The first test executed is using the Latency-IO micro-benchmark developed as part of the latency test suite [8], which is a micro-benchmark executing either individual or collective I/O operations. Initially, we compare the performance obtained with the blocking version of the dynamic segmentation algorithm vs. a sequence of NBC_File_iwrite_all followed by NBC_Wait. Table 1 presents the bandwidth achieved in both scenarios for 64 and 128 MPI processes when using 32 aggregator processes and a 4 MB cycle buffer size. The overall file size written were 63 GB and 125 GB respectively (1000MB per process). All tests have been executed three times, and we present the average bandwidth obtained over all three runs. Note that the variation in the individual performance numbers between different runs very fairly small. The results indicate a small overhead for the 64 processes test case of the nonblocking implementation, which achieved 94% of the bandwidth obtained with the blocking version. For 128 processes the nonblocking version slightly outperformed the blocking version, which we attribute however to measurement jitter. All-in-all, the conclusion drawn from this analysis is that nonblocking implementation does not impose a significant, fundamental overhead compared to the blocking version.

In the second test we evaluate the ability to overlap collective I/O operations with compute operations. For this, the same benchmark is executing a compute

Table 1. Performance comparison of blocking vs. nonblocking collective I/O algorithm

No. of processes	Blocking Bandwidth	Nonblocking Bandwidth
64	703 MB/s	660 MB/s
128	574 MB/s	577 MB/s

Table 2. Evaluating the overlap potential of nonblocking collective I/O operations

No. of processes	I/O only time	Overlapping time	Time spent in computation
64	85.69 sec	85.80 sec	85.69 sec
128	205.39 sec	205.91 sec	205.39 sec

function after posting the nonblocking collective write operation. The compute operation is configured to take the equal amount of time as the I/O operation. Thus, we expect to observe an overall execution time equal or larger than the time required to perform the I/O operation only for the according scenario, with the upper bound being twice the amount of time required for the same test without the compute operation in case I/O and computation cannot be overlapped. Table 2 presents the results achieved for the same test cases as outlined above, the first column being the time spent in the I/O test without overlap, the second column representing the time spent in writing the same amount of data and performing an equally expensive compute operation, and the third column showing the time spent in the compute operation for the overlap test.

The results in this section indicate the ability to entirely hide the I/O operation under optimal circumstances. These optimal circumstances are represented by the ability of libNBC to progress the operation either through a progress thread or through inserting regularly NBC_Test function calls into the compute operation. Within the context of this analysis, we choose the second approach. Moreover, we also identified that the frequency and number of calls to NBC_Test have a tremendous influence on the overlap performance: calling it too often will introduce an additional overhead, if there are to few calls to this function, the library will not be able to progress the function. In our experimental results we identified the time required to execute one cycle in the dynamic segmentation algorithm as the optimal interval between two subsequent calls to NBC_Test.

3.2   An Application Scenario

Further tests have been executed with a parallel image processing application. This application is used to analyze smear sample from fine needle aspiration cytology, with the overall goal being to assist medical doctors in identifying cancer cells [9]. The challenge imposed by this application is due to the high resolution of the microscopes and the fact that images are captured at various wave-length to identify different chemical properties of the cells. For a $1cm \times 1cm$ sample with 31 spectral channels the image can contain overall up to 50GB of raw data. The MPI version of the code has furthermore the option to write the texture data into output files to facilitate future processing steps in realizing a complete computer aided diagnosis (CAD) solution. This makes the application compute and I/O intensive.

For the following tests, we focus on the code section which writes the texture data into files. This code sequence contains a loop in which texture data for each of the twelve Gabor filters is calculated and then written to a separate file. The computational part within this loop consists of two parallel fast-fourier transforms (FFTs), which are implemented using the FFTW library [10] version 2.1.5, and a convolution operation. For the version using the non-blocking collective I/O functions, writing the texture data in one iteration is overlapped with the execution of the FFTs and the convolution of the next iteration. Progress of the non-blocking collective I/O operation is implemented in two ways. The first one uses NBC_Test function calls in-between each FFT and the convolution operation. The second code version uses a patched version of the FFTW library which contains further function calls to NBC_Test. Note, that the initial reading of the image and final writing of the cluster assignments have not been modified and still use the blocking collective MPI I/O version.

For evaluation purposes we used two separate images. The first image has 8192 × 8192 pixels and 21 spectral channels, writing 12 times 256 MB of texture data (3 GB total) . The second image has 12281 × 12281 pixels and also 21 spectral channels, writing 12 times 576 MB of data (6.75 GB total). Tests have been executed with 64 and 96 processes on the same cluster and file system as in the previous section. Figure 1 show the times spent in I/O operations for each test case. We present again the average obtained over three separate runs. Note, that we ensured that both blocking and non-blocking collective I/O operation use the same algorithm, with the same number of aggregator processes and the same cycle buffer size.

The results indicate that the version of the code which uses the FFTW library as a 'black box', i.e. without any NBC_Test function calls inserted, offers only little benefit compared to the original version of the code which uses blocking, collective MPI I/O operation. The main problem is the limited ability to progress the non-blocking operations without a progress thread and with a very small number of calls to NBC_Test. On the other hand, using the patched version of the FFTW library ensures more progress and demonstrates significant benefits of the non-blocking collective I/O operations. The benefit is more obvious for the 64 processes test cases compared to the 96 processes test cases due to the increased execution time of the FFTs and the convolution for the 64 process test cases, which offer therefore more potential for overlapping computation and I/O

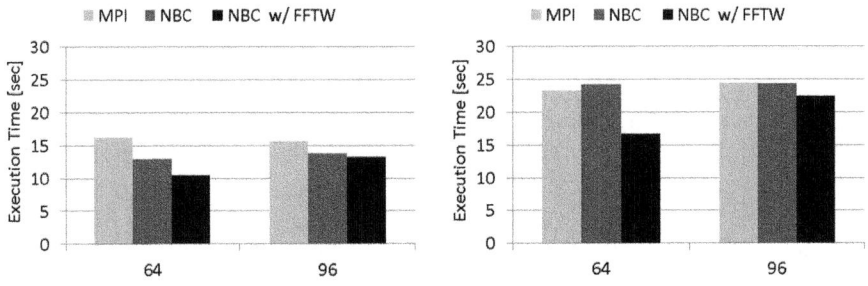

Fig. 1. Comparison of I/O times for 8k × 8k image (left) and 12k × 12k image (right) for 64 and 96 MPI processes

operations. Hiding the entire costs of the I/O operations for a real application is however very difficult, since i) the application has to have compute intensive sections that can be used for overlapping computation and I/O operations and ii) the timespan between two subsequent calls to NBC_Test can not be controlled in the similar manner as for the micro-benchmark. Nevertheless, with some efforts we were able to reduce the time spent in I/O operation by up to 35% – which can be highly significant for large scale applications.

4 Discussion

To mitigate potential performance bottlenecks, MPI added support for nonblocking file routines. However, collective MPI file operations can only be expressed with the limited split collective interface. The main limitations are that (1) there must only be a single split collective active on a file handle at any time and (2) no other collective file I/O operations can be issued on a file handle when a split collective is active. The first limitation prevents optimization techniques such as pipelined communications for communication/communication overlap [11] and the second limitation reduces programmability. The MPI-2.2 standard also allows to perform a global synchronization in the begin call of a split collective. This limits certain usage patterns.

Our nonblocking collective I/O framework allows to offer two additional features: (1) explicit progress and (2) multiple outstanding operations.

Thus, we propose to extend the MPI standard similar to nonblocking collective operations, i.e., to add immediate versions of all split collective operations, e.g., MPI_File_iread_all(..., MPI_Request req) and adding a request as last parameter. For file operations, the file pointer is advanced within the immediate function call, so that following calls operate on the right offset. We omit a list of all functions for space reasons.

The new functions can be used like nonblocking point-to-point and collective operations and the returned requests can be tested and waited on for completion with the usual functions (e.g., MPI_Test). Implicit progress can be problematic under certain circumstances while explicit progress puts a higher burden on the user [12]. Our interface proposal allows the implementation to offer both choices to the user. In addition, having multiple outstanding operations allows to employ pipelining techniques for overlapping communication and computation.

5 Conclusion

In this paper we discussed the challenges associated with non-blocking collective I/O operations. We present a framework which provides non-blocking versions of the collective read and write operations by extending the libNBC library. The performance of write operation has been evaluated using a micro benchmark and parallel image processing application. The results indicate the potential to actually overlap computation and I/O operations using these functions. However, the main challenge is how to ensure progress of the non-blocking collective I/O operations in the absence of a progress thread. The currently ongoing work includes multiple domains. First, we plan to extend the analysis to collective read operations. Second, we plan to perform a similar set of analysis as shown in this paper on different file systems, specifically on a large scale Lustre installation.

References

1. Brightwell, R., Underwood, K.D.: An analysis of the impact of MPI overlap and independent progress. In: ICS 2004: Proceedings of the 18th Annual International Conference on Supercomputing, pp. 298–305. ACM Press, New York (2004)
2. Baude, F., Caromel, D., Furmento, N., Sagnol, D.: Optimizing metacomputing with communication-computation overlap. In: Malyshkin, V.E. (ed.) PaCT 2001. LNCS, vol. 2127, pp. 190–204. Springer, Heidelberg (2001)
3. Hoefler, T., Gottschling, P., Lumsdaine, A., Rehm, W.: Optimizing a Conjugate Gradient Solver with Non-Blocking Collective Operations. Elsevier Journal of Parallel Computing (PARCO) 33(9), 624–633 (2007)
4. Hoefler, T., Lumsdaine, A., Rehm, W.: Implementation and Performance Analysis of Non-Blocking Collective Operations for MPI. In: Proc. of the 2007 Intl. Conf. on High Perf. Comp., Networking, Storage and Analysis, SC 2007, IEEE Computer Society/ACM (November 2007)
5. Kothe, D., Kendall, R.: Computational science requirements for leadership computing. Technical report, ORNL/TM-2007/44 (2007)
6. Chaarawi, M., Chandok, S., Gabriel, E.: Performance Evaluation of Collective Write Algorithms in MPI I/O. In: Allen, G., Nabrzyski, J., Seidel, E., van Albada, G.D., Dongarra, J., Sloot, P.M.A. (eds.) ICCS 2009. LNCS, vol. 5544, pp. 185–194. Springer, Heidelberg (2009)
7. Chaarawi, M., Gabriel, E., Keller, R., Graham, R.L., Bosilca, G., Dongarra, J.J.: OMPIO: A Modular Software Architecture for MPI I/O. In: Cotronis, Y., et al. (eds.) EuroMPI 2011. LNCS, vol. 6960, pp. 81–89. Springer, Heidelberg (2011)
8. Gabriel, E., Fagg, G.E., Dongarra, J.J.: Evaluating dynamic communicators and one-sided operations for current MPI libraries. International Journal of High Performance Computing Applications 19(1), 67–79 (2005)
9. Gabriel, E., Venkatesan, V., Shah, S.: Towards high performance cell segmentation in multispectral fine needle aspiration cytology of thyroid lesions. Computational Methods and Programs in Biomedicine 98(3), 231–240 (2009)
10. Frigo, M., Johnson, S.G.: The Design and Implementation of FFTW3. Proceedings of IEEE 93(2), 216–231 (2005); Special issue on Program Generation, Optimization, and Platform Adaptation
11. Bell, C., Bonachea, D., Cote, Y., Duell, J., Hargrove, P., Husbands, P., Iancu, C., Welcome, M., Yelick, K.: An evaluation of current high-performance networks. In: Proc. of the 17th Int. Symp. on Par. and Distr. Proc., p. 28.1 (2003)
12. Hoefler, T., Lumsdaine, A.: Message Progression in Parallel Computing - To Thread or not to Thread?. In: Proceedings of the 2008 IEEE International Conference on Cluster Computing. IEEE Computer Society, Los Alamitos (2008)

Optimizing MPI One Sided Communication on Multi-core InfiniBand Clusters Using Shared Memory Backed Windows

Sreeram Potluri, Hao Wang, Vijay Dhanraj, Sayantan Sur, and Dhabaleswar K. Panda

Department of Computer Science and Engineering, The Ohio State University
{potluri,wangh,dhanraj,surs,panda}@cse.ohio-state.edu

Abstract. The Message Passing Interface (MPI) has been very popular for programming parallel scientific applications. As the multi-core architectures have become prevalent, a major question that has emerged is about the use of MPI within a compute node and its impact on communication costs. The one-sided communication interface in MPI provides a mechanism to reduce communication costs by removing matching requirements of the send/receive model. The MPI standard provides the flexibility to allocate memory windows backed by shared memory. However, state-of-the-art open-source MPI libraries do not leverage this optimization opportunity for commodity clusters. In this paper, we present a design and implementation of intra-node MPI one-sided interface using shared memory backed windows on multi-core clusters. We use MVAPICH2 MPI library for design, implementation and evaluation. Micro-benchmark evaluation shows that the new design can bring up to 85% improvement in Put, Get and Accumulate latencies, with passive synchronization mode. The bandwidth performance of Put and Get improves by 64% and 42%, respectively. Splash LU benchmark shows an improvement of up to 55% with the new design on 32 core Magny-cours node. It shows similar improvement on a 12 core Westmere node. The mean BFS time in Graph500 reduces by 39% and 77% on Magny-cours and Westmere nodes, respectively.

Keywords: MPI, shared memory, one-sided communication.

1 Introduction

Message Passing has been the most popular model for developing parallel scientific applications. However, with the advent of multi-core processors, researchers have questioned the use of message passing for intra-node communication, since the sender-receiver interaction and tag matching pose a significant overhead. MPI One Sided Communication provides a better alternative by avoiding these overheads. In this model, the origin process can independently initiate and complete transfers from remote memory without requiring any involvement from the process owning it. Traditionally, many MPI libraries have implemented intra-node one sided communication calls over the two-sided model. This prevents them from achieving the true potential of one-sided communication within a node.

Y. Cotronis et al. (Eds.): EuroMPI 2011, LNCS 6960, pp. 99–109, 2011.

1.1 Motivation

The one sided communication model provides the flexibility to create windows in shared memory, thus providing true one sided intra-node transfers with minimal overheads. MPI implementations on shared memory machines and on platforms with the SHM protocol [4,13] have taken advantage of this to achieve good performance. However, their usage is limited to the particular platforms or a single node. MVAPICH2 supports the use of kernel modules to implement one sided communication [6]. The usability of kernel-assisted methods are in general limited due to the requirement for kernel compatibility, system administrator support in installation, and overheads for small messages. This has motivated the work in this paper to design shared memory backed windows in MVAPICH2 for use on multi-core InfiniBand clusters. Figure 1 depicts the design choice presented in this paper.

1.2 Contributions

In this paper, we make the following key contributions:

1. We design and implement shared memory backed windows in MVAPICH2.
2. We discuss the interactions of this design with the existing designs for inter-node communication on InfiniBand clusters.
3. Using micro-benchmarks and application-level benchmarks, we show the improved performance with our new design.

Micro-benchmark evaluation shows that the design using shared memory backed windows can achieve up to 85% improvment in latency for Put, Get and Accumulate, in passive mode. Bandwidth for Put and Get improves up to 64% and 42%, respectively. Splash LU benchmark, and Graph 500 benchmark can get up to 55%, and 39% improvement, respectively, on a 32 core AMD Magny-Cours node. They show 55% and 77% improvement respectively, on a 12 core Intel Westmere node.

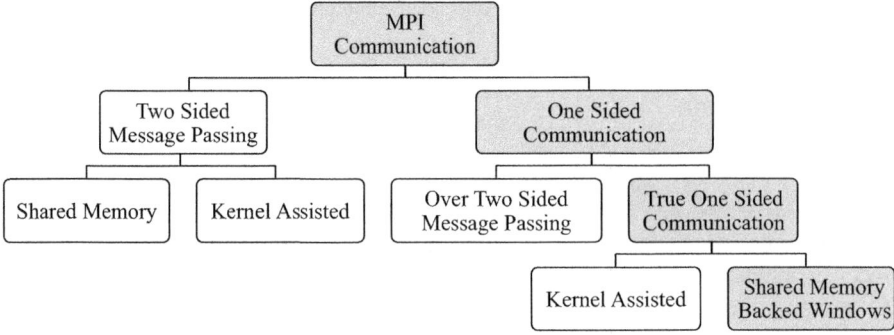

Fig. 1. MPI Intra-Node Communication Design (design choice in this paper is colored)

2   Background and Related Work

In this section, we provide an overview of the one-sided interface in MPI-2 and the extensions in MPI-3 Draft [2] that are relevant to our work. We also discuss the existing implementations for one-sided communication within a node.

2.1   MPI One-Sided Communication Model

MPI one-sided interface enables direct remote memory access through the concept of a "window". Window is a region in memory that each process exposes to others through a collective operation: MPI_Win_create. After that, each process can directly read or update data in the windows at all other processes in the communicator. Semantically, this differs from two-sided communication in that, the origin specifies all the parameters required for a communication (including remote address and datatype). In MPI-2, the user has to allocate the buffer and pass it onto the MPI_Win_create function. The standard suggests the use of MPI_Alloc_mem to allocate window memory as this allows the MPI libraries to allocate memory in way to achieve high performance. The MPI-3 draft introduces a new function: MPI_Win_allocate which lets the library to both allocate a buffer and create the window in one call. MPI-2 provides three communication operations all of which are non-blocking: MPI_Put (write), MPI_Get (read) and MPI_Accumulate (update). Access to windows and start/completion of communication operations is managed through synchronization calls. Synchronization modes provided by MPI-2 RMA can be classified into passive (no explicit participation from the target) and active (involves both origin and target). In the passive mode, an epoch (period in which a window can be accessed) is bounded by calls to MPI_Win_lock and MPI_Win_unlock. As the name suggests, this does not require any participation from the remote process. MPI-2 one-sided interface provides two modes of active synchronization: a) a collective synchronization on the communicator: MPI_Win_fence; and b) a group based synchronization: MPI_Win_Post-MPI_Win_Wait and MPI Win start-MPI_Win_complete. In the later mode, the origin process uses MPI_Win_start and MPI_Win_complete to specify access epoch for a group of target processes, and a target process calls MPI_Win_post and MPI_Win_wait to specify exposure epoch for a group of origin processes to access its window. The communication operations issued by an origin can execute only after the target has called post, and the target can complete an epoch only when all the origins in the post group have called complete on the window. Normally multiple RMA operations are issued in an epoch to amortize the synchronization overhead. The Draft MPI-3 standard adds several new calls for communication and synchronization. However, we focus our discussions in this paper to the functions discussed above.

2.2   Existing Implementation for Intra-node MPI One-sided Communication

One-sided communication within a node can be designed in several ways. Most libraries implement them over Send/Receive calls using shared memory buffers. The origin process copies the data in and the destination process copies it out. It cannot provide asynchronous progress that is desired in one-sided communication. Also, for large messages, the copies into and out of shared memory buffers become a huge overhead. Some kernel assisted techniques like LIMIC2 [5] and KNEM [3] have been proposed to avoid these extra copies. The source and destination buffers are mapped onto the kernel's address space and the data is copied directly between the two buffers, through a kernel module. However, this still does not solve the problem of asynchronous progress. More recent work [6] has proposed designs for direct implementation of one-sided communication using kernel assistance. This is supported in MVAPICH2 from version 1.6. The window buffers are mapped into the kernel address space which enables zero-copy transfers. The usability of kernel-assisted methods are in general limited due to the requirement for kernel compatibility, system administrator support in installation, and overheads for small messages. One-sided communication has been efficiently implemented over shared memory systems [13,4] where memory is globally accessible. However these designs are platform dependent or are restricted to a single node. In this paper we propose the use of shared memory backed windows for intra-node one-sided communication on modern multi-core clusters.

3   Design and Implementation

In this section, we describe in detail, the design of the One Sided Communication and Synchronization semantics using shared memory backed windows. We also discuss the interaction of the proposed intra-node design with existing implementations of inter-node communication.

3.1   Window Creation

In this section, we present the design of shared memory backed windows in the context of two window creation mechanisms: MPI_Win_create (MPI-2 and draft MPI-3) and MPI_Win_allocate (draft MPI-3).

Using MPI_Alloc_mem: In our design, memory regions created using MPI_Alloc_mem are allocated as shared memory files. They are mapped and maintained as a list sorted by their starting addresses. When the application calls MPI_Win_create with a buffer, we parse the shared memory file list to check if it was allocated in shared memory or not. If window memory at all the processes on a given node was not allocated in shared memory, the processes fall-back to the default implementation that uses Send/Receive calls. Otherwise, each process checks if it has already mapped the corresponding shared memory file. This check is required as two or more windows can be created from memory in the same file/buffer. If not, the process maps the shared memory files onto its virtual address space and generates the base addresses.

Using MPI_Win_allocate: Window allocation in shared memory becomes easier with the new semantics. Each buffer will have only one window associated with it. The steps of window creation are similar to the earlier case except that the information about the shared memory files can associated with and stored within the window structure. No external data structure is required.

3.2 Communication

Implementation of communication operations over shared memory backed windows requires two simple steps. The process initiating the call calculates the target address using the base address of the window (mapped from the shared memory file) and the displacement specified in the operation. It can then directly copy or compute on the data at the target address.

3.3 Synchronization

Lock-Unlock: MPI_Win_lock and MPI_Win_unlock can be implemented over Send/Receive using lock request and lock grant messages. In the case when there is just one communication operation between the lock and unlock calls, the sequence of lock-op-unlock operations can be accomplished through just one message. This implementation applies for both inter-node and intra-node scenarios, in the current implementation of MVAPICH2.

In the new design, each intra-node lock is implemented as two shared memory counters allocated during window creation. One counter has the information about the kind (exclusive/shared) of lock any process holds on the window. The second counter contains the number of processes holding the lock when the type is shared. When a process is handling a request it received from an off-node process, it checks the shared memory lock counters for any existing conflicting locks. While granting a lock, it sets its shared memory lock counters appropriately so that no intra-node peer can acquire a conflicting lock.

Locks can be implemented using InfiniBand RDMA atomics as described in [9,7]. However, InfiniBand does not provide atomicity between CPU initiated lock operations and network-initiated atomic operations. Therefore, we implement locks between intra-node processes using loop-back network atomic operations as suggested in [9].

Post-Wait/Start-Complete: In a Send/Receive based implementation of Post-Wait/Start-Complete, an MPI_Win_Post call at the target converts to a message being sent to each process in the specified group. On the other end, an origin process calling MPI_Win_start will wait for post messages to arrive from all the processes in the specified group, before issuing communication calls. A similar interaction happens in the opposite direction when MPI_Win_complete and MPI_Win_wait calls are called. When communication operations are deferred until the second set of synchronization calls, complete messages can be piggybacked onto the last data message being sent from a source to a target.

We implement Post-Wait/Start-complete synchronization within a node using shared memory counters. Each process has a Post and a Complete flag per window for every other process on the same node. Processes can directly set these counters to signal Post and Complete operations instead of sending messages. The processes calling Start and Complete poll on these counters. A similar intra-node design works in conjunction with RDMA-based implementation where the signaling happens through RDMA writes to pre-allocated counters.

Fence: MPI_Win_Fence is a collective operation. In one call, it marks the end of the previous epoch, ensuring the completion of all the operations issued before it and also marks the beginning of the next epoch. In Send/Receive based implementations, each process gets the count of operations that were issued in the epoch with its window as the target. MPI_Reduce_Scatter is used for this exchange. Then each process waits for the expected number of messages to arrive from every other process. An MPI_Barrier is called to ensure completion at all the processes before starting the next epoch. In our design, intra-node communication operations are blocking and can be considered complete when they return. So no additional handling is required during the Fence call. The existing Barrier ensures the required synchronization.

Several design choices exist for implementing Fence using RDMA operations. One can use counters like in the case of Post-Wait/Start-Complete, maintaining Post and Complete counters for all the processes in the communicator. More optimized schemes can be designed using RDMA Write with Immediate Data as described in [11]. In either case, the intra-node synchronization can still be handled by calling a Barrier on the shared memory (intra-node) communicator.

4   Experimental Results

In this section we describe our experiments and give an in-depth analysis of the results. We have carried out all micro-benchmark experiments on the Intel Westmere platform. Each node has 8 cores (4 per socket) running at 2.67GHz with 12GB of DDR3 RAM. The operating system is RHEL Server 5.4. We use MVAPICH2 1.6 version and LIMIC2 0.5.4 kernel module. We have evaluated the designs with application benchmarks on two platforms. One is Trestles at San Diego Supercomputing Center. Each node contains four sockets, each with a 8-core AMD Magny-Cours processor, and 64 GB memory per node. The other is Lonestar at Texas Advance Computing Center. Each node has two sockets, each with a 6-core Intel Westmere processor, and 24GB memory per node. Both use CentOS 5.5.

4.1   Micro-Benchmark Evaluation

In order to avoid effects of compiler and system optimizations on the memory to memory copies, we have modified OSU Micro Benchmark (OMB) [8] to use a different buffer in each transfer of a given message size and we modify the contents of the buffers after each message size. All point-to-point experiments were run across cores on different sockets with no shared cache. "MV2" refers to the version that uses Send/Recv based implementation of One Sided Communication. "MV2-LIMIC" refers to the designs using LIMIC2 discussed in [6]. "MV2-SWIN" represents the designs presented in this paper, using shared memory backed windows.

Latency: Figure 2 shows latency performance for MPI_Put, MPI_Get and MPI_Accumulate. These benchmarks use Post-Wait/Start-Complete mode of synchronization. MVAPICH2 enables LIMIC2 in one sided communication for message sizes greater than 4Kbytes. For smaller messages, the overhead from LIMIC2 is greater than the

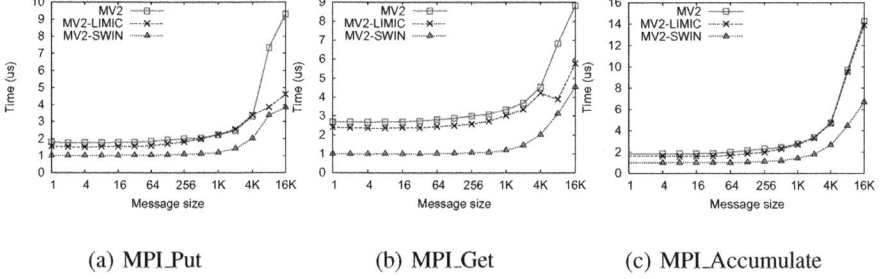

(a) MPI_Put   (b) MPI_Get   (c) MPI_Accumulate

Fig. 2. Latency Performance

benefits it gives. The slight improvement in "MV2-LIMIC" over "MV2" for small messages is from a shared-memory based synchronization. Compared with "MV2-LIMIC", "MV2-SWIN" does not have the overhead of the kernel module and can be used for all message sizes. It performs consistently better than "MV2-LIMIC" for small messages. For larger messages, the overhead from the kernel module becomes negligible, so "MV2-SWIN" and "MV2-LIMIC" converge as the message size increases. At 16 KB, "MV2-SWIN" performs 16% better than "MV2-LIMIC" and 59% better than "MV2". MPI_Accumulate does not use the kernel assisted design. Hence we see similar performance for "MV2" and "MV2-LIMIC". We see 53% improvement in Accumulate latency with "MV2-SWIN" for 8Kbyte messages.

Bandwidth: The bandwidth performance is shown in Figure 3. "MV2-SWIN" clearly out-performs the existing techniques and achieves close to system peak performance. We see up to 64% improvement in bandwidth for Put and 42% for Get compared with "MV2-LIMIC". "MV2-LIMIC" and "MV2" designs queue operations until the synchronization phase. These queuing overheads are removed in "MV2-SWIN" where operations are executed as soon as they are issued.

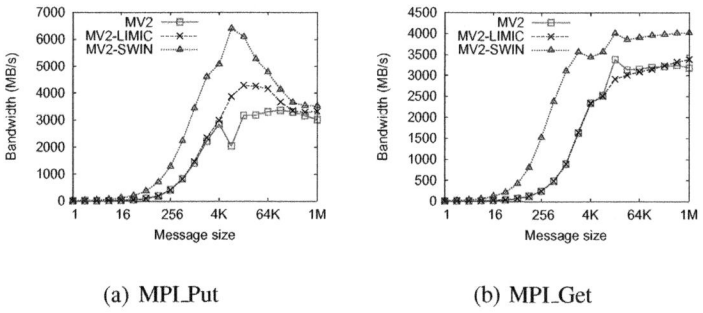

(a) MPI_Put   (b) MPI_Get

Fig. 3. Bandwidth Performance

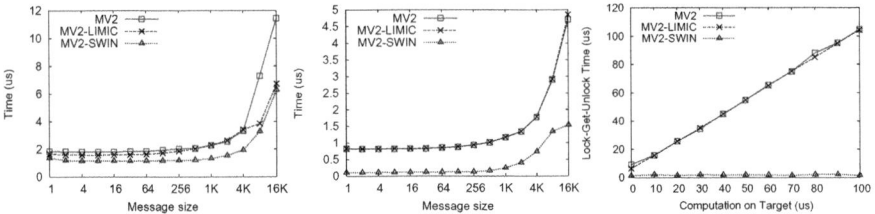

(a) Off-cache MPI_Put Latency (b) Passive MPI_Put Latency (c) MPI_Get Latency w/ Busy Target

Fig. 4. Performance with Off-cache data, Passive synchronization and Busy target

Latency Performance with Off-Cache Data: In Figure 4(a), we have modified Put latency benchmark to ensure the data is off-cache, by accessing data from a different page in each iteration and by flushing the cache after each message size. We believe this experiment will show the performance behavior of the designs in real-world applications. For 4Kbyte messages "MV2-SWIN" shows 43% improvement compared to "MV2-LIMIC".

Latency with Passive Synchronization: Figure 4(b) shows Put latency with passive mode of synchronization. One Sided Communication in this mode has not been optimized with LIMIC2 kernel module. Hence we observe similar performance for "MV2" and "MV2-LIMIC" up to 16K. We see more than 85% improvement using "MV2-SWIN".

Passive Latency with Busy Target: This experiment shows the asynchronous progress that can be achieved using the shared memory backed windows. We measure the latency of Lock-Put-Unlock when the remote process is busy in computation for an increasing amount of time. The results are shown in Figure 4(c). For "MV2-SWIN", where the completion of passive operations does not require any remote process intervention, we see that the time remains constant.

Multi-pair Bandwidth and Message Rate: We use all 8 cores on the Westmere node with 4 pairs of MPI processes. Figure 5(a) shows the multi-pair bandwidth performance using MPI_Put. "MV2-SWIN" gives 69% better bandwidth than "MV2-LIMIC" for 4Kbyte messages. For messages beyond 4Kbytes, the two designs perform similarly. Figure 5(b) shows the multi-pair message rate performance of the different designs. "MV2-SWIN" achieves nearly 4.5 times improvement in message rate compared to other designs for 4byte messages.

4.2   Application Benchmark Evaluation

In this section we evaluate our new design with application benchmarks. The application benchmarks use communication paths (passive synchronization and accumulate operations) that were not optimized using LIMIC2. So we present results comparing

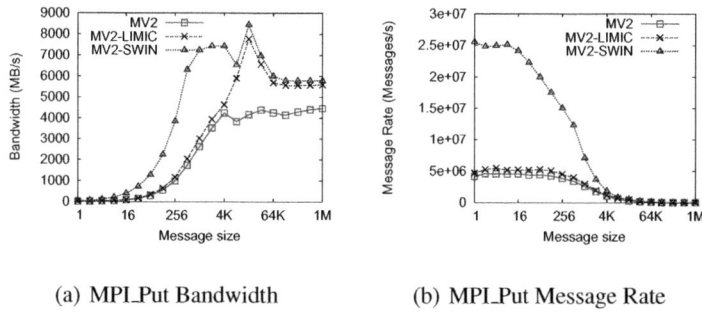

(a) MPI_Put Bandwidth   (b) MPI_Put Message Rate

Fig. 5. Multi-Pair Put Bandwidth and Message Rate

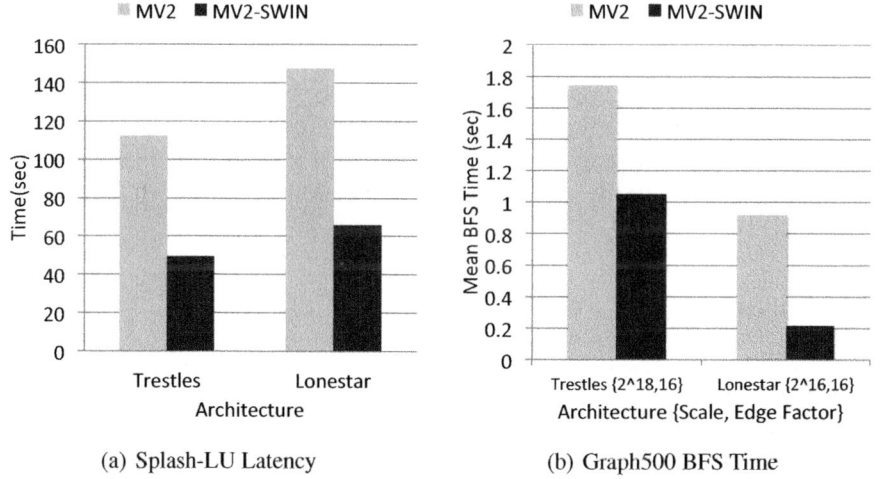

(a) Splash-LU Latency   (b) Graph500 BFS Time

Fig. 6. Splash LU and Graph500

"MV2-SWIN" with the default design "MV2". All the experiments were run on a single node i.e. with 32 cores on Trestles and 12 cores on Lonestar.

Splash LU: We use Splash LU benchmark [12] modified to use MPI-2 communication calls. The design of this benchmark is outlined in [10] and uses the passive synchronization and MPI_Get calls. We run the test on a matrix of size 16Kx16K matrix of doubles and a block size of 128x128. From results shown in 6(a), the new design "MV2-SWIN" shows a 55% improvement compared to "MV2" on Trestles. It shows a similar improvement on Lonestar.

Graph500: Now, we compare the performance of the Breadth First Search(BFS) kernel from the Graph500 [1] benchmark suite using "MV2" and "MV2-SWIN". This kernel uses Fence synchronization and Accumulate operations. Graphs of size 2^18 and

$2^{\wedge}16$ nodes were used on Trestles and Lonestar respectively. Results in 6(b) show that "MV2-SWIN" achieves 39% lower mean BFS time compared to "MV2", on Trestles. It achieves a 77% improvement on Lonestar.

5   Conclusion

Two-sided message passing leads to significant overhead for intra-node communication due to requirement for sender-receiver interactions and tag matching. The one-sided communication interface in MPI provides a better alternative. In this paper, we design one-sided communication using shared memory backed windows for use on multi-core InfiniBand clusters. Experimental evaluation using micro-benchmarks and application benchmarks has shown that our design performs significantly better than the existing kernel-assisted and send/receive-based options. We have implemented our design in MVAPICH2 and will make it publicly available through a release in the near future.

Acknowledgments. This research is supported in part by U.S. Department of Energy grants #DE-FC02-06ER25749 and #DE-PFC02-06ER25755; National Science Foundation grants #CCF-0833169, #CCF-0916302, #OCI-0926691 and #CCF-0937842; grants from Intel, Mellanox, Cisco, QLogic, and Sun Microsystems; Equipment donations from Intel, Mellanox, AMD, Appro, Chelsio, Dell, Microway, QLogic, and Sun Microsystems.

References

1. Graph500, http://www.graph500.org/
2. MPI-3 RMA, https://svn.mpi-forum.org/trac/mpi-forum-web/raw-attachment/wiki/mpi3-rma-proposal1/one-side-2.pdf
3. Barrett, B.W., Shipman, G.M., Lumsdaine, A.: Analysis of implementation options for MPI-2 one-sided. In: Cappello, F., Herault, T., Dongarra, J. (eds.) PVM/MPI 2007. LNCS, vol. 4757, pp. 242–250. Springer, Heidelberg (2007)
4. Booth, S., Mourao, E.: Single sided MPI implementations for SUN MPI. In: Proceedings of the ACM/IEEE Conference on Supercomputing, p. 2 (2000)
5. Jin, H.W., Sur, S., Chai, L., Panda, D.K.: Lightweight kernel-level primitives for high-performance MPI intra-node communication over multi-core systems. In: Proceedings of IEEE International Conference on Cluster Computing, pp. 446–451 (2007)
6. Lai, P., Sur, S., Panda, D.K.: Designing Truly One-Sided MPI-2 RMA Intra-node Communication on Multi-core Systems. In: Proceedings of International Supercomputing Conference (ISC), vol. 25, pp. 3–14 (2010)
7. Narravula, S., Mamidala, A., Vishnu, A., Vaidyanathan, K., Jin, H.W., Panda, D.K.: High Performance Distributed Lock Management Services using Network-based Remote Atomic Operations. In: Proceedings of IEEE/ACM International Symposium on Cluster Computing and the Grid (CCGrid), pp. 583–590 (2007)
8. OSU Microbenchmarks: http://mvapich.cse.ohio-state.edu/benchmarks/
9. Santhanaraman, G., Balaji, P., Gopalakrishnan, K., Thakur, R., Gropp, W., Panda, D.K.: Natively Supporting True One-Sided Communication in MPI on Multi-core Systems with Infini-Band. In: Proceedings of the 9th IEEE/ACM International Symposium on Cluster Computing and the Grid (CCGrid), pp. 380–387 (2009)

10. Santhanaraman, G., Narravula, S., Panda, D.K.: Designing Passive Synchronization for MPI-2 One-Sided Communication to Maximize Overlap. In: Proceedings of International Parallel and Distributed Processing Symposium (IPDPS), pp. 1–11 (2008)
11. Santhanaraman, G., Gangadharappa, T., Narravula, S., Mamidala, A., Panda, D.K.: Design Alternatives for Implementing Fence Synchronization in MPI-2 One-sided Communication on InfiniBand Clusters. In: Proceedings of IEEE Cluster, pp. 1–9 (2009)
12. Singh, J.P., Weber, W., Gupta, A.: Splash: Stanford parallel applications for shared-memory. Tech. rep., Stanford, CA, USA (1991)
13. Thakur, R., Gropp, W., Toonen, B.: Optimizing the Synchronization Operations in MPI One-Sided Communication. In: International Journal of High Performance Computing Applications (IJHPCA), pp. 119–128 (2005)

A uGNI-Based MPICH2 Nemesis Network Module for the Cray XE[*]

Howard Pritchard[1], Igor Gorodetsky[1], and Darius Buntinas[2]

[1] Cray Inc.
[2] Argonne National Laboratory

Abstract. Recent versions of MPICH2 have featured Nemesis – a scalable, high-performance, multi-network communication subsystem. Nemesis provides a framework for developing Network Modules (Netmods) for interfacing the Nemesis subsystem to various high speed network protocols. Cray has developed a user-level Generic Network Interface (uGNI) for interfacing MPI implementations to the internal high speed network of Cray XE and follow-on computer systems. This paper describes the design of a uGNI Netmod for the MPICH2 nemesis subsystem. MPICH2 performance data on the Cray XE are presented.

1 Introduction

The Cray XE represents a fundamental change in network architecture from its predecessor XT systems [1]. The Cray XE *Gemini* network provides user-space applications with a low-overhead, programmed I/O (PIO) mechanism for accessing memory on remote nodes in a true one-sided fashion. Termed *Fast Memory Access* (FMA), this hardware supports remote direct memory access (RDMA) read, write, and atomic memory operations (AMOs) to memory at remote nodes. The Gemini also has a Block Transfer Engine (BTE) to offload RDMA read and write operations from the host processor. In addition, Gemini was designed with fault-tolerance related features that allow software to recover from various network errors more reliably than on the predecessor systems.

In this paper, we present the design and implementation of the MPICH2 uGNI network module along with a brief overview of the Cray XE network's Generic Network Interface (GNI) API. We also present performance evaluation of the new network module.

The paper is organized as follows. A summary of the Generic Network Interface (GNI) is given in Section 2. An overview of MPICH2 Nemesis and its Network Module framework follows in Section 3. Details of the uGNI Network Module and related support software are presented in Section 4. Some basic performance results on Cray XE are presented in Section 5. The paper concludes with a discussion of future work in Section 6.

---

[*] This material is based upon work supported by the Defense Advanced Research Projects Agency under its Agreement No. HR0011-07-9-0001. Any opinions, findings and conclusions or recommendations expressed in this material are those of the author(s) and do not necessarily reflect the views of the Defense Advanced Research Projects Agency. This work was supported in part by the Office of Advanced Scientific Computing Research, Office of Science, U.S. Department of Energy, under Contract DE-AC02-06CH11357.

Y. Cotronis et al. (Eds.): EuroMPI 2011, LNCS 6960, pp. 110–119, 2011.

2   Generic Network Interface

The Generic Network Interface (GNI) [3] provides a low-level API for network middleware to efficiently utilize the Cray XE network. GNI is primarily intended for user-space and kernel-space network applications whose communication patterns are message-based in nature, and where the ability to recover from network faults is of importance. The API is not intended to be used to develop end-user applications, but rather to be used by library developers to develop message passing libraries, such as MPI.

A layered approach was taken in designing GNI. A lowest level Generic Hardware Abstraction Layer (GHAL) is used to interface to particular implementations of Gemini. This layer is used to mask implementation details, such as specific details on hardware registers, from the upper level components. Other components of the GNI stack include *kGNI* - the device driver which also implements the kernel level API and a *uGNI* library which implements the API for user-space applications. Figure 1 depicts the layered view of the software stack with two sample clients: the MPICH2 MPI implementation in user-space, and Luster's LNET layer, (GNILND) in kernel space. Elements of the API and characteristics of RDMA transactions and messaging are briefly described below.

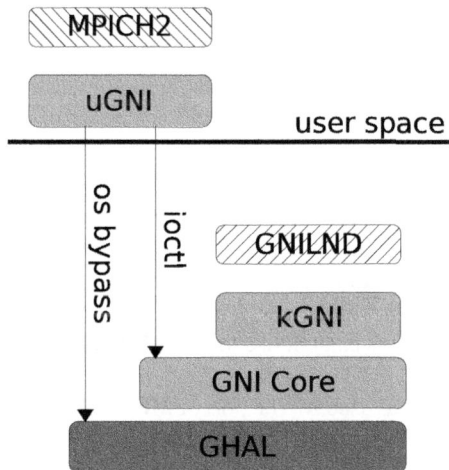

Fig. 1. GNI Software Stack with MPICH2 and LNET(GNILND) example clients

2.1   Elements of the API

The GNI API uses a number of software constructs to leverage the underlying Gemini hardware in a way that provides flexibility to developers using the API, while at the same time providing sufficient abstraction to map efficiently to future instantiations of Gemini. Details of the API can be found in [3]. Important concepts of the API include *Communication Domains* (CDM), *End Points* (EP), *Completion Queues* (CQ) and *Memory Handles* (MH). A CDM is a group of processes (i.e., *peers*) that form a hardware protection domain within the network. An EP is a software construct used to manage data exchange between peers within a CDM. A MH is a handle to a region of memory used for RDMA operations. A CQ delivers *Completion Queue Events* (CQEs) that provide a light weight notification mechanism for tracking network transactions. CQEs provide indication of global completion of network transactions, including any associated errors, and can be used to provide notification that messages have arrived within a memory region associated with a MH.

2.2   Remote Direct Memory Access Transactions and Messaging

GNI provides an interface for initiating RDMA transactions using either the BTE or FMA units. RDMA writes, reads and AMOs are supported. An application typically requests a global CQE for every RDMA or messaging transaction. The Gemini provides a specialized RDMA Write with remote notification operation which is used by GNI to support two messaging methods. The GNI *Short Message* (SMSG) facility provides the highest performance in terms of latency and short messages rates, but comes at the expense of memory usage, which grows linearly with the number of peer-to-peer connections. A *Message Queue* (MSGQ) facility is also available and is much more scalable in terms of memory usage than SMSG channels, but has some additional performance overhead. MSGQ memory usage scales with the number of *nodes* in the job rather than peers. Both facilities provide reliable, in-order message delivery by exploiting features of the Gemini write-with-remote-notification hardware.

3   MPICH2 Nemesis

MPICH2 is a widely used, open-source implementation of MPI developed and maintained by Argonne National Laboratory (ANL) [6]. Vendors have several options for porting MPICH2 to their custom interconnect. The vendor can choose to create a *device*, *channel* or Nemesis network module (*netmod*), depending on their need for flexibility vs. implementation effort. Figure 2 shows the different layers of MPICH2.

Network hardware vendors can choose to port MPICH2 to a custom interconnect by implementing *device* to the ADI3 interface. This option, which gives the greatest flexibility in implementation, was used for the port of MPICH2 to the Cray XT. However, developing and maintaining a complete, custom ADI3 device can be quite expensive and frequently leads to redundant development that contributes little in the way of differentiation of a custom interconnect. Additionally, the pace of development of MPICH2 at Argonne has accelerated recently, partially driven by the desire to implement proposed MPI-3 extensions to MPI.

An alternative to developing a full ADI3 device is to implement a CH3 *channel*. MPICH2 includes a default device implementation called CH3. CH3 provides the *CH3 interface* to allow vendors to implement channels. In principle this interface requires significantly less development effort to code to, while still delivering reasonable performance. However, as the availability of commodity RDMA capable networks increased, and thus the interest in porting MPICH2 to networks using protocols other than TCP sockets, deficiencies with the channel API became apparent. Vendors and other organizations more often than not just reworked the CH3 ADI3 device for a particular network, rather than using the channel interface.

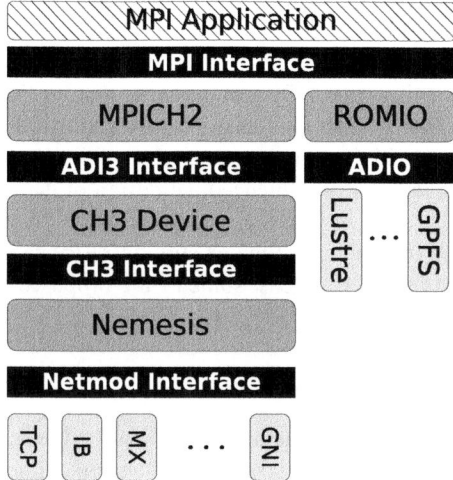

Fig. 2. MPICH2 and Nemesis CH3 channel software stack with sample Network Modules

This was one of the motivations for the introduction of a new *Nemesis* channel to the CH3 ADI3 device. The goals of the Nemesis channel, as stated by the authors, are scalability, high performance intranode communication, high performance internode communication, and multi-network internode communication [2]. Although the ultimate goal is to make Nemesis a self-standing ADI3 device, it was found sufficient at the time to make modifications to the CH3 device itself to better support the Nemesis package as a channel. Note that although the figure shows multiple software layers, the layers do not add significant overhead due to the use of function pointers and up-calls directly to the upper layers.

The major components of Nemesis are a highly optimized on-node messaging system and a multi-method capable framework for implementing *network modules* (Netmods) within Nemesis. The framework is flexible and can be used for a variety of interconnects as evidenced by existing modules such as Myrinet MX and GM and a recent IB module available in the MVAPICH2 version of MPICH2 [8]. The basic function of a Netmod is to move control messages, which can be application messages, and data across a network. The upper components of Nemesis implement the MPI portion, e.g. message matching, handling of unexpected messages, etc. There are hooks in Nemesis to support Netmods that have MPI-awareness such as hardware message matching in their networks.

Nemesis features a *Long Message Transfer* (LMT) protocol that facilitates implementation of zero-copy transfers for Netmods interfacing to networks that support these types of operations. When the LMT path is used, only short control messages actually move through the Nemesis stack itself. The bulk message data can be transferred in a zero-copy fashion from the application's send buffer into the application's receiver buffer. Note there are some exceptions to when the LMT is used, even when the message size is greater than the eager message size. Ready send messages do not use the LMT path. Messages generated from MPI-2 RMA functions do not currently use this path.

4   The uGNI Netmod

An important goal in the design of the uGNI Netmod was to exploit features of Gemini specifically designed to improve MPI performance, such as the FMA hardware. This hardware enables the MPICH2 implementation on the Cray XE to realize much lower latencies and significantly higher message rates than with the predecessor XT systems. Other factors which significantly influenced the design of the uGNI Netmod were the requirement to be able to handle transient network faults, interoperability with other program models, reuse of as much of the existing infrastructure in MPICH2 as possible, and extensibility to support at least some of the proposed MPI-3 Fault Tolerance features [4].

4.1   Initialization and Connection Setup

The uGNI Netmod's initialization method is invoked by Nemesis as part of the overall MPI initialization procedure that takes place when an application calls *MPI_Init* or *MPI_Init_thread*. A CDM is created using the *ptag* value supplied by the ALPS process manager. The Netmod then attaches the CDM to all available Gemini NICs. The Netmod next initializes a registration cache (see Section 4.4), CQs are created using the NIC handles, DMA buffers are registered with the NIC handles, and a freelist of transaction management structures is created. An initial block of SMSG *mailboxes* is created and registered with the NIC handles. A set of EPs are created in order to post *wildcard* session management datagrams [3] with each of the NIC handles.

By default, SMSG channels are only established when a given rank in the job needs to send a message to another rank. If a channel has not been established yet, the sender allocates an SMSG mailbox, and prepares a channel establishment message describing the mailbox location within the pool of registered memory from which the mailbox was allocated. The GNI session managment protocol is used to set up the SMSG channel. The sender then delivers the original application message using the SMSG channel.

4.2   Eager Message Path

Owing to the relatively short messages that can be delivered by GNI SMSG, the eager path in the GNI Netmod actually uses two paths. If the application message data and internal MPICH2 CH3 header is under the maximum size message possible for the SMSG mailbox, then the message is delivered using this path alone. If the message is larger than can be delivered using GNI SMSG, an RDMA read path is used. Owing to semantics of Nemesis, arbitrarily large messages may actually be sent using this RDMA read path.

On the receive side, dequeuing of incoming messages is driven by the CQ associated with the SMSG mailboxes (see Section 4.1). The receiver polls the CQ to determine which SMSG *mailboxes* have messages to dequeue. As messages are received off the network, either directly via SMSG, or via completion of RDMA reads, they are handed off in the order received to Nemesis using the *MPID_nem_handle_pkt* function. The fact that Nemesis can handle processing of partial packets of a message significantly simplified this push/pull model for handling eager messages.

Table 1. SMSG Maximum Message and Mailbox Size

Job Size	Max. Msg. Size including CH3 hdr	Mailbox Size (bytes) per channel
≤ 1024	1024	4672
> 1024 ≤ 16384	512	2624
> 16384	256	1088

The maximum size message that can be sent using SMSG varies with the job size, with smaller mailboxes being used as the job size increases (see Table 1). An upcoming release of MPICH2 for Cray XE will be able to optionally use the MSGQ facility. For MSGQs the memory requirements are typically about 50 KB/node for each inter-node connection. Thus for a job spanning 10,000 nodes, about 500 MB is required on each node for the MSGQ.

4.3   Rendezvous Message Path

The Nemesis LMT path is used for delivering messages exceeding the eager message size threshold. As described on the Nemesis API wiki [7], the LMT path supports read, write, and cooperative data transfer mechanisms. The uGNI Netmod employs a read method for smaller LMT transfers and a cooperative, RDMA write-based method for longer transfers. The short control messages Nemesis uses for steering an application's MPI messages through the LMT procedure all use the SMSG path for eager messages described above in Section 4.2.

This path utilizes a memory registration cache (see Section 4.4). The bandwidth achieved using the LMT path is sensitive to the efficiency with which the registration cache is being utilized. The efficiency of the RDMA read path is also sensitive to the alignment of the send and receive buffers. RDMA writes are much less sensitive to alignment.

4.4   uDREG Library and Memory Registration

A registration cache library (*uDREG*) was implemented to reduce the overhead of memory registration for large message transfers. There are well known pitfalls to using a user-space memory registration cache in the context of the GNU/Linux environment [9]. To avoid these problems, a device driver was developed which utilizes the Linux MMU Notifier facility to inform uDREG when virtual memory (VM) activity by a process has resulted in invalidation of entries in the registration cache. VM issues attributable to *fork* operations are handled by kGNI.

4.5   Network Fault Tolerance

As discussed in Section 2.2, the GNI SMSG and MSGQ facilities guarantee reliable delivery of messages between two EPs. However, GNI does not deal with failed FMA or BTE initiated RDMA transactions. The Netmod implements fault tolerance with respect to transient network errors by using replayable RDMA transactions for bulk data delivery. The error code in the CQE associated with a transaction is used to determine whether it should be replayed. Notification messages go exclusively over the reliable channels made available by SMSG. The Netmod is only one component of the Cray XE network fault tolerance/fault recovery strategy. A complete description of the mechanism is beyond the scope of this paper.

5   Basic Performance Characteristics

The intent of this section is to provide basic performance data relevant to the uGNI Netmod and to explain how the data relates to both to the internal operation of the Netmod as well as the Gemini NIC and the Cray XE node architecture. A basic knowledge of the node architecture is assumed in these discussions. For reference, a depiction of the Cray XE node using AMD *Magny Cours* 12-core sockets is shown in Figure 3. All performance results were obtained on a Cray XE with Magny Cours 12-core socket nodes running at 2.0 GHz. The operating system was Cray Linux Env. (CLE) 3.1.61 and the MPICH2 packaged in MPT 5.3.0.5. Large pages were not used in any of the tests. Unless explicitly mentioned, default MPICH2 environment variables were used.

5.1   Message Rate and Latency

The OSU 3.3 MPI latency test was used to measure the latency for MPICH2. The 8-byte message latency between processes pinned to CPU 0 (Die 0) on nodes connected by adjacent Geminis for these test conditions was measured to be a little over 1.3 μsecs.

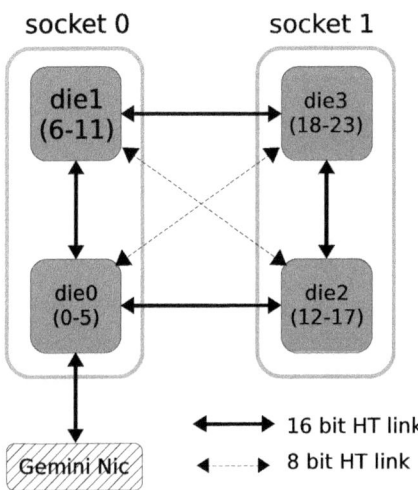

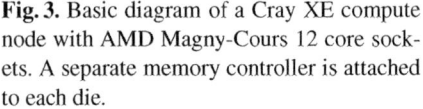

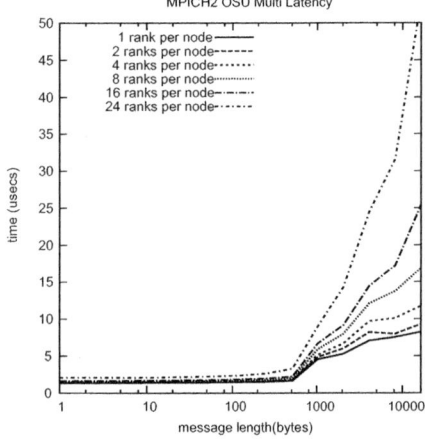

Fig. 3. Basic diagram of a Cray XE compute node with AMD Magny-Cours 12 core sockets. A separate memory controller is attached to each die.

Fig. 4.   MPICH2   Latency   for   multiple sender/receiver pairs per node

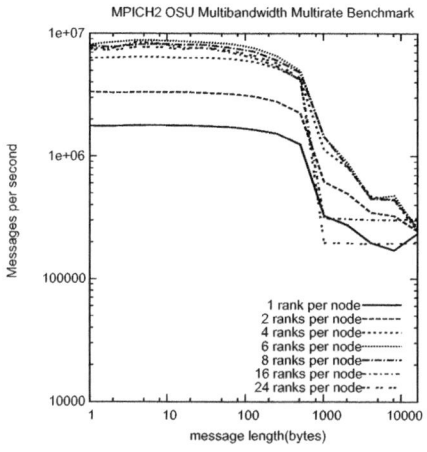

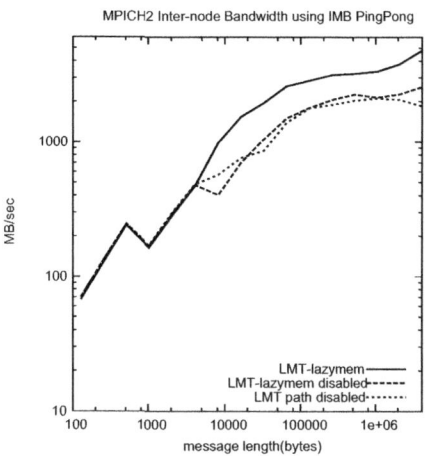

Fig. 5. MPICH2 message rate measured using the OSU mbw_mr test with different numbers of MPI ranks per node

Fig. 6. MPICH2 Inter-node IMB PingPong Bandwidth using various options for handling long messages

The cost of a network hop for an MPI message is about 150–200 nsecs. The one-way cost of the intra-node hop from one of the cores not adjacent to the Gemini NIC was measured to be about 90 nsecs.

Although the MPI latency for a single sender/receiver pair is useful to know, a more important metric for applications which are typically run using multiple MPI ranks per node is the latency when multiple sender/receivers are trying to exchange messages across a network interface. Figure 4 shows the results from the OSU 3.3 multi-latency (mult_lat) test. The test was run between two adjacent Gemini NICs. To improve the throughput for medium size messages, the MPICH_GNI_RDMA_THRESHOLD environment variable was set to 16 KiB for this test. The MPICH_GNI_MBOX_PLACEMENT environment variable was set to specify *NIC* placement for the SMSG mailboxes and CQs. This results in the GNI Netmod placing the SMSG mailboxes and CQs on the memory of Die 0 (see Figure 3). This gives much better performance than if the mailboxes and CQs are placed local to the MPI ranks. One observes that very good latency is observed for small messages even when there are 24 ranks per node up to 1024 bytes. It is at this point that the MPICH2 switches to the RMDA read eager protocol. At the largest message lengths shown in the figure, the latency is beginning to be dominated by the serializing effect of the BTE.

The aggregate message rate for short and medium size MPI messages is shown in Figure 5. These measurements were made also made with the MPICH_GNI_MBOX_PLACEMENT environment variable set to specify *NIC* placement. The MPICH_GNI_RDMA_THRESHOLD environment variable was not set for these measurements. The maximum message rate realized with this placement option, and using 2.0 GHz processors, is about 8 million MPI messages/sec. Rates of 9.8 MM/sec can be attained with 2.4 GHz processors. The drop off in message rate at 1024 bytes is due to the switch to the RDMA read eager protocol (Section 4.2).

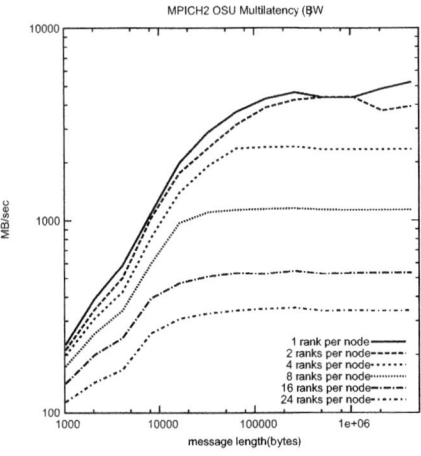
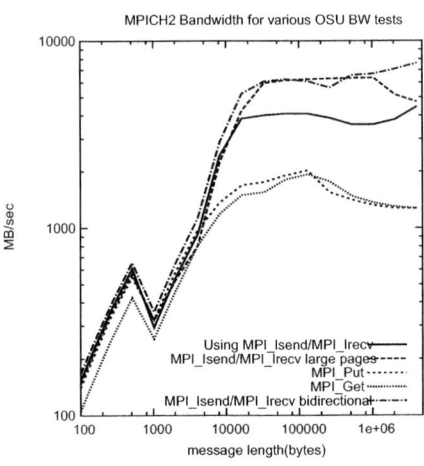

Fig. 7. MPICH2 bandwidth per rank for multiple ranks per node as derived from the latency measurements obtained using the OSU multi_lat test

Fig. 8. Comparison of realized bandwidth for MPI_Send/MPI_Recv and MPI_Put and MPI_Get. Also shown is bidirectional bandwidth obtained using the OSU bibw test.

5.2 Bandwidth

Bandwidth measurements were made using the IMB 3.2.2 PingPong test and various OSU 3.3 bandwidth tests. Unless otherwise mentioned, all tests were run between adjacent Gemini NICs.

Results of the IMB PingPong test are shown in Figure 6 for various ways of handling large messages. As shown in the figure, the best bandwidth is obtained when using the LMT path described in Section 4.3 and also using *lazy* memory deregistration for the registration cache. The bandwidth drops significantly if the lazy memory registration policy is not used. Disabling the LMT path at all has a similar effect on the bandwidth for large messages. The drop in bandwidth between 512 and 1024 bytes is again due to the switch to the RDMA read protocol in the eager path. The differences in bandwidth for the longer transfers methods only appear at 8 KiB and above because that is the default threshold for switching from the eager to the rendezvous protocol.

Since many MPI applications are typically run with multiple processes per node, bandwidth results when using multiple MPI send/receive pairs are shown in Figure 7. For this test, the MPICH_GNI_RDMA_THRESHOLD environment variable was again set to 16 KiB. The bandwidths are derived from the latencies obtained using the OSU 3.3 multi_lat test. These are the results in bandwidth rather than latency, for messages longer than those shown in Figure 4. At transfer sizes beyond 16 KiB bytes, the available bandwidth per rank is dominated by the effects of sharing the BTE between the ranks for transferring the message data.

Figure 8 is included to show effects of the MPICH2 Nemesis design on the bandwidth realized using different MPI methods and page sizes for transferring data, and also to show results of the OSU bidirectional bandwidth test. Significantly better bandwidth is obtained when using large pages. The Nemesis device currently does not use the LMT path for MPI-2 RMA transfers. Thus, the realized bandwidth for MPI RMA operations is similar to that for long MPI_Send messages with the LMT path disabled (Figure 6).

6   Future Work

A main area for enhancement of the uGNI Netmod is providing better support for independent progress of the state-engine, thus allowing for better overlap of computation with communication. Approaches being investigated include enhancing of the existing asynchronous-thread infrastructure within MPICH2, as well as more complex approaches (e.g. [5]). Longer-term, work on the Netmod will include adding support for MPI-3 features such Fault Tolerance and extended MPI-3 RMA functionality.

References

1. Alverson, R., Roweth, D., Kaplan, L.: The Gemini System Interconnect. In: Symposium on High-Performance Interconnects, vol. 0, pp. 83–87 (2010)
2. Buntinas, D., Mercier, G., Gropp, W.: Design and Evaluation of Nemesis, a Scalable, Low-Latency, Message-Passing Communication Subsystem. In: CCGRID 2006, pp. 521–530 (2006)
3. Cray, Inc.: Cray Software Document S-2446-3103: Using the GNI and DMAPP APIs (March 2011)
4. Fault Tolerance Working Group: Run-though Stabilization Interfaces and Semantics, `svn.mpi-forum.org/trac/mpi-forum-web/wiki/ft/ run_through_stabilization`
5. Lai, P., Balaji, P., Thakur, R., Panda, D.K.: ProOnE: a General-purpose Protocol Onload Engine for Multi- and Many-core Architectures. Computer Science - R&D, 133–142 (2009)
6. MPICH2: `www.mcs.anl.gov/research/projects/mpich2/`
7. MPICH2–Nemesis: Nemesis Network Module API, `wiki.mcs.anl.gov/mpich2/index.php/Nemesis_Network_Module_API`
8. Network–Based Computing Laboratory: MVAPICH: MPI over Infiniband, 10GigE/iWARP and RoCE, `mvapich.cse.ohio-state.edu/overview/mvapich2`
9. Wyckoff, P., Wu, J.: Memory Registration Caching Correctness. In: Proceedings of CCGrid 2005. IEEE Computer Society, Los Alamitos (2005)

Using Triggered Operations to Offload Rendezvous Messages

Brian W. Barrett[1], Ron Brightwell[1], K. Scott Hemmert[1],
Kyle B. Wheeler[1], and Keith D. Underwood[2]

[1] Sandia National Laboratries*
P.O. Box 5800, MS-1319
Albuquerque, NM, 87185-1319
{kshemme,bwbarre,kbwheel}@sandia.gov
[2] Intel Corporation
Hillsboro, OR, USA
keith.d.underwood@intel.com

Abstract. Historically, MPI implementations have had to choose between eager messaging protocols that require buffering and rendezvous protocols that sacrifice overlap and strong independent progress in some scenarios. The typical choice is to use an eager protocol for short messages and switch to a rendezvous protocol for long messages. If overlap and progress are desired, some implementations offer the option of using a thread. We propose an approach that leverages *triggered operations* to implement a long message rendezvous protocol that provides strong progress guarantees. The results indicate that a triggered operation based rendezvous can achieve better overlap than a traditional rendezvous implementation and less wasted bandwidth than an eager long protocol.

1 Introduction

Many MPI-based science and engineering applications use large messages for bulk data transfer. As the increases in processor performance rapidly outstrips the performance improvement of the network, it becomes increasingly important to maximize the overlap of these messages with computation to improve network efficiency. It is critical for the MPI implementation to provide support for overlapping long message transfers with computation.

MPI implementations traditionally implement one of two protocols for delivering large messages: the message may be sent eagerly [2], which presumes that the receive for the message has already been posted, or the message may be transferred as part of a rendezvous protocol [6], sending a header followed by a bulk transfer of the body after matching. Because eager-long messages require

---

* Sandia National Laboratories is a multi-program laboratory managed and operated by Sandia Corporation, a wholly owned subsidiary of Lockheed Martin Corporation, for the U.S. Department of Energy's National Nuclear Security Administration under contract DE-AC04-94AL85000.

Y. Cotronis et al. (Eds.): EuroMPI 2011, LNCS 6960, pp. 120–129, 2011.

a retransmit of the message body when the message is unexpected [3], large messages are typically sent with a rendezvous protocol of some kind.

Traditional rendezvous protocols require the application to enter the MPI library to progress communication for expected long messages. For unexpected messages, the transfer can be initiated with an RDMA get issued by the target when the receive is posted [15]. However, supporting overlap usually implies that the receive is posted before the "work" begins. In these cases, the message often arrives after the receive and the MPI library cannot initiate the bulk data transfer until the application work completes and the MPI_Wait() is called.

We propose a rendezvous protocol leveraging the triggered operations recently introduced in Portals 4. Using these simple building blocks, we are able to provide a rendezvous implementation that can achieve overlap without requiring a host level thread or the use of eager sends for long messages. Furthermore, the implementation of these constructs is more straight-forward than a full NIC based rendezvous protocol.

2   Related Work

Most MPI implementations employ some form of a rendezvous protocol for transferring large messages. Many strategies have been explored for optimizing the transfer of data, overlapping communication with computation, and progressing communication independently of the application. Rendezvous protocol optimizations generally fall into two categories: host-based and network-based.

Host-based rendezvous optimizations include performing the rendezvous solely inside the MPI library. In this case, message delivery is only progressed when the application makes MPI library calls and the internal progress engine is engaged. Many have attempted to optimize rendezvous inside the MPI library using remote DMA (RDMA) operations [8,11,12,15]. The effectiveness of this approach in enabling overlap for large messages is limited by the rate at which the application makes MPI library calls.

Another host-based approach dedicates a user-level thread to running the MPI progress engine. Most current MPI implementations support this option, and some HPC networks [4,9] use this approach inside their own communication library to provide progress to MPI. Using a progress thread avoids depending on the application to make frequent MPI library calls, but can add significant complexity in terms of scheduling and coordination.

Timer- and network-based interrupts have been used to provide progress [5], but this approach has given way to using threads. Finally, we previously explored the use of eager long messages followed by remote read in the case when the long message was unexpected [1,3]. This approach provides independent progress and overlap for expected messages, but retransmits the entire message buffer when a long message is unexpected.

Network-based approaches for providing progress and overlap for large MPI messages have also been explored. The Quadrics [10] network supported running a user-level thread directly on the network interface hardware. This thread performed MPI matching and could issue remote read requests directly from the network interface without any interaction with the host. This approach adds significant complexity to the hardware design of the network interface, but maximizes the ability to overlap computation and communication and provides a very elegant solution to ensuring independent progress.

Our strategy using triggered operations described in this paper has the advantage of providing simple building blocks on the network interface that can be used to implement a rendezvous protocol for point-to-point messages, but can also be used for other capabilities, such as MPI collective operations [7]. These building blocks are less complex from a network interface hardware and software standpoint when compared to the infrastructure needed to provide the ability to run a user-level thread on the network interface. Using triggered operations provides the same desired round-trip delay reduction by offloading the remote read operation to the network, but it does so with a relatively simple and flexible mechanism.

3   Triggered Operations in Portals 4

Triggered operations and counting events were introduced into Portals 4 [13] as semantic building blocks for collective communication offload. Triggered operations provide a mechanism through which an application can schedule message operations that initiate when a counting event reaches a threshold. Triggered versions of each of the Portals data movement operations were added (e.g., `PtlTriggeredPut()`, `PtlTriggeredGet()`, and `PtlTriggeredAtomic()`) by extending the argument list to include a counting event on which the operation will trigger and a threshold at which it triggers. In turn, counting events are the lightweight semantic provided to track the completion of network operations. Counting events are opaque objects containing an integer that can be allocated, set to a value, or incremented by a value through the Portals API. In addition, they can be attached to various Portals structures and configured to count a variety of network operations, such as the local or remote completion of a message as well as the completion of incoming operations on a buffer (e.g., the completion of a `PtlPut()` or `PtlAtomic()` to a local buffer).

Through careful use of counting events and triggered operations, an almost arbitrary sequence of network operations can be setup by the application and then allowed to progress asynchronously. A discussion of how collective operations can be implemented using triggered operations is presented in [7].

4   Evaluation Methodology

The Structural Simulation Toolkit (SST) v2.0 [14] was used to simulate a NIC offload implementation of Portals 4. SST provides both cycle-accurate and event-based simulation capabilities, and these simulations utilized a cycle-approximate router model combined with an event driven model of the network interface and the host. Message injection rates, data copy delays to and within the NIC, and memory copy delays were modeled as interrelated occupancies in a queuing model. The timings used are described in [7]. The timings model a 1 μs back-to-back zero-byte latency; routing and data copy overheads result in a 1.4 μs latency for 1-byte messages. Additionally, the simulated network achieves 2.7 GB/s peak payload bandwidth after network overheads.

Three long message protocols were examined in the simulator: an eager long protocol [2], a host-based rendezvous protocol, and a triggered rendezvous protocol. In all three cases, the same eager protocol is used for short messages, which are either delivered directly into the user's receive buffer or delivered into a bounce buffer and copied when the receiver posts a matching receive.

4.1 Eager Protocol

The eager protocol sends messages of all sizes eagerly (Fig. 1). If a message matches a pre-posted receive, it is delivered directly into the user's receive buffer and an ack is automatically generated to notify the sender the message was successfully delivered. If the message is unexpected, the payload is discarded with only header data kept by the receiver. Before a long message transfer is initiated, a match list entry covering the send buffer is created and matching information included in the message allows the receiver to issue a get request to retrieve the data if the initial message is discarded. The protocol ensures asynchronous progress in both cases: the receive is either posted before incoming data and the message is asynchronously delivered in the user buffer or the receive is posted after the incoming data begins arriving and the get request is issued before the receive call returns. However, the protocol results in wasted bandwidth for unexpected messages, which may result in further unexpected messages.

4.2 Host-Based Rendezvous Protocol

The host-based rendezvous protocol only sends a piece of the message, up to the threshold between eager and rendezvous messages, as shown in Fig. 2. The

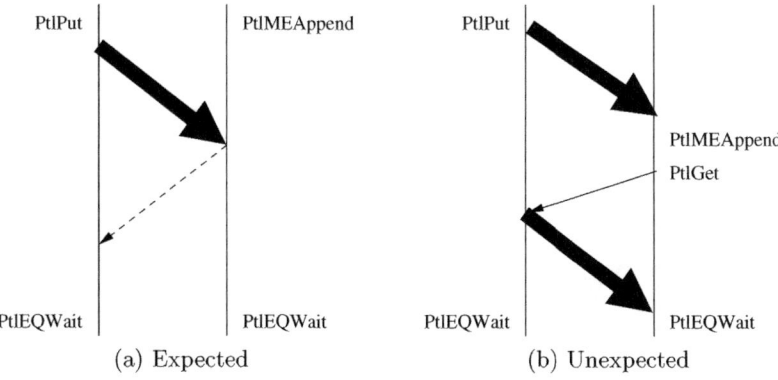

Fig. 1. Communication pattern for eager message protocol with both expected and unexpected messages

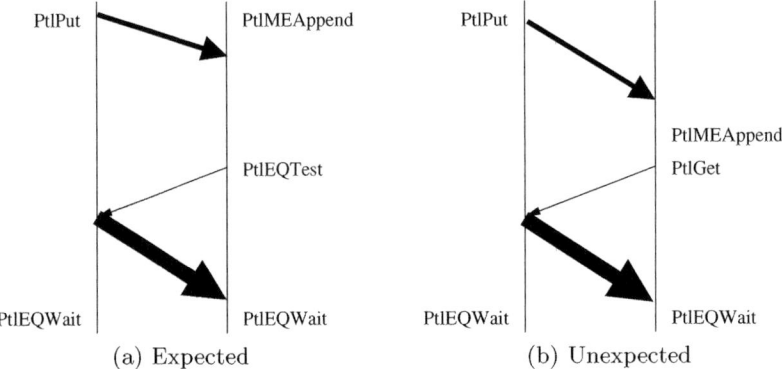

Fig. 2. Communication pattern for host-based rendezvous protocol with both expected and unexpected messages

message includes both the size of the message and sufficient information in the header to allow the receiver to issue a get to retrieve the message when the receive is posted. If the message is expected, the first part of the message is delivered directly into the receive buffer, otherwise it is delivered into bounce buffers. The protocol ensures asynchronous progress for unexpected messages, as the header data is immediately available when the receive is posted. However, the protocol does not ensure asynchronous progress for expected messages, as the receiver must enter the library after the header arrives to issue the get request.

4.3   Triggered Rendezvous Protocol

The triggered rendezvous protocol utilizes Portals triggered operations to issue the receiver-side get request without involving the host application (Fig. 3). The first *eager_limit* + 1 bytes of the message are sent to the receiver when the send

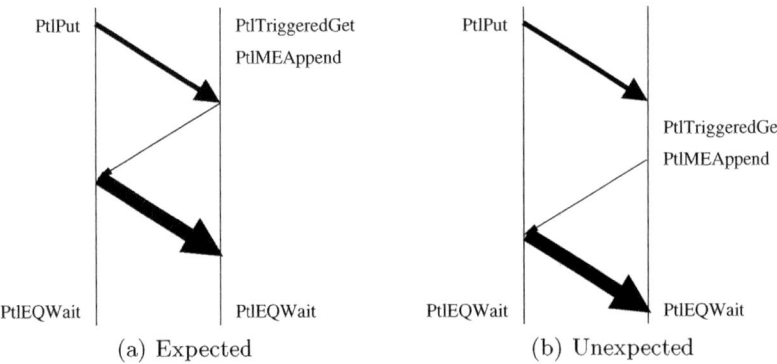

Fig. 3. Communication pattern for triggered rendezvous protocol

is posted. If the message is expected, the first part of the message is delivered directly into the receive buffer, otherwise it is delivered into bounce buffers. A counting event which counts bytes delivered is attached to the receive buffer, and a triggered get is scheduled to execute when a message larger than the eager limit arrives. The counting event is modified whether the message is expected or unexpected, so the protocol provides asynchronous progress in either case.

Portals triggered operations require all arguments to be set when the triggered operation is scheduled, including target, match information, and message size. `MPI_ANY_SOURCE` complicates the protocol, as the sender (the target of the get operation) is not known until the matching message header arrives. When a long receive with `MPI_ANY_SOURCE` is posted, the triggered rendezvous protocol falls back to the host-based rendezvous protocol. When `MPI_ANY_SOURCE` receives have completed, the triggered rendezvous protocol resumes.

Matching information for the get must also be pre-calculated, rather than retrieved from the header data as in the other protocols. Each rank maintains two sets of counters: the number of messages it has sent to each peer and the number of messages it has received from each peer. The matching information for the get is the current message count between the peers. While the array of counters is non-scalable, a 16 bit counter should be sufficient, leading to memory usage of only 4 MiB per process for a million rank application.

The triggered get operation assumes the message being transferred is the same size as the posted receive. The MPI standard allows the send size to be smaller than the receive size and defines an error class for the case of a message larger than the posted receive buffer. The larger receive buffer case presents an issue for the triggered get operation, as the get request will be larger than the send buffer. Portals includes the ability to truncate any data transfer request (put or get) to the size of the target-side match list entry, allowing the get to be truncated by the sender. Sends that are larger than the posted receive are handled during event completion, by comparing the size of the send request included in the initial send meta data with the size of the posted receive. The message is delivered up to the posted receive size and an error is raised.

5   Results

An eager long protocol has several advantages, but it also has one significant disadvantage, as illustrated in Fig. 4. The six lines presented represent the bandwidth of an eager long protocol when various percentages of the messages are unexpected. After crossing from an eager short protocol to an eager long protocol, the bandwidth is reduced in direct proportion to the fraction of unexpected messages. When all of the messages are unexpected, all of the messages are transmitted eagerly, dropped, and retransmitted when the receive is posted. This yields only 50% of the networks potential bandwidth. While the results are simulated, a similar result was seen in practice with early software releases on the Cray XT3 platform.

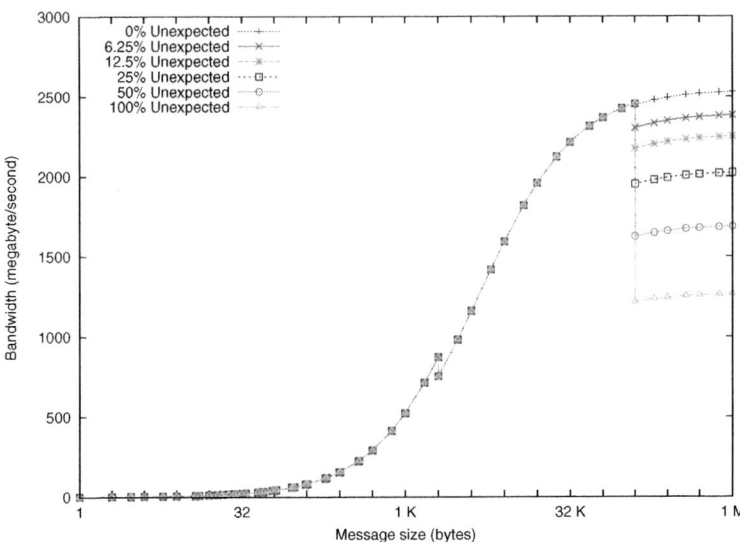

Fig. 4. Ping-pong bandwidth for eager protocol with varying percentage of unexpected messages

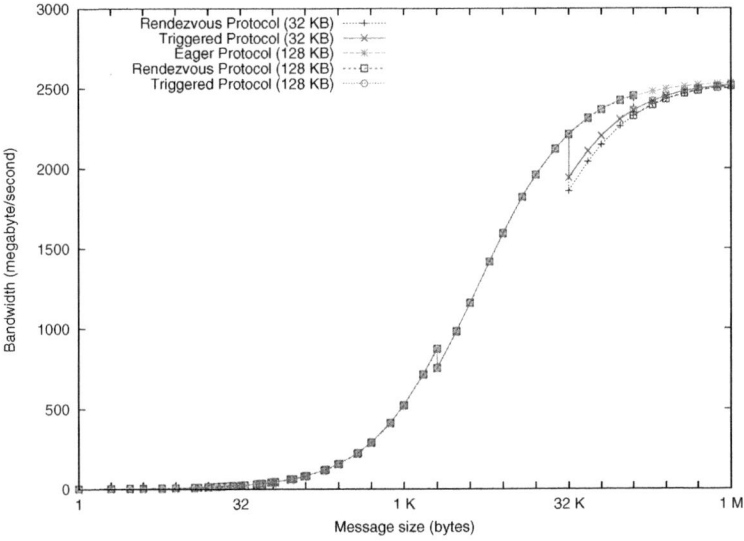

Fig. 5. Ping-pong bandwidth using three different long message protocols with blocking receives

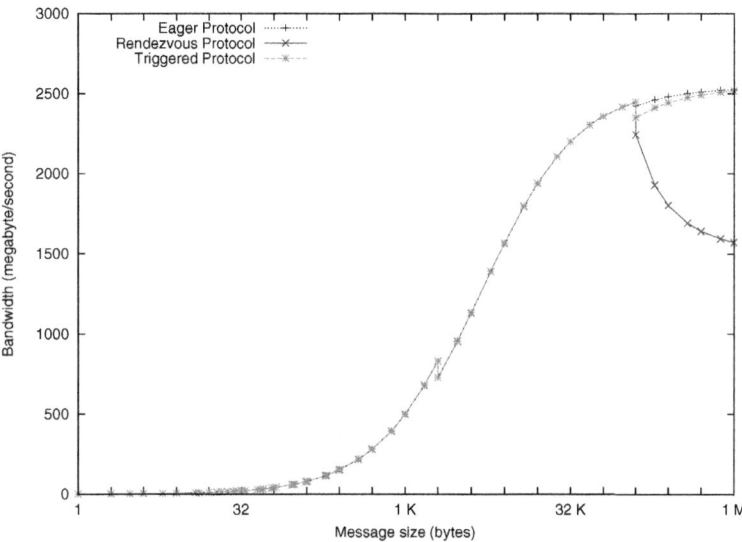

Fig. 6. Ping-pong bandwidth using three different long message protocols with non-blocking receives and work proportional to total message transfer time inserted between posting receive and waiting for message completion

With blocking receives and expected messages, the eager long protocol has a slight edge over the rendezvous protocols because it does not incur the round-trip delay to perform the rendezvous (Fig. 5). It is also important to note that the triggered rendezvous protocol has a slight edge over the host based rendezvous protocol, because the round-trip delay is reduced by having the triggered rendezvous get request released by the counting operation on the NIC rather than using the host processor. The real advantage of the triggered rendezvous protocol relative to the host based rendezvous protocol is seen in Fig 6. In this case, a non-blocking receive is posted before the send arrives and the application enters a work loop that is proportional to the size of the message (i.e. the work delay equals the communication time delay) before re-entering the MPI library. This means that the host based long message rendezvous cannot achieve overlap; however, the triggered based rendezvous still achieves full overlap and nearly matches the performance of a pure eager protocol.

6   Conclusions

This paper demonstrates how triggered operations can be leveraged to implement a rendezvous protocol under host software control. Because triggered operations are a relatively simple building block (simply defer an outgoing message until a condition is met), they are an easier target for NIC offload than a full rendezvous protocol. Nonetheless, they still offer many of the performance advantages of offloading a rendezvous protocol: reduced latency of issuing the rendezvous get and full overlap through independent progress. Simulation results are used to illustrate both of these advantages.

References

1. Brightwell, R.: A Comparison of Three MPI Implementations for Red Storm. In: Di Martino, B., Kranzlmüller, D., Dongarra, J. (eds.) EuroPVM/MPI 2005. LNCS, vol. 3666, pp. 425–432. Springer, Heidelberg (2005)
2. Brightwell, R., Maccabe, A.B., Riesen, R.: Design, implementation, and performance of MPI on Portals 3.0. International Journal of High Performance Computing Applications 17(1), 7–20 (2003)
3. Brightwell, R., Underwood, K.D.: Evaluation of an eager protocol optimization for MPI. In: Dongarra, J., Laforenza, D., Orlando, S. (eds.) EuroPVM/MPI 2003. LNCS, vol. 2840, pp. 327–334. Springer, Heidelberg (2003)
4. Dickman, L., Lindahl, G., Olson, D., Rubin, J., Broughton, J.: PathScale Infini-Path: A First Look. In: Proceedings of the 13th Symposium on High Performance Interconnects (HOTI 2005) (August 2005)
5. Franke, H., Wu, C.E., Riviere, M., Pattnaik, P., Snir, M.: MPI programming environment for IBM SP1/SP2. In: Proceedings of the 15th International Conference on Distributed Computing Systems (1995)
6. Gropp, W., Lusk, E., Doss, N., Skjellum, A.: A high-performance, portable implementation of the MPI message passing interface standard. Parallel Computing 22(6), 789–828 (1996)
7. Hemmert, K.S., Barrett, B., Underwood, K.D.: Using triggered operations to offload collective communication operations. In: Keller, R., Gabriel, E., Resch, M., Dongarra, J. (eds.) EuroMPI 2010. LNCS, vol. 6305, pp. 249–256. Springer, Heidelberg (2010)
8. Kumar, R., Mamidala, A.R., Koop, M.J., Santhanaraman, G., Panda, D.K.: Lock-Free Asynchronous Rendezvous Design for MPI Point-to-Point Communication. In: Lastovetsky, A., Kechadi, T., Dongarra, J. (eds.) EuroPVM/MPI 2008. LNCS, vol. 5205, pp. 185–193. Springer, Heidelberg (2008)
9. Myricom, Inc.: Myrinet Express (MX): A high performance, low-level, message-passing interface for Myrinet (July 2003),
   http://www.myri.com/scs/MX/doc/mx.pdf
10. Petrini, F., chun Feng, W., Hoisie, A., Coll, S., Frachtenberg, E.: The Quadrics Network: High-Performance Clustering Technology. IEEE Micro 22(1), 46–57 (2002)
11. Rashti, M.J., Afsahi, A.: Improving Communication Progress and Overlap in MPI Rendezvous Protocol over RDMA-enabled Interconnects. In: Proceedings of the 22nd International Symposium on High Performance Computing Systems, pp. 95–101 (June 2008)
12. Rashti, M.J., Afsahi, A.: A speculative and adaptive MPI rendezvous protocol over RDMA-enabled interconnects. International Journal of Parallel Programming 37, 223–246 (2009)
13. Riesen, R.E., Pedretti, K.T., Brightwell, R., Barrett, B.W., Underwood, K.D., Hudson, T.B., Maccabe, A.B.: The Portals 4.0 message passing interface. Tech. Rep. SAND2008-2639, Sandia National Laboratories (April 2008)

14. Rodrigues, A.F., Hemmert, K.S., Barrett, B.W., Kersey, C., Oldfield, R., Weston, M., Risen, R., Cook, J., Rosenfeld, P., CooperBalls, E., Jacob, B.: The structural simulation toolkit. SIGMETRICS Perform. Eval. Rev. 38, 37–42 (2011), http://doi.acm.org/10.1145/1964218.1964225
15. Sur, S., Jin, H.W., Chai, L., Panda, D.K.: Rdma read based rendezvous protocol for mpi over infiniband: design alternatives and benefits. In: Proceedings of the eleventh ACM SIGPLAN Symposium on Principles and Practice of Parallel Programming, PPoPP 2006, pp. 32–39. ACM, New York (2006), http://doi.acm.org/10.1145/1122971.1122978

pupyMPI - MPI Implemented in Pure Python

Rune Bromer[1], Frederik Hantho[1], and Brian Vinter[2]

[1] Computer Science Institute, University of Copenhagen
[2] Niels Bohr Institute, University of Copenhagen

Abstract. As distributed memory systems have become common, the
de facto standard for communication is still the Message Passing Interface (MPI). pupyMPI is a pure Python implementation of a broad
subset of the MPI 1.3 specifications that allows Python programmers
to utilize multiple CPUs with datatypes and memory handled transparently. pupyMPI also implements a few non-standard extensions such as
non-blocking collectives and the option of suspending, migrating and re-
suming the distributed computation of a pupyMPI program. This paper
introduces pupyMPI and presents benchmarks against C implementations of MPI, which show acceptable performance.

Keywords: MPI, Python, scientific computing, parallel computing, high-
level languages.

1 Introduction

The Message Passing Interface (MPI) is a widely used model for expressing
parallelism. The success of MPI has many causes, chief among these are its
portability, completeness and performance. Supplanting a variety of incompati-
ble options in parallel communication libraries, MPI as a standard has greatly
improved the portability of HPC applications. However, MPI is generally not
considered simple, nor easy to program which is a problem since many potential
users are not skilled programmers, but rather researchers who are not expert at
expressing their algorithms efficiently in a deterministic logical framework.

Traditionally HPC applications are written in C/C++ or Fortran, both of
which are statically typed, compiled, low-level languages with excellent performance characteristics. In more recent years high-level interpreted languages,
most notably MATLAB and Python, have gained popularity. Both MATLAB
and Python are easily extended which means that performance critical parts
of the code can be written in eg. C or Fortran resulting in much faster execution. By using the correct libraries and datastructures Python, mixed with C
and Fortran, has been shown to be on par with pure C for scientific computing
including message passing [3]. With the increasing maturity of libraries for scientific computing, such as NumPy and SciPy, Python is now often used to do the
computationally heavy brunt work, e.g. GPAW [2] a large quantum mechanics
code that has been shown to scale to thousands of CPUs [12].

Y. Cotronis et al. (Eds.): EuroMPI 2011, LNCS 6960, pp. 130–139, 2011.

1.1   MPI with Python

Several projects have taken advantage of the relative ease with which you can wrap C code in Python, to provide MPI functionality in the form of bindings to existing MPI implementations, such as Open MPI or MPICH. Using Python bindings on top of a mature MPI implementation in C has the advantages that the communication layer can be assumed to perform very well and of course that a great deal of MPI internal functionality can be reused as is. The drawbacks are that the bindings need to conform to the specific MPI implementations supported and change with them. And of course - from an system administrator's point of view - the C MPI library used and any dependencies need to be maintained alongside the Python bindings. Another problem is portability, since Python bindings need compilation and linking against the MPI library, this requires the appropriate toolchain and significant effort for every installation needed.

An important reason for a Python version of MPI is to have a "pythonic" API , i.e. remove the information on datatypes from the users perspective. This paper describes pupyMPI (pure Python MPI) which is a fully functional implementation of a large subset of the MPI 1.3 specification in Python. Implementing MPI in pure Python instead of merely layering bindings on top of an MPI implementation in C has several advantages, one is portability; Python is available for many platforms, indeed it is included with most common operating systems (Microsoft Windows being the exception). By relying only on Python we eliminate a host of dependencies and greatly simplify the task of maintaining pupyMPI. The second major advantage is that developing in a high level language is efficient in terms of developer resources. As a case in point using Python has meant that in around 1.5 man-years we have produced a fully functional MPI implementation and still have had time for experimentation with MPI internals and different design strategies.

1.2   Related Work

The project Numerical Python (NumPy) provides a library written in C for numerical operations to minimize the overhead associated with standard interpreted Python code.

Projects such as Pypar, pyMPI, myMPI and MPI for Python (mpi4py) offer MPI functionality in Python through bindings with various degrees of completeness. The best developed of these is mpi4py [4] with both excellent coverage of the MPI-2 specification and a sensible Pythonic MPI syntax. mpi4py had automatic datatype discovery added in late 2009. However users are still required to distinguish between sending objects that expose the buffer interface and those that do not.

MPI has been implemented in languages besides C, such as Java and OCaml. To the best of our knowledge no other Python implementation than pupyMPI is available.

2   Overview of pupyMPI

A detailed description of pupyMPI is beyond the scope of this text. Instead we describe selected details of the design and implementation in the following.

The goal of message passing is low latency and high throughput, so that time can be spent on the user's computations. In addition we strive to make pupyMPI easy and intuitive to use. Ease of use means that we embrace the dynamic nature of Python where type and memory management are handled for the user, as well as the programming paradigms - most notably object orientation - supported by Python. The target for pupyMPI are scientists who are not experts at MPI or C and we have struck a balance between the expectations of veteran MPI users and new users unfamiliar with the syntax laid out in the MPI specifications.

2.1   Concurrency in Python

High performance MPI programs require that communication takes place simultaneously with computation. pupyMPI has a threaded architecture with a thread doing raw network input or output and a thread doing internal command and control (the MPI thread) along with the user's thread executing the pupyMPI program. The threading is transparant to the user although the user is not prohibited from calling the same pupyMPI instance from multiple threads since pupyMPI is in effect thread-safe.

True concurrency exploiting multiple cores in the standard version of Python (ie. the CPython interpreter) is hindered by the existence of the *Global Interpreter Lock* (GIL). A running thread must generally hold the GIL which in effect serializes the execution of multiple threads. However in many situations the GIL will be released while a thread is still executing valuable work. This is the case for I/O, most Numpy operations, and some built-in Python operations that are implemented in C. The thread-architecture of pupyMPI means that multiple cores can be utilized in most cases, and ideally the only competition for the GIL is between the user thread and the MPI thread when the user thread is not doing heavy computations that would release the GIL.

In practice the overhead of GIL contention manifests itself when all cores on a node are mapped to a pupyMPI process. This is a problem since pupyMPI need to be able to restrict itself to running on one core to be performance competitive with regular MPI libraries. It is important to recognize that the problems are not caused by only having one core available but rather how vulnerable threading in Python is to contention for the GIL.

The CPython community have to some degree adressed the problems of GIL contention in Python 3.2 [7] [1] which we hope to be able to test in the future. At the same time we are working on different ways of architecting the concurrency in pupyMPI to minimize the overhead of the library.

2.2   Supporting Numpy

Numpy arrays expose the *buffer interface* which allows direct access to the memory segment that contains the array. Thus it is possible to send the bytestream directly to another process avoiding serialization which would create an extra copy of the array contents and waste CPU resources. The header of a message that contains Numpy data allows the receiving process to reconstruct the bytestream into a Numpy array with proper type and dimensions instead of resorting to Python's standard deserialization. When reductions are performed on Numpy arrays optimized Numpy functions replace the reduce operation if possible.

2.3   Caching Socket Connections

Establishing connections between all processes in an MPI program on initialization incurs considerable overhead, strain limited OS resources for large systems and is often unneccessary due to sparse communication patterns. Instead every pupyMPI process has a socketpool caching connections to other processes as they are established. To guard against a proliferation of open connections there is a default maximum size of the socketpool that can be changed by the user. A *second chance FIFO* cache algorithm decides which connections to throw out if the socketpool has run out of space.

It is of course possible to set the size of the socket pool to the size of `MPI_COMM_WORLD` in which case no connection will ever be thrown out. The user can also specify that connections between all processes should be made during start up.

3   pupyMPI API

The interface differs from traditional MPI in two ways: most calls are methods on a communicator object and type management is eliminated. pupyMPI exposes almost all the operations from MPI 1.3 that make sense in an dynamically typed and non-memory managed language. That is operations like `MPI_Bsend`, `MPI_Sendrecv_replace`, `MPI_Type_*` are left out. The collective v-operations are also left out since handling varying datasizes is easily done with the general collective operations.

3.1   General Operations

pupyMPI is initialized by importing the module and instantiating the `MPI` class. Hereafter the default world communicator is available as `MPI_COMM_WORLD` supporting the usual communicator operations. pupyMPI is finalized by invoking `finalize` on the MPI object:

```
from mpi import MPI
pupy = MPI()
world = pupy.MPI_COMM_WORLD
print "I am rank %i of %i processes" % (world.rank(), world.size())
pupy.finalize()
```

3.2   Point-to-Point Operations

For point-to-point operations tags can be used but if none are specified pupyMPI defaults to `MPI_TAG_ANY` for both send and receive operations. A send operation to a receiving process with rank 1 is as simple as: `world.send(1, "Test message")`. A non-blocking receive operation using a tag:

```
MY_RED_TAG = 9
handle = world.irecv(1, MY_RED_TAG)
msg = handle.wait()
```

3.3   Collective Operations

The collective operations default to `root=0`, so `gather` to rank 0 is simply:

```
res = world.gather(my_part) # if not root res will be None
```

Operations that work on vectors (sequences in Python) will partition based on the size of the communicator:

```
# for NP=4 root will get the string "ab"
part = world.scatter("abcdefgh")
# for a sequence reduction is elementwise
res = world.allreduce([1,2,3,4], MPI_sum)
```

3.4   Other Operations on Communicators

The group operations are supported and can be used to e.g. split a communicator into even ranks and odd ranks

```
size = world.size()
world_g = world.group()                    # First a world group
even_g = world_g.incl(range(0,size,2))   # Two target groups
odd_g = world_g.incl(range(1,size,2))    #    called even and odd
even_comm = world.comm_create(even_g)    # Split communicators
odd_comm = world.comm_create(odd_g)      #    from world
```

3.5   A Working Example

The following Python function is the main part of a 2D stencil solver using pupyMPI and Numpy. Initialization of MPI and obvious variables such as `rank` along with distribution and reassembly of global state happen outside the function:

```
def stencil_solver(local,epsilon):
    W, H = local.shape
    maxrank = np - 1
```

```
work = numpy.zeros((W-2,H-2)) # temp workspace

cells = local[1:-1, 1:-1] # interior
left  = local[1:-1, 0:-2]
up    = local[0:-2, 1:-1]
down  = local[2:  , 1:-1]
right = local[1:-1, 2:  ]

delta = epsilon+1
while epsilon<delta:
    if rank != 0:
        local[0,:] = world.sendrecv(local[1,:], dest=rank-1)
    if rank != maxrank:
        local[-1,:] = world.sendrecv(local[-2,:], dest=rank+1)
    work[:] = (cells+up+left+right+down)*0.2
    delta = world.allreduce(numpy.sum(numpy.abs(cells-work)),
            MPI_sum)
    cells[:] = work
```

4 Collective Operations

Collective operations are useful abstractions and their implementations are vital
for performance. The framework supporting collective operations in pupyMPI is
too complex to examine here but three areas bear mentioning.

4.1 Topology Reordering

The MPI specification defines both Cartesian and Graph topologies even through
a Cartesion is a subset of the very general Graph toplogy. This is done since a
Cartesian grid is a very common partitioning and easier specified directly than
via its corresponding graph. pupyMPI extends virtual topologies with another
subset of Graph namely Tree. We consider trees a very practical communication structure and provide easy topology creation and an API with expected
functionality like parent(), children() and descendants(). Also internally
pupyMPI defaults to a Tree topology (normally a binomial tree) for distributing communication during most collective operations.

Whenever this topology is created[1] the ranks mapping to tree nodes are logically reordered so that processes close to each other in the physical network
are mapped to nodes in the tree that are expected to communicate heavily. For
instance a scatter operation will have the root sending the greatest amount of
data to the child that is itself root in the biggest subtree since all data to any
leaf in this subtree will have to go through that child.

---

[1] Any topology created by the collective operations is cached so the creation penalty
will only occur once.

Many reordering schemes exist to take advantage of various physical network topologies [6,11]. We currently only reorder to place processes on the same network node but are working on extending this to topology-hierarchies of levels deeper than 2. To suit the audience for pupyMPI this should be done transparently requiring little or no knowledge of the underlying network architecture from the user.

4.2   Algorithm Selection

Based on the information that is avilable when the collective operation starts, a suitable algorithm is chosen. The information currently used to select an algorithm is *data size* and *communicator size*. As a special case several situations can use the same algorithm but change the topology based on information about the input data. Further optimizations are possible by extending the selection logic [5,14] and by using portfolio algorithms [13].

4.3   Non-blocking Collective Operations

Regular collective operations are blocking which can limit performance. A common way to avoid this is to create a seperate user thread that performs the collective operation [9].

Instead we support non-blocking collective operations as proposed in [8] using the same syntax as for point to point operations. The proposal have already been voted in for the MPI-3 standard, so we consider this as part of covering the MPI standard.

```
data = world.allreduce(42, sum)     # A blocking allreduce
handle = world.iallreduce(42, sum) # A non blocking call
# (overlap calculations here)
data = handle.wait()
```

5   pupyMPI User's Toolset

pupyMPI comes with a set of experimental user tools that allows users to accomplish tasks including profiling of pupyMPI programs with visualization of communication/computation patterns and inspection of the state of running programs.

A special feature worth mentioning allows the entire distributed state of a running pupyMPI program to be packed into a file. The program can be resumed on another (or the same) set of hosts or even copied and rerun several times. The user specifies desired checkpoints with @checkpoint decorators in the code.

6   Benchmarks

We have ported a large part of the IMB [10] testsuite to pupyMPI for benchmarking along with a few non-synthetic benchmark applications. The pupyMark

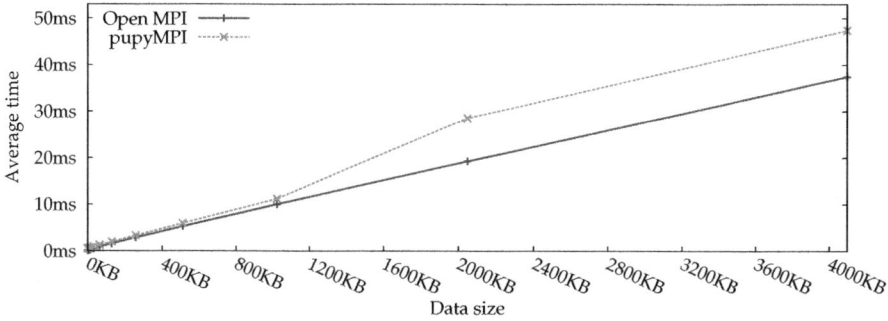

Fig. 1. Benchmark for PingPing

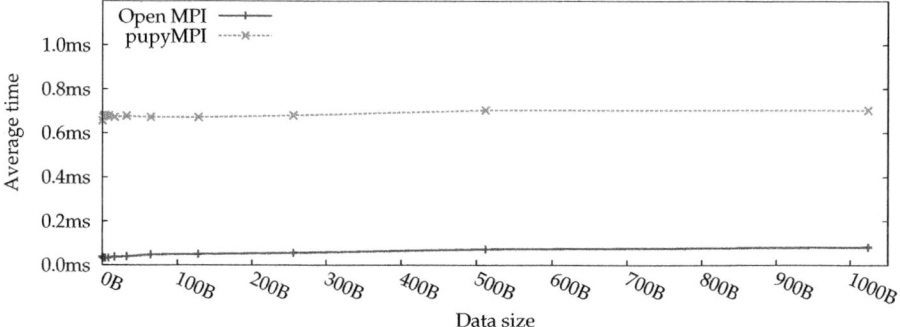

Fig. 2. Benchmark for PingPong

benchmarking tools are included with the pupyMPI library for easy reproduction of our results.

Shown here are a few results of benchmarking the latest stable version of pupyMPI versus Open MPI version 1.5.3. The platform used was Linux 2.6.28, with Python 2.6.2 and Numpy 1.2.1. Benchmarks were run on a cluster of 8 nodes with Intel Q9400 processors (4 cores) and Intel 82567LM-3 Gigabit Ethernet NIC connected via a D-Link DGS-1016D 16 port Gigabit Switch.

It is evident from the PingPing and PingPong (Fig 1 and 2) results that while pupyMPI can almost keep up with Open MPI in raw throughput there is an order of magnitude difference in latency. This is most likely due to threading issues where the GIL inhibits a fast hand-off from user thread via MPI thread to I/O thread. Surprisingly the test of allgather in Fig. 3 shows almost no performance difference - other collective operations show a much greater advantage in Open MPIs favor. This we attribute to a less optimized implementation in Open MPI whereas pupyMPI uses a *dissemination allgather* algorithm.

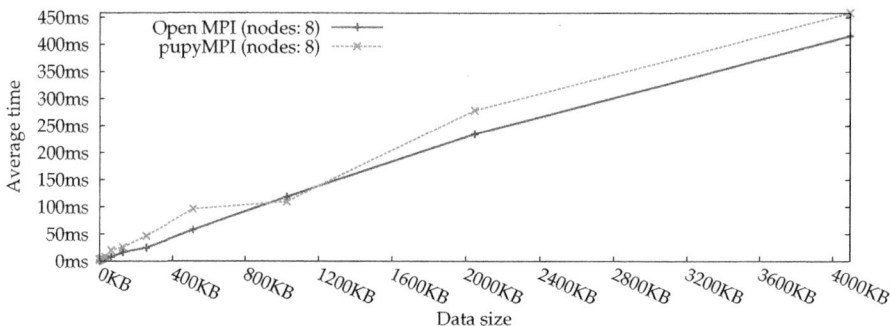

Fig. 3. Benchmark for Allgather

7   Conclusion and Future Work

pupyMPI provides a pythonic MPI for parallel computing without management of types or memory. Performance is as should be expected for an interpreted language since the network is the worst bottle-neck. A young project and still under intense development, pupyMPI has already seen succesful use in an academic setting at Copenhagen University. It has been used by students at a class in cluster computing 3 times and for parallel computations in an research project on text recognition. The appr. 1.5 man-years of development that has gone into pupyMPI is testament to the efficiency of programming in Python.

For the near future we plan to focus especially on the collective operations where performance gains are still be had. The current solution for selecting the fastest collective operation is based solely on static information available at startup of the operation. A smarter solution is be to introduce dynamically adaptive operations [15]. We also plan to introduce a utility script that would run several tests on a given network and determine the optimal settings for collective operations.

Acknowledgements. We wish to thank former members of the pupyMPI development team Jan Wiberg and Asser Femø for their hard work and inspiration. Also we are indebted to the reviewers for their encouragement and thorough critique. We owe gratitude to the three classes of students who learned about MPI using our library and helped improve the quality of the implementation. Lastly we thank Josva Kleist of Aalborg University for letting us use the Fyrkat cluster and evaluating our work at multiple occasions.

References

1. Beazley, D.: Convoy effect with I/O bound threads and New GIL (2010),
   http://bugs.python.org/issue7946
2. Blochl, P.E.: Projector augmented-wave method. Phys. Rev. B 50(24), 17953–17979 (1994)

3. Cai, X., Langtangen, H., Moe, H.: On the performance of the Python programming language for serial and parallel scientific computations. Scientific Programming 13(1), 31–56 (2005)
4. Dalcin, L., Paz, R., Storti, M.: MPI for python. Journal of Parallel and Distributed Computing 65(9), 1108–1115 (2005)
5. Faraj, A., Yuan, X., Lowenthal, D.: STAR-MPI: self tuned adaptive routines for MPI collective operations. In: Proceedings of the 20th Annual International Conference on Supercomputing, ICS 2006, pp. 199–208. ACM, New York (2006), http://doi.acm.org/10.1145/1183401.1183431
6. Hatazaki, T.: Rank reordering strategy for mpi topology creation functions. In: Recent Advances in Parallel Virtual Machine and Message Passing Interface, pp. 188–195 (1998)
7. Hettinger, R.: What's new in python 3.2 (2011), http://docs.python.org/dev/whatsnew/3.2.html#multi-threading
8. Hoefler, T., Kambadur, P., Graham, R., Shipman, G., Lumsdaine, A.: A Case for Standard Non-Blocking Collective Operations. In: Recent Advances in Parallel Virtual Machine and Message Passing Interface, pp. 125–134 (2007)
9. Hoefler, T., Lumsdaine, A., Rehm, W.: Implementation and performance analysis of non-blocking collective operations for mpi. In: Proceedings of the 2007 ACM/IEEE Conference on Supercomputing, p. 52. ACM, New York (2007)
10. Intel, M.: Intel® MPI Benchmarks 3.2.2 (2008), http://software.intel.com/sites/products/mpi-benchmarks/IMB_3.2.2.tgz (accessed May 2, 2011)
11. Karonis, N.T., De Supinski, B.R., Foster, I., Gropp, W., Lusk, E., Bresnahan, J.: Exploiting hierarchy in parallel computer networks to optimize collective operation performance. In: Proceedings of 14th International Parallel and Distributed Processing Symposium, IPDPS 2000, pp. 377–384. IEEE, Los Alamitos (2000)
12. Kristensen, M.R.B., Happe, H.H., Vinter, B.: GPAW optimized for Blue Gene/P using hybrid programming. In: International Parallel and Distributed Processing Symposium, vol. 0, pp. 1–6 (2009)
13. Leyton-Brown, K., Nudelman, E., Andrew, G., McFadden, J., Shoham, Y.: A portfolio approach to algorithm selection. In: International Joint Conference on Artificial Intelligence, Citeseer, vol. 18, pp. 1542–1543 (2003)
14. Thakur, R., Gropp, W.D.: Improving the performance of collective operations in MPICH. In: Dongarra, J., Laforenza, D., Orlando, S. (eds.) EuroPVM/MPI 2003. LNCS, vol. 2840, pp. 257–267. Springer, Heidelberg (2003)
15. Vadhiyar, S., Fagg, G., Dongarra, J.: Automatically tuned collective communications. In: Proceedings of the 2000 ACM/IEEE Conference on Supercomputing (CDROM), p. 3. IEEE Computer Society, Los Alamitos (2000)

Scalable Memory Use in MPI: A Case Study with MPICH2

David Goodell[1], William Gropp[2], Xin Zhao[2], and Rajeev Thakur[1]

[1] Argonne National Lab., Argonne, IL 60439, USA
{goodell,thakur}@mcs.anl.gov
[2] Univ. of Illinois, Urbana, IL 61801, USA
{wgropp,xinzhao3}@illinois.edu

Abstract. One of the factors that can limit the scalability of MPI to exascale is the amount of memory consumed by the MPI implementation. In fact, some researchers believe that existing MPI implementations, if used unchanged, will themselves consume a large fraction of the available system memory at exascale. To investigate and address this issue, we undertook a study of the memory consumed by the MPICH2 implementation of MPI, with a focus on identifying parts of the code where the memory consumed per process scales linearly with the total number of processes. We report on the findings of this study and discuss ways to avoid the linear growth in memory consumption. We also describe specific optimizations that we implemented in MPICH2 to avoid this linear growth and present experimental results demonstrating the memory savings achieved and the impact on performance.

1 Introduction

We have already reached an era where the largest parallel machines in the world have a few hundred thousand cores and are soon approaching an era of million-core systems. For example, an IBM Blue Gene/Q system (Sequoia) to be deployed at Lawrence Livermore National Laboratory in 2012 will have more than 1.5 million cores and a peak speed of 20 petaflops. Roadmaps for future systems indicate that we can expect systems with many millions of cores over the next 5–10 years. For example, a DOE technology and architecture roadmap for exascale envisions a 1 exaflop/s machine by 2018 with 1 *billion* cores [8]. Another significant trend is that although the number of cores is increasing rapidly, the amount of memory available per core is not increasing.

As systems grow to these sizes, many researchers and users wonder whether MPI will scale to such large systems. Scalability of performance is not the only concern; an often-cited concern is the potential memory consumption of MPI at scale. It is generally believed that as the system size grows, the memory consumed by MPI on each process also grows linearly. Given the limited amount of memory per core, it is believed that, unless steps are taken, MPI itself will consume a large fraction of available memory on exascale systems.

Anecdotal evidence exists of isolated examples indicating memory consumption issues in some functions in some MPI implementations, often reflecting a

Y. Cotronis et al. (Eds.): EuroMPI 2011, LNCS 6960, pp. 140–149, 2011.

bug in the code or oversight on the part of the developers. At small scale, developers tend to make assumptions or take shortcuts that need to be fixed at scales several orders of magnitude higher. However, quantitative data on MPI memory consumption at scale is hard to find. Particularly lacking is information about what aspects are merely bugs that need to be fixed and what are more intrinsic problems that require a redesign or rethinking of conventional ways of implementing MPI, including changes that may incur a performance penalty at small scale but are necessary for the code to even run at large scale.

To investigate these aspects and address potential problems, we undertook a study of the memory consumed by MPICH2 [7], an MPI implementation that is widely used on many of the largest machines in the Top500 list. We focused particularly on parts of the code where the memory consumption increases linearly with system size. We report on the findings of this study and discuss ways in which such linear growth in memory can be avoided. We also designed and implemented specific optimizations in MPICH2 to avoid this linear memory growth. We describe these optimizations and present experimental results demonstrating the memory savings achieved and the negligible impact on performance.

Related Work. Balaji et al. [3] discuss issues related to scaling MPI to millions of cores, in terms of what is needed both in the MPI specification and in MPI implementations. The authors consider implementation issues in general, not specific to any particular MPI implementation. In this paper, on the other hand, we focus on identifying and fixing memory scalability issues in the MPICH2 implementation of MPI. Other researchers have also explored memory-space related optimizations for MPI implementations, such as the memory required for storing communicators and groups [5,6,9].

2   Apparent Nonscalable Memory Use in MPI

At first glance, MPI appears to have a number of areas where it must store $\mathcal{O}(p)$ data on each MPI process, where p is the number of processes in the MPI program. In this section, we discuss some of these areas and comment on what MPI really requires for them. For simplicity and to match the behavior of most MPI implementations on large systems, we assume that all processes are in MPI_COMM_WORLD (e.g., no dynamic processes).

> **Group Representation.** An MPI *group* describes a collection of processes. The obvious implementation is an enumeration of processes by some identifier, such as rank in MPI_COMM_WORLD or an IP address and process ID. However, MPI only requires that this information be available, not the form in which it is stored. Lossless compression of the data is permitted; for example, for MPI_COMM_WORLD, the group can be represented as simply 0:p-1 (all ranks from 0 to $p-1$, requiring only a few words of storage).
>
> **Connections and Message Buffers.** MPI allows a process to communicate directly with all other processes. It is sometimes alleged that this feature requires MPI to maintain $\mathcal{O}(p)$ data for such connections and to allocate

significant buffer space to each possible connection. For example, providing only 16 KB for each connection for eager message delivery would require 16 GB on each process for a million-process MPI program (a total of 16 petabytes of memory). However, MPI does not specify when connections are established or how buffer memory is allocated and associated with connections; in fact, MPI does not even define "connections." For example, an MPI implementation may instantiate a connection only when needed and dynamically associate buffer memory to active connections. For a scalable application (which by definition cannot communicate with $\mathcal{O}(p)$ other processes), only a small number of such connections can be active.

RMA Windows. Each MPI RMA window is created with its own displacement value, start address, size, and info object for hints. Because RMA is for one-sided operations, it is natural to store information about the remote windows locally, where the information can be quickly accessed. However, locally storing the information for all ranks is not required by MPI. Other options include using a cache strategy for such data, acquiring it on first use, or even fully distributing the data and using one-sided operations to acquire the data.

Nonscalable Arguments. Some MPI routines have array parameters of size p. These are nonscalable routines and simply cannot be used in a scalable application. They do not reflect a problem in an MPI implementation.

In all of these cases, allocating memory for each of the $\mathcal{O}(p)$ items both simplifies the implementation and may be (slightly) faster. However, $\mathcal{O}(p)$ memory is not required, and we argue that the performance cost is often negligible.

3 Memory Usage in MPICH2

MPICH2 has been carefully designed and developed to be adaptable to environments with a paucity of memory resources. The current design is parsimonious with memory in certain areas, such as the usage and representation of MPI groups. In other areas of the code, decisions were consciously made to trade increased memory consumption to obtain decreased algorithmic running times. In a severely memory-constrained environment, some of these decisions could be revisited and potentially altered when such a change would be beneficial. Yet unsurprisingly, several memory inefficiencies remain in the current code. We discuss these strengths and weaknesses of the current stable version of MPICH2 in this section.

3.1 Link-Time Program Text Size Savings

MPICH2 was designed from the beginning to be highly modular. Less-used code is organized so that the code and the associated data structures are included (by the linker) only when actually used by the application. For example, the buffered send code is included only if the user references one of the buffered send routines. The code for each of the MPI collectives is another example. This reduces the size of the executable code, which is good for both very large systems and ones where use of dynamically loaded code from shared libraries may stress the I/O system, such as nodes without local or nearby disks.

3.2   One-Sided Communication

The current implementation of MPICH2 stores a copy of the window start address, size, and displacement unit of all processes locally on each process for easy lookup. This clearly requires $\mathcal{O}(p)$ space on each process, which is nonscalable. Possible approaches to fix this problem are outlined in Section 2, and we will consider them as part of our future work.

3.3   MPI Groups

Within every MPI process, the process assigns to each other process to which it is connected[1] a local process ID, or *LPID*, in the range $[0, p)$, where p is the number of connected processes. Note that LPIDs are not unique across processes; they exist as a purely local concept to simplify process-related bookkeeping operations.

An MPI group is a totally ordered set of processes in which each process is indexed by an integer rank in the range $[0, p_g)$, where p_g is the size of the group. This set is currently stored as a dense int array of LPIDs, where element i in the array stores the LPID of the process corresponding to rank i in the group. This information is sufficient, though nonoptimal, to be able to correctly implement all local MPI_Group_ operations.

As a performance optimization, a list of indices sorted by increasing LPID order can also be constructed and stored in the group object, which significantly improves the performance of MPI_Group_translate_ranks, MPI_Group_compare, and MPI_Group_union. In order to conserve memory (and list construction time) in codes that do not use these routines, this sorted LPID list is constructed lazily only when these routines are first invoked. Constructing this list requires an $\mathcal{O}(p_g \log p_g)$ time sorting step and roughly doubles the size of the group object for nontrivial values of p_g.

3.4   Virtual Connections

In most practical MPI implementations, each process must maintain at least a modicum of state for each other process with which it is communicating. In MPICH2, this state is kept in a *virtual connection* object (or *VC*) associated with the remote process. MPICH2's current implementation creates one of these objects for each other process in the system, on every process. That is, across an entire MPI application these VC objects consume $\mathcal{O}(p^2)$ memory.

This obvious scalability issue is addressed in Section 4. However, even the current design is more scalable than a naïve implementation. Many buffers that are

---

[1] See MPI-2.2, § 10.5.4, for a formal definition of "connected" in this context.

attached to the VC object are not created until communication actually occurs with the process corresponding to that VC. For connection-oriented communication substrates such as TCP, these connections are not created until communication actually occurs, thereby conserving operating system resources.

The VC implementation provides another example of a location where additional space is consumed in exchange for reduced access time. Each VC contains a "scratch pad" area that may be used by lower-level code to store per-VC information. In order to decouple lower layers from the upper layers of MPICH2, such storage space must exist. However, it would also be sufficient for this space to be just large enough to hold a pointer, such that the lower-level code could allocate a separate object and store a pointer to it in this minimal scratch pad region. This approach would require an additional pointer dereference for the lower-level code to access its own VC-specific data. Instead, by making the scratch pad region larger, this additional pointer dereference is saved for latency-sensitive data accesses that can be fit into the scratch pad. Of course, tuning the size of this scratch pad becomes critical for large p.

3.5   Communicator and Topology Information

Among many responsibilities, MPI communicators are responsible for storing enough data in order determine which underlying process corresponds to a given rank in that communicator. For example, when the user calls `MPI_Send(...,` `5,...,comm)`, the implementation must be able to determine that rank 5 in comm will result in communication with a particular process on a particular network host. More concretely in the case of MPICH2, this means that given a communicator and a rank, the implementation must be able to produce a VC object. This translation is currently supported by a *virtual connection reference table* (or *VCRT*).

VCRTs consist of a dense array of *VC references* (or *VCRs*), indexed by communicator rank. The VCR is an opaque type, but because of practical details of the interface, it must typically be implemented as a pointer to the underlying VC. Each communicator stores a pointer to its VCRT and manipulates reference counts inside that VCRT. This reference counting permits shallow copies of the VCRT for the common case of `MPI_Comm_dup`, reducing memory consumption. However, besides sharing a VCRT between two communicators, VCRTs themselves have only $\mathcal{O}(p)$ per communicator space scalability in the general case.

MPICH2 stores additional information on a per process and per communicator basis in order to support hierarchical collective communication algorithms. For each connected process a *node ID* is stored, consuming $\mathcal{O}(p)$ memory on each process. This approach enables creating two internal subcommunicators for each user-created communicator: one that contains only "node leaders" and another that contains only processes on the same node. For a top-level communicator of size p that is spread evenly over k nodes, the node-leaders communicator will contain k members, while the node-local communicator will contain p/k members. Every process will be a member of a node-local communicator, but only the node leaders will be a member of the leader communicator. In MPICH2's current implementation these communicators will consume $\mathcal{O}(p^2/k + pk) \Rightarrow \mathcal{O}(p^2)$ memory across the whole system (assuming constant k).

4   Steps to Reduce MPICH2 Memory Consumption

The current deficiencies in MPICH2 memory usage mentioned above can be addressed in several ways. We detail here solutions we have implemented in an experimental version of MPICH2, and we outline several additional solutions we intend to implement in the near future.

4.1   Implemented Solutions

The most serious scalability problem discussed earlier is the $\mathcal{O}(p^2)$ memory consumption by VCs (across the whole system) even when communication takes place with zero or few partners. To rectify this issue, we have developed a prototype version of MPICH2 that substantially overhauls the way VCs are managed.

Under the new scheme, entire VC objects are created lazily only as needed instead of statically at `MPI_Init` time. This change required a fundamental shift in the way VCs are stored and accessed. The per communicator VCRTs discussed in Section 3.5 have been eliminated and replaced with a similar, yet more efficient concept: the *LPID mapping* (or *LPM*). These objects perform a similar role; but rather than mapping communicator ranks to VCs directly and always via a dense array mechanism, the LPM maps communicator ranks to LPIDs. This mapping decouples the upper-level code, for example MPI collective routines, from any notion of VCs that exist only at the lower level.

Unlike VCRTs, LPMs are truly opaque objects that are accessed only via function calls and macros. This design provides the opportunity to encode the communicator representation in the most succinct, memory-efficient manner possible. Examples include using compression techniques that take advantage of domain-specific knowledge [9] or more general compression methods [4]. Another example of domain-specific compression is supporting identity mappings, wherein the LPID is always equal to the communicator rank. Implementing this identity mapping is trivial, given the new interface, and reduces per process memory consumption from $\mathcal{O}(p)$ to $\mathcal{O}(1)$ for communicators for which this mapping holds (such as `MPI_COMM_WORLD`).

Conveniently, the LPM concept and interface also permitted us to unify the representation of groups and the representation of communicator VC contents. Future improvements to this common LPM facility will yield dividends in both the group and communicator subsystems of MPICH2.

At a lower level, VCs are obtained only via APIs that refer to them by their LPIDs. This design permits true lazy instantiation and storage of VC objects, such as in a hash table, since upper-level code no longer holds pointers to all VCs. This hash-based approach is exactly what we implemented, with a run-time environment variable to select between the hash table and a dense, fully populated array.

For convenience and robustness, we used the open source `uthash` package [10]. Additional constant factor time and memory savings may be possible with an alternative implementation.

4.2 Proposed Solutions

Though we implemented several space-saving techniques in our experimental version of MPICH2, there remain many that we did not have time to implement. For example, data on the same SMP node can be shared, such as information about `MPI_Win` objects. Data-caching strategies can be employed, particularly if efficient remote memory access is available. We leave these approaches to future work.

5 Results

In this section we provide experimental evidence that an MPI implementation can limit the use of memory for scalable applications without a significant performance impact. We first look at some simple benchmarks, including ping-pong performance microbenchmarks, and then evaluate several application-based benchmarks.

All results were gathered on the "Fusion" cluster at ANL. Each node consists of two Intel Xeon X5550 quad-core processors, and the nodes are connected by QDR Infiniband. MPICH2 was configured as `--with-device=ch3:nemesis:tcp` and `--enable-fast`.

5.1 Scalable Memory Use

To validate the expected memory consumption of the prototype, we crafted three microbenchmarks that isolate basic communication behavior from more sophisticated application MPI usage. These microbenchmarks respectively perform no communication, scalable communication (a single `MPI_Allreduce`), and non-scalable communication (pairwise communication between all processes). Furthermore, the MPI library was instrumented to permit memory consumption measurements to be taken. The results from running these simple programs respectively provide minimum, modest, and maximum memory consumption baselines that are harder to observe as clearly in applications with more sophisticated communication patterns.

Figure 1 shows the results of running these experiments with the lazy initialization prototype code enabled. As expected, the "no communication" and "allreduce" benchmarks consumed essentially no additional memory per process as the job size was increased, while the "all communication" benchmark showed per process memory consumption increasing linearly with job size. This increase indicates an $\mathcal{O}(p^2)$ systemwide memory consumption scalability problem, one that our technique has addressed for programs with a scalable communication pattern.

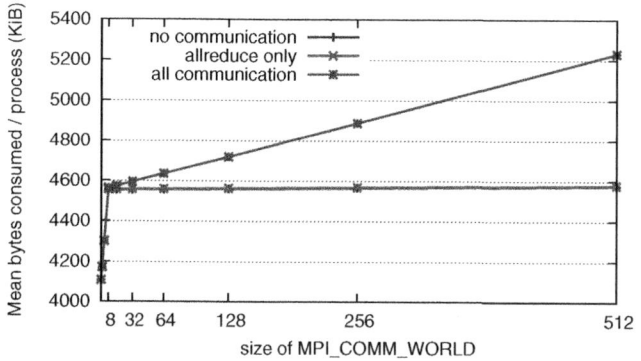

Fig. 1. Per process memory consumption in the prototype for three microbenchmarks

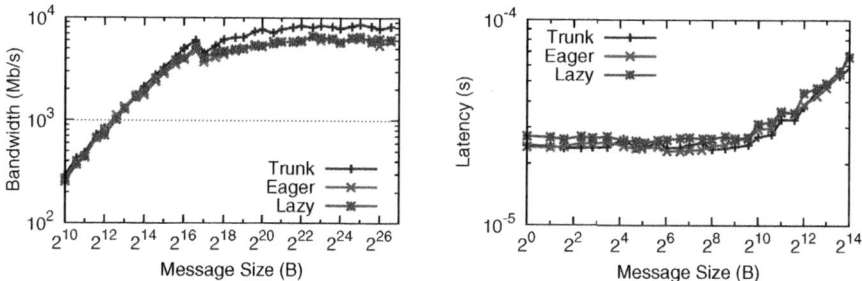

Fig. 2. Netpipe ping-pong performance results (log-log plot for relevant message sizes)

5.2   Performance Impact

The techniques discussed in Section 4 are expected to at least slightly impact performance. Figure 2 shows MPI-level bandwidth and one-way latency numbers for the stable ("Trunk") version of MPICH2 as a reliable baseline, as well as the prototype configured to use an eagerly constructed dense array ("Eager") or lazily constructed sparse hash table ("Lazy") for VC storage. Both a slight decrease in large-message bandwidth and a slight increase in small-message latency can be seen. We emphasize, however, that the prototype code has not been tuned to any noteworthy extent; we expect to eliminate most of this performance gap with further effort.

5.3   Application Impact

We measured the impact of our changes on scalable applications by examining the performance and memory consumption behavior of certain NAS Parallel Benchmarks [2] and the Sequoia AMG benchmark that are representative of application behavior. All of these benchmarks exhibit fairly scalable communication patterns; that is, the number of communication partners remains flat

Table 1. Performance of selected NAS Parallel Benchmarks, version 3.3 run with 512 processes

Benchmark	MPI	Time (s)		Memory/Process (kiB)	
	Trunk	536.77		5,149.2	
cg.D.512	Eager	520.55	(-3.02%)	5,144.7	(-0.09%)
	Lazy	556.82	$(+3.74\%)$	4,588.2	(-10.89%)
	Trunk	18.69		5,154.2	
mg.D.512	Eager	19.19	$(+2.68\%)$	5,154.3	$(+0.00\%)$
	Lazy	19.49	$(+4.28\%)$	4,602.3	(-10.71%)

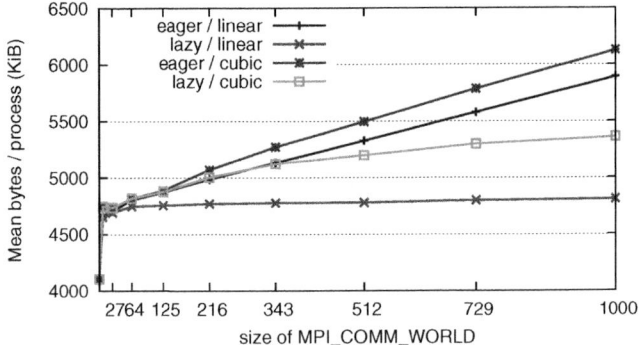

Fig. 3. Per process memory consumption in the prototype for the Sequoia AMG benchmark

or increases slowly as job size increases. These codes are also well known and commonly used to represent the behavior of many real-world MPI numerical applications.

Table 1 lists the performance impact and average per-process memory consumption of our techniques when applied to the CG and MG class D NAS Parallel Benchmark. The benchmarks were run with the same three configurations from Figure 2. At this modest scale MPI memory consumption is reduced in the Lazy approach by approximately 550 bytes per process ($\approx 11\%$), at a cost of less than 5% in performance. We did observe variability in the run times, despite great consistency in the memory consumption numbers, which we attribute to noise from the shared Infiniband network on this system.

Figure 3 shows per process memory consumption versus job size when running the Sequoia AMG benchmark [1] on the prototype with both the eager and lazy VC initialization strategies. The benchmark was configured to solve a Laplace-type problem[2] with two different three-dimensional processor layouts. The first layout was cubic (e.g., 36 processes organized as $P_x \times P_y \times P_z = 6 \times 6 \times 6$). The

---

[2] AMG was run with the following options: `-laplace -n 25 25 25 -solver 4`.

plot clearly shows a substantially slower-growing memory consumption curve for this case when lazy VC initialization is used. The second layout was entirely linear (e.g., $36 \times 1 \times 1$). Although unrealistic as a choice of typical application parameters, this layout has far fewer communication partners, which yields the expected almost entirely flat per-process memory consumption curve.

6   Conclusions

We have shown that an MPI implementation can be constructed so that memory use grows slowly as the number of processes increase and that the performance cost for a real application is low.

Acknowledgments. This work was supported by the U.S. Department of Energy, under Contract DE-AC02-06CH11357 and Award DE-FG02-08ER25835, and by the National Science Foundation Grant #0702182.

References

1. ASC Sequoia Benchmark Codes: AMG,
   https://asc.llnl.gov/sequoia/benchmarks/#amg (May 2011)
2. Bailey, D., Harris, T., Saphir, W., Van Der Wijngaart, R., Woo, A., Yarrow, M.: The NAS Parallel Benchmarks 2.0. NAS Technical Report NAS-95-020, NASA Ames Research Center, Moffett Field, CA (1995)
3. Balaji, P., Buntinas, D., Goodell, D., Gropp, W., Hoefler, T., Kumar, S., Lusk, E., Thakur, R., Träff, J.L.: MPI on millions of cores. Parallel Processing Letters 21(1), 45–60 (2011)
4. Barbay, J., Navarro, G.: Compressed representations of permutations, and applications. In: Proc. of 26th Int'l Symposium on Theoretical Aspects of Computer Science (STACS), pp. 111–122 (2009)
5. Chaarawi, M., Gabriel, E.: Evaluating sparse data storage techniques for MPI groups and communicators. In: Bubak, M., van Albada, G.D., Dongarra, J., Sloot, P.M.A. (eds.) ICCS 2008, Part I. LNCS, vol. 5101, pp. 297–306. Springer, Heidelberg (2008)
6. Kamal, H., Mirtaheri, S.M., Wagner, A.: Scalability of communicators and groups in MPI. In: Proc. of the ACM International Symposium on High Performance Distributed Computing, HPDC (2010)
7. MPICH2, http://www.mcs.anl.gov/mpi/mpich2
8. Stevens, R., White, A.: Report of the workshop on architectures and technologies for extreme scale computing (December 2009),
   http://extremecomputing.labworks.org/hardware/report.stm
9. Träff, J.L.: Compact and efficient implementation of the MPI group operations. In: Keller, R., Gabriel, E., Resch, M., Dongarra, J. (eds.) EuroMPI 2010. LNCS, vol. 6305, pp. 170–178. Springer, Heidelberg (2010)
10. uthash (May 2011), http://uthash.sourceforge.net/

Performance Expectations and Guidelines for MPI Derived Datatypes

William Gropp[1], Torsten Hoefler[1], Rajeev Thakur[2], and Jesper Larsson Träff[3]

[1] University of Illinois, Urbana, IL 61801, USA
{wgropp,htor}@illinois.edu
[2] Argonne National Laboratory, Argonne, IL 60439, USA
thakur@mcs.anl.gov
[3] University of Vienna, Austria
traff@par.univie.ac.at

Abstract. MPI's derived datatypes provide a powerful mechanism for concisely describing arbitrary, noncontiguous layouts of user data for use in MPI communication. This paper formulates *self-consistent performance guidelines* for derived datatypes. Such guidelines make performance expectations for derived datatypes explicit and suggest relevant optimizations to MPI implementers. We also identify self-consistent guidelines that are too strict to enforce, because they entail NP-hard optimization problems. Enforced self-consistent guidelines assure the user that certain manual datatype optimizations cannot lead to performance improvements, which in turn contributes to performance portability between MPI implementations that behave in accordance with the guidelines. We present results of tests with several MPI implementations, which indicate that many of them violate the guidelines.

1 Introduction

Self-consistent performance requirements for MPI are an invitation to MPI implementers to ensure consistent performance among interrelated functionalities. In addition to guarding against unpleasant performance surprises, such guidelines can support performance portability among MPI implementations: They avoid the need for hand optimizations to compensate for unsatisfactory performance of specific functions in specific contexts, systems, or MPI implementations, which could also be counterproductive on other systems, implementations, or circumstances. Self-consistent MPI performance guidelines can be construed as performance expectations for application programmers, recommendations for MPI implementers, or even requirements that would be desirable to fulfill.

Performance expectations for MPI communication functions were formulated in [11] and for MPI-IO in [3]. This paper proposes performance expectations and guidelines for the derived datatype mechanism in MPI. We identify a number of guidelines for the performance of the MPI datatype mechanism that an MPI implementation should meet so as to enable and encourage performance-portable programming. We also present the results of simple experiments to validate MPI

Y. Cotronis et al. (Eds.): EuroMPI 2011, LNCS 6960, pp. 150–159, 2011.

implementations. Our measurement results for several implementations indicate that many of them violate the performance guidelines, which can lead to unpleasant surprises for users. This result should serve as an encouragement for further research and implementation work on improving the handling of MPI derived datatypes.

1.1   Related Work

The derived datatype mechanism is one of the central concepts of the MPI standard. It separates communication operations from the structure of data being communicated [7, Chapter 4] and is vital for the MPI-IO specification for distributed file structures [7, Chapter 13]. The generality and expressive power of the derived datatype mechanism is one feature that sets MPI apart from other interfaces with similar intentions and scope. Describing complex, local data layouts by derived datatypes makes it possible for the MPI implementation to handle such structures by efficient packing and unpacking mechanisms that interact closely with (pipelined) communication algorithms or by exploiting available hardware support for noncontiguous data communication. Achieving similar or better effects by hand is tedious and in many cases non-portable performance wise. The ultimate goal of an efficient MPI implementation of the datatype mechanism is, in some loose sense, never to be worse than what the application programmer can achieve by hand packing/unpacking and communicating the packed buffers. This paper is an attempt toward defining this goal more precisely.

Providing efficient implementations of MPI datatypes has therefore been the focus of several groups [2,4,9,10,12], and much progress has been achieved, although there are still situations where datatype performance is less satisfactory as we discuss in Section 7. The use of MPI datatypes to provide better performance within applications has been explored in several studies, e.g., [1,5,6]. Benchmarks for datatypes focusing on the complexity of the different constructors were defined in [8]. We are not aware of any work directly addressing performance expectations and guidelines for MPI datatypes.

2   Derived Datatype Constructors

MPI derived datatypes can be thought of as concise descriptions of layouts of data in process memory. MPI derived datatypes are described in [7, Chapter 4], which the reader should consult for precise definitions (constructors, type signatures and maps). There are five main MPI functions for constructing new datatypes out of old ones. Let n be the value of the count argument supplied to the constructors. We omit all arguments that are not essential for the discussion.

1. MPI_Type_contiguous(n,T): n successive blocks of type T, denoted as contig(n, T)
2. MPI_Type_vector(n,m,T): n strided blocks of m instances of type T, denoted as vector(n, m, T)

3. MPI_Type_create_indexed_block(n,m,T): n blocks of m instances of type T each with own displacement, denoted as index_block(n, m, T)
4. MPI_Type_indexed(n,m_n,T): n blocks of type T each with own count m_i ($0 \leq i < n$) and displacement, denoted as index(n, m_n, T). The total number of blocks is $\sum_{i=0}^{n} m_i$.
5. MPI_Type_create_struct(n, m_n, T_n): n blocks of types T_i each with own count m_i ($0 \leq i < n$) and displacement, denoted as struct(n, m_n, T_n)

The constructors can be applied recursively, so T can be a primitive, basic datatype or a previously constructed, derived datatype. In addition, there are convenience functions for creating datatypes representing subarrays and distributed arrays. Another special constructor makes it possible to change the extent of a (derived) datatype, which is important when using nested type constructors, see for instance [1].

A first benchmark measures the basic communication performance for strided layouts described by each of the five constructors. The benchmark can be parameterized in type T (here we use only the basic MPI_DOUBLE type), stride s and number of blocks n. Communication performance is measured by point-to-point ping-pong communication in order to be able to focus as far as possible on the datatype component.

Benchmark 1. *The same strided layout of a n repetitions of type T with stride s described by the five different type constructors. Communication time for the five types as a function of number of repetitions n.*

On a given architecture the layout of the data elements in memory eventually determines the performance of communication operations involving the derived datatype. Alignment of the basic datatypes might be good or bad, the basic datatypes may be blocked, or strided or otherwise regularly spaced which might be advantageous for some architectures, there might be special hardware that can exploit certain structures in the layout, etc. For these reasons it is not possible to pose *absolute* performance requirements on MPI operations involving datatypes. A natural user expectation, however, would be that hardware support, e.g., for strided memory access or communication, bulk transfers etc. be utilized wherever possible by the MPI library.

However, what can be done, and this is the key point, is to relate the many different ways that a *given* type map can be described by the derived datatype mechanism (e.g., as in Benchmark 1). A self-consistent MPI performance guideline for datatypes would state that the performance of an MPI communication operation with some datatype T describing the *given* (non-contiguous) layout should be no worse than the same operation with any other datatype T' that describes the *same* layout. Otherwise, the user could improve performance by possibly tedious and non-portable redefinitions of the datatype description of the application data.

Another user expectation which we discuss in more detail in Section 5 is that MPI operations with datatypes perform at least as well as manually packing the data into a contiguous buffer before the MPI operation.

Each of the type constructors describes a sequence of n blocks. The contiguous and vector types do so with constant extra information, the indexed block requires one array for the displacements, the indexed one extra array for the block lengths, and the structure yet one more array for the datatypes of the blocks. We formalize this by associating the *penalties* $0, 0, 1n, 2n, 3n$, plus some constant $O(1)$ overhead, with the five constructors, accordingly. The total penalty of a datatype is defined recursively as the penalty of the top-level constructor times the penalty of the subtype, or, for the struct constructor, the sum of the penalties of the subtypes. The intuition is that in order to process a layout described by a datatype with some total penalty h, $\Omega(h)$ operations are required just to parse the type map. The strictest, self-consistent performance guideline then says that the performance of an MPI function with datatype T should be no worse than the performance with a datatype T' that has minimal total (considering possibly recursive type specifications) penalty.

3   Trivial Expectations

We will use the following notation to express performance expectations and guidelines: $\mathsf{MPI\_A}(n, T_{A'}) \preceq \mathsf{MPI\_B}(n, T_{B'})$ shall mean that MPI function A operating on n elements as described by datatype $T_{A'}$ is not slower than MPI function B with type $T_{B'}$ for almost all n, all other things, including in particular the type map of the datatypes $T_{A'}$ and $T_{B'}$, being equal.

Expectation (1) comes directly from the MPI standard which states that a call to a communication function with a count and a datatype argument is functionally equivalent to the same call where the count and the datatype have been encapsulated in a contiguous datatype [7, Section 4.1.11]. It would be sensible to expect that these two equivalent call forms would also perform similarly:

$$\mathsf{MPI\_A}(1, \mathrm{contig}(n, T)) \approx \mathsf{MPI\_A}(n, T) \tag{1}$$

This should hold for any type T. Exhaustive verification is of course not possible, but a simple benchmark will indicate whether the expectation is reasonably fulfilled.

Benchmark 2. *Six basetypes $T_0 = T$, $T_1 = \mathrm{contig}(k, T)$, $T_2 = \mathrm{vector}(k, T)$, $T_3 = \mathrm{index\_block}(k, T)$, $T_4 = \mathrm{index}(k, T)$, and $T_5 = \mathrm{struct}(k, T)$ with repetition count n, versus $T_0, \ldots, T_5$ encapsulated in a contiguous type with count n; for T_0 repetition count is kn so as to have the same number of element in all six cases. Communication performance with the two versions for the six types.*

This benchmark measures both sides of Equation 1. It should be extended with more irregular layouts, e.g., from the following benchmarks.

The five constructors are able to express more and more irregular layouts of data in memory, but at an increasing penalty (more parameters for displacements/indices, block lengths, and datatypes). For a given, regularly strided layout that can be expressed with all five constructors, it is therefore natural to expect, for any function $f(n) \leq n$ (e.g., constant), that

$$
\begin{aligned}
\mathsf{MPI\_A}(n, \mathsf{contig}(f(n), T)) \preceq\ & \mathsf{MPI\_A}(n, \mathsf{vector}(f(n), T)) \\
\preceq\ & \mathsf{MPI\_A}(n, \mathsf{index\_block}(f(n), T)) \\
\preceq\ & \mathsf{MPI\_A}(n, \mathsf{index}(f(n), T)) \\
\preceq\ & \mathsf{MPI\_A}(n, \mathsf{struct}(f(n), T))
\end{aligned}
\tag{2}
$$

Guideline (2) says that if a given layout can be expressed with fewer parameters (less penalty), then for any MPI function A this should perform no worse, ideally better, than expressing this layout with a datatype constructor with higher penalty. It is a (trivial) self-consistent performance requirement: if the higher penalty datatype constructor would perform better in some context, the user could obtain this performance by manually rewriting his code to use the better performing constructor. With Benchmark 1 Expectation (2) can be checked for non-nested instances of the five constructors, and we discuss this in Section 7.

4   Non-trivial Guidelines

Non-trivial guidelines either constrain or impose requirements on an MPI implementation. Not all MPI libraries may fulfill them, but for performance portability reasons it is beneficial for implementations to adhere to them. This saves the user from the temptation to look for the best performing constructor, and let him focus instead on the most convenient, close-to-the-application-logic description.

The self-consistent principle would seem to require that MPI libraries do *type normalization* of any user-defined datatype to the "most efficient" representation that could be expressed by other datatype constructors. The MPI_Type_commit function is the point where MPI libraries can do such normalization. For instance, a $\mathsf{struct}(n, m_n, T)$ where all n blocks have the same basetype could trivially be converted into an indexed type which has penalty $2n$ instead of $3n$. Or an $\mathsf{index}(n, m_n, T)$ where all blocks have the same size could be converted into an $\mathsf{index\_block}(n, m, T)$, again with less penalty. If in addition the indices are regularly strided the $\mathsf{index\_block}(n, m, T)$ could be converted into a $\mathsf{vector}(n, m, T)$, now with constant penalty, and if the stride is equal to the block length, this could also be expressed as a $\mathsf{contig}(n, T)$. This is stated as guidelines/requirements of the form

$$
\mathsf{MPI\_A}(n, \mathsf{struct}(n', m_{n'}, T_{n'})) \approx \mathsf{MPI\_A}(n, \mathsf{index}(n', m_{n'}, T))
\tag{3}
$$

for indexed layouts where all indexed elements have the same basetype $T_i = T$. From such requirements it would follow that communication with a datatype T whose type map consists of consecutive, basic datatypes in increasing offset order should be no worse than communication with a basic datatype alone, that is

$$
\mathsf{MPI\_A}(n) \approx \mathsf{MPI\_A}(n, T)
\tag{4}
$$

In general, self-consistency would require MPI implementations to solve the following problem.

Definition 1. *The* datatype normalization problem *is the problem of finding, for given layout, the derived datatype with the lowest penalty describing the same layout.*

At the top level use of a constructor, type normalization is easy and looks sensible. A simple scan through index, blocksize and type lists can easily discover whether a type constructor with high penalty can be expressed in terms of a more regular constructor with lower penalty. However, for nested types, normalization is not trivial. A datatype layout described by the MPI constructors can be described by a tree with repetition counts and displacement/type lists at the nodes, and there are many trees describing the same layout. Finding the one with least penalty is similar to hard optimization problems on trees, and the presence of repetition counts makes the problem particularly difficult. We conjecture that the type normalization problem is NP-hard.   If this conjecture is true, it is not reasonable to *require* that an MPI implementation performs optimum type normalization in all cases.

The next benchmark is intended to test whether slightly non-trivial normalizations are performed. It is parameterized in a type T.

Benchmark 3. *a) A strided layout where the ith element is placed at position $is + (i \bmod 2)$ described with the MPI_Type_create_indexed_block constructor (cannot be normalized to a one level vector type) versus a two level vector of $n/2$ blocks of a two element vector with stride $s + 1$ and extent $2s$. The first description has penalty n, the second penalty $O(1)$.*

b) A layout of two elements, a stride, three elements, a stride, and a single element is repeated $n/6$ times. This layout described with the MPI_Type_indexed constructor versus description as two elements followed by a vector of $n/3 - 1$ blocks of three elements, followed by a single element. The latter description has penalty $O(1)$, the former penalty $2n$.

Communication performance with the two versions of the layouts.

5 Packing

MPI provides functionality for packing any layout described by a derived datatype into a contiguous buffer. It is reasonable to expect that in communication functions this is done internally as necessary, such that first packing and then communicating the consecutive buffer does not make sense, performance wise. This is an example of a self-consistent performance requirement in which an MPI functionality (namely, any communication function with a non-contiguous layout) is implemented (by the application programmer) in terms of other MPI functionality [11].

$$\text{MPI\_A}(n, T) \preceq \text{MPI\_Pack}(n, T, B) + \text{MPI\_A}(B) \tag{5}$$

where B is the intermediate packed buffer.

Benchmark 4. *The previous benchmarks in two versions: communication with datatypes directly in the communication functions, and with a pack/unpack to/-from contiguous buffers before/after communication. Also pack time is measured stand alone.*

As n grows large, a reasonable MPI implementation should be able to do pipelining to overlap any internal packing that may be necessary with other operations. For very small data, explicit packing with MPI_Pack could make sense, but should make no difference.

Again, by self-consistency recursive application of pack should not lead to an improvement [11]. Pack for basetypes should be comparable to memcpy; otherwise, the user would be tempted to do this optimization by hand. This implies that packing by hand in the sequence implied by the datatype constructors will not make sense. Hand-packing can lead only to an improvement if non-trivial tricks or domain knowledge is exploited. This can be expressed as

$$\mathsf{MPI\_Pack}(n, T, B) \preceq \mathsf{Userpack}(n, T, B) \tag{6}$$

Note that user-provided code for pack and unpack operations range from very simple loops to complex, memory-hierarchy-aware codes using deep application knowledge. A natural user expectation is that the MPI operations perform at least as well as "simple" user code implemented by straightforward loops over and recursive decomposition of the datatype T.

Benchmark 5. *Packing time versus user packing time with a* simple *pack loop for the datatypes of the previous benchmarks.*

6   Datatype Preprocessing and Commit

It appears difficult to pose self-consistent or absolute performance requirements for the type constructors and the MPI_Type_commit function. For the constructors at least all parameter lists must be read (and unfortunately copied, because the user may change the buffers after the creation call), so the time is $\Omega(n)$ where n is the total size of parameters in the call, and possibly $\Omega(m)$ where m is the penalty of the constituent datatypes (here it probably suffices to go through the normalized subtypes). The MPI_Type_commit function may for trivial library implementations do nothing and take constant time otherwise an expectation may be that no more than linear time (in either penalty or total size of parameters) be taken.

Benchmark 6. *Type construction and commit times are measured for the datatypes of previous benchmarks.*

7   Initial Experimental Results

We have implemented a first datatype expectation benchmark program incorporating some of Benchmarks 1-6. The benchmark creates datatypes for describing

Table 1. Results for Stride-1 in μs, $n = 32768$. **Bold** entries indicate violation of a performance guideline.

Library	Phase	User	Contig	Resized	Vector	BIdx	Idx	Struct
MPICH2	Pack	74	65	65	65	94	**180**	**262**
MPICH2	Send	486	459	463	**544**	480	460	457
Open MPI	Pack	74	66	66	66	**375**	**370**	**279**
Open MPI	Send	428	428	428	428	428	428	428
BG/P	Pack	386	148	148	**409**	149	149	149
BG/P	Send	238	238	238	238	238	238	238
POE	Pack	224	195	196	196	195	197	198
POE	Send	368	362	363	362	351	361	373

strided layouts of single basetype elements by means of the five basic constructors. A basic experiment compares the performance with a ping-pong benchmark. Likewise, packing by MPI_Pack can be performed. The benchmark also measures the construction time and the commit time.

We here present some of the benchmark results for communicating n MPI_DOUBLE values with stride 1 (contiguous) and stride 16 (vector) for different MPI implementations communicating in shared memory . We used Open MPI 1.4.3 and MPICH2 1.3.2p1 on a 1GHz Quad Core Opteron 270 HE system at Indiana University, IBM's BG/P MPI on Intrepid at Argonne National Laboratory, and POE MPI 5.1 on a 16 core POWER5+ system at the University of Illinois at Urbana-Champaign (we also have results for POE on POWER7 under Linux; they are qualitatively similar to the POWER5+ results and are omitted).

We compare a simple pack loop (**User**) with types constructed with MPI_Type_contiguous (**Contig**, only stride 1), MPI_Type_create_resized (**Resized**, the extent of the type is used to generate the correct stride), MPI_Type_vector (**Vector**), MPI_Type_indexed_block (**BIdx**), MPI_Type_indexed (**Idx**), and MPI_Type_struct (**Struct**). The combination Send/User means that the data is sent directly from the user buffer (this is only possible in the contiguous stride-1 case).

Table 1 shows the results for different specifications of stride-1 data access. This can be considered the simplest case (a Benchmark 0), and provides both a basis for comparing non-unit strides in Table 2 and for identifying which datatypes the MPI implementation simplifies to a more efficient internal representation. Our experiments show that, for stride-1 data, MPI_Pack is generally faster than a pack loop. We also observed that sending datatypes directly was generally faster than combining packing and sending manually. Our results show that almost all libraries fail to detect the contiguous data pattern reliably. Bold entries in Table 1 show where the self-consistency requirements are violated because the requirement of Equation (3) is not met. Note that these timing results have some uncertainty and small differences are not significant.

Table 2 shows the results for different specifications of stride-16 data access. Our experiments show that, for stride-16 data, MPI_Pack is often slower than

Table 2. Results for Stride-16 in μs, $n = 32768$. **Bold** entries indicate violation of a performance guideline.

Library	Phase	User	Resized	Vector	BIdx	Idx	Struct
MPICH2	Pack	592	1035	1034	**843**	**704**	**797**
MPICH2	Send	-	2917	3045	3015	3013	3036
Open MPI	Pack	600	1494	1490	**2773**	**2769**	**2717**
Open MPI	Send	-	3086	3060	**5281**	**5279**	**5269**
BG/P	Pack	2049	2116	2115	**2218**	**2292**	**2368**
BG/P	Send	-	6402	6402	6414	6414	6412
POE	Pack	563	631	623	**2056**	**2064**	**2072**
POE	Send	-	1658	1694	**6203**	**6263**	**6296**

a pack loop. One notable exception is MPICH2 where the pack performance is slightly better. We also observed that sending datatypes directly was generally faster than combining packing and sending manually. Our results show that almost all libraries fail to detect the vector pattern reliably. Bold entries in Table 2 show where the self-consistency requirements are violated because the requirement of Equation (3) is not met.

8   Conclusion

By identifying self-consistently motivated performance guidelines and performance expectations for the MPI derived datatype mechanism first steps were taken toward a benchmark for testing aspects of datatype performance. The datatype normalization problem was formalized in terms of penalties, and we conjecture that this problem is NP-hard. This limits the amount of type normalization that an MPI library can be expected to do, and therefore the user still needs to be careful how data layouts are described. Our experiments on a selection of platforms and MPI libraries showed unpleasant performance surprises, indicating for instance that very little type normalization is performed, even for cases where this would be trivially possible. The experiments also clearly showed large performance differences depending on the way a given layout is described, thus more normalization could well make sense in MPI implementations.

Acknowledgments. This work was supported in part by the Office of Advanced Scientific Computing Research, Office of Science, U.S. Department of Energy, under contract DE-AC02-06CH11357 and DE-FG02-08ER25835, and by the Blue Waters sustained-petascale computing project, which is supported by the National Science Foundation (award number OCI 07-25070) and the state of Illinois.

References

1. Bajrović, E., Träff, J.L.: Using MPI derived datatypes in numerical libraries. In: Cotronis, Y., et al. (eds.) EuroMPI 2011. LNCS, vol. 6960, pp. 29–38. Springer, Heidelberg (2011)
2. Byna, S., Sun, X.-H., Thakur, R., Gropp, W.D.: Automatic memory optimizations for improving MPI derived datatype performance. In: Mohr, B., Träff, J.L., Worringen, J., Dongarra, J. (eds.) PVM/MPI 2006. LNCS, vol. 4192, pp. 238–246. Springer, Heidelberg (2006)
3. Gropp, W.D., Kimpe, D., Ross, R., Thakur, R., Träff, J.L.: Self-consistent MPI-IO performance requirements and expectations. In: Lastovetsky, A., Kechadi, T., Dongarra, J. (eds.) EuroPVM/MPI 2008. LNCS, vol. 5205, pp. 167–176. Springer, Heidelberg (2008)
4. Gropp, W.D., Lusk, E., Swider, D.: Improving the performance of MPI derived datatypes. In: Proceedings of the Third MPI Developer's and User's Conference, pp. 25–30. MPI Software Technology Press (1999)
5. Hoefler, T., Gottlieb, S.: Parallel zero-copy algorithms for fast fourier transform and conjugate gradient using MPI datatypes. In: Keller, R., Gabriel, E., Resch, M., Dongarra, J. (eds.) EuroMPI 2010. LNCS, vol. 6305, pp. 132–141. Springer, Heidelberg (2010)
6. Lu, Q., Wu, J., Panda, D.K., Sadayappan, P.: Applying MPI derived datatypes to the NAS benchmarks: A case study. In: 33rd International Conference on Parallel Processing Workshops (ICPP 2004 Workshops), pp. 538–545. IEEE Computer Society, Los Alamitos (2004)
7. MPI Forum: MPI: A Message-Passing Interface Standard. Version 2.2 (September 4, 2009), http://www.mpi-forum.org
8. Reussner, R., Träff, J.L., Hunzelmann, G.: A benchmark for MPI derived datatypes. In: Dongarra, J., Kacsuk, P., Podhorszki, N. (eds.) PVM/MPI 2000. LNCS, vol. 1908, pp. 10–17. Springer, Heidelberg (2000)
9. Ross, R., Miller, N., Gropp, W.D.: Implementing fast and reusable datatype processing. In: Dongarra, J., Laforenza, D., Orlando, S. (eds.) EuroPVM/MPI 2003. LNCS, vol. 2840, pp. 404–413. Springer, Heidelberg (2003)
10. Santhanaraman, G., Wu, J., Huang, W., Panda, D.K.: Designing zero-copy message passing interface derived datatype communication over Infiniband: Alternative approaches and performance evaluation. International Journal on High Performance Computing Applications 19(2), 129–142 (2005)
11. Träff, J.L., Gropp, W.D., Thakur, R.: Self-consistent MPI performance guidelines. IEEE Transactions on Parallel and Distributed Systems 21(5), 698–709 (2010)
12. Träff, J.L., Hempel, R., Ritzdorf, H., Zimmermann, F.: Flattening on the fly: Efficient handling of MPI derived datatypes. In: Margalef, T., Dongarra, J., Luque, E. (eds.) PVM/MPI 1999. LNCS, vol. 1697, pp. 109–116. Springer, Heidelberg (1999)

The Analysis of Cluster Interconnect with the Network_Tests2 Toolkit

Alexey Salnikov, Dmitry Andreev, and Roman Lebedev

Lomonosov Moscow State University, Dorodnicyn Computing Centre of RAS,
Moscow National research nuclear university "MEPhI"
salnikov@cs.msu.ru, andreevd@cs.msu.su,
rmn.lebedev@gmail.com

Abstract. The article discusses MPI-2 tools for benchmarking and extracting information on features of interconnect in HPC clusters. Authors develop a toolkit named "network_tests2". This toolkit highlights hidden cluster's topology, illuminates the so-called "jump points" in latency during message transfer, allows user to search defective cluster nodes and so on. The toolkit consists of several programs. The first one is an MPI-program that performs message transfer in several modes to provide certain communication activity or benchmarking of a chosen MPI-function and collects some statistics. The output of this program is a set of communicative matrices which are stored as a NetCDF file. The toolkit includes programs that perform data clustering and provide GUI for visualisation and comparison of results obtained from different clusters. This article touches some results obtained from Russian supercomputers such as Lomonosov T500 system. We also present data on Infiniband Mellanox and Blue Gene/P interconnect technologies.

1   Introduction

Nowadays, there is significant amount of cluster systems with number of processors greater than 1000. Some examples of such systems in Russia are MVS-100K (JSSC RAS), Chebyshev MSU , BlueGene/P (CMC department of MSU) and Lomonosov MSU. The developer uses functions from one of the library implementation of MPI standard (Message Passing Interface). Message passing delays during MPI-messages transfer over the cluster interconnect is a substantial problem for writing parallel applications. Experiments show that the observed delays become more and more varied with increase of the number of processors in cluster. This situation is caused by dramatical complexity growth of communication subsystem in High Performance Computational (HPC) cluster with increase of the number of processor units.

Often vendors provide information about communication environment in multiprocessor system, specifying its topology. The modern popular topologies are 2-dimensional torus, 3-dimensional torus, grid or 3-dimensional cube. For these topologies it is possible to divide the cluster nodes into a set of "neighbour nodes" and a set of other nodes. Physically, several neighbour nodes can be located in

Y. Cotronis et al. (Eds.): EuroMPI 2011, LNCS 6960, pp. 160–169, 2011.

different server racks what is reasoning for differences in messages passing delays. It is entirely possible that full specification of the actual architecture of HPC cluster's interconnect isn't available for end user. This makes solving problem of apriory delay prediction in communications during messages passing between processors extremely difficult.

Nowadays, in the World some teams are actively developing a set of MPI-programs which performs artificial communication activity by functions of the MPI library (so called "synthetic tests"). Such tests are intended for revealing features of communications behaviour and estimating performance of computational clusters. The results of cluster testing with such MPI-programs are available for usage by system administrators for detecting over all problems in cluster, for example to locate problematic nodes in cluster. The authors have developed the software (toolkit) of such type. The software is intended for testing and analysing communicational environment in computational clusters by means of the set of certain "synthetic tests" and data visualisation.

There are several tools which perform testing of communications based on sending messages: MPIbenchsuit[1], NetPIPE[2], SKaMPI[6], MPIBlib[4], Intel MPI Benchmarks[2]. Also, we found some works where people concentrate on comparison of communications of different cluster system using MPI, for example article [5].

Most of these tools have following shortcomings

- Often they lack a complex modes of measurement delays, which suggest a given behaviour not just measured MPI-processes, but also others MPI-processes which are not involved in the measurement. These modes are important for acquiring holistic information about features of communicational environment in HPC cluster.
- It is possible, that visualising and analysing tools aren't included into testing software. Availability of such instruments becomes essentially important if we are testing computational clusters with a large number of processors, where amount of generated by tools data doesn't allow possibility of manual analysis, and requires special software for processing it.

In order to diminish these shortcomings, as a part of PARUS[7] project, we have developing our own toolkit for analyzing communication bandwidth of computational cluster. This toolkit is called *network_tests2* and is available for download in source codes from the SourceForge site http://parus.sf.net.

Some group of authors in USA "Sandia National Laboratories" maintain free software project "Cbench"[3]. The goal of the Cbench project is uniting already existing testing tools into a uniform system which makes possible to extract additional information about computational cluster as a whole. However, unlike *network_tests2* in PARUS project, Cbench isn't concentrated on features of

---

[1] MPIbenchsuit is available for download from http://parallel.ru/testmpi

[2] Description of the Intel MPI Benchmarks is available at
http://software.intel.com/en-us/articles/intel-mpi-benchmarks

[3] Project Cbench website: http://sourceforge.net/apps/trac/cbench

communication environment, and it seems that Cbench doesn't have its own visualising tools. It is wise to integrate *network_tests2* into the Cbench as one of its component in the future. Let us look closer to the *network_tests2* software.

2   The Components Description

The *network_tests2* toolkit consists of several components.

- The *network_test* is a testing application which supports many testing modes.
- Useful converters, that perform transformation of testing results into three formats: "plain text representation", "non-clustered NetCDF representation", "clustered NetCDF representation".
- The implementation of clustering algorithm which performs a data compression.
- Programs for visualizing the testing results. There are two programs, first one is written in Java, and the second one in C++ using Qt4.5 framework.

2.1   Description of the Method for Cluster Interconnect Testing

Authors have developed program the *network_test*, which uses functions of MPI-2 standard [3] to perform data transmission between single processor units of computational cluster. Bandwidth of communication environment is computed by multiple repeated measurements of delay values. Values of delay are measured for one of the MPI functions. The name of MPI function is determined by the testing mode its name is given in the parameters of *network_test* program. On the basis of performed repeated measurements the software determines such characteristics as minimum latency, average latency, median latency and the standard deviation of the delay values.

The application starts working by setting length of message to *begin*, then increases this length by *step* after each iteration, and stops when limit *end* is reached. The application consider all possible pairs of MPI-processes. Measurements of delays in MPI-functions are performed for each pair (i, j) independently from other pairs of MPI-processes, but according to the testing mode. Measurements are performed for one fixed length of message. The result is a matrix of measured delays.

Thus, vector of matrices of delay values is created during the each step of *current_length*. This vector length is determined by number of iterations for each step by message length. Statistical characteristics are computed for each such vector and their values are stored to the files which collect results. Thus, statistical characteristics are stored in file. Characteristics stored as elements of three-dimensional space with coordinates $(i, j, current\_length)$, where $begin \leq current\_length \leq end$.

We design NetCDF file format for storing testing results. The following text describes CDL-header for this format:

```
netcdf info {
dimensions:
    x = <number of processors> ;
    y = <number of processors> ;
    n = UNLIMITED ;
variables:
    int proc_num ;            /*
    int test_type ;           * information about test
    int data_type ;           * and its parameters
    int begin_mes_length ;    */
    int end_mes_length ;
    int step_length ;
    int noise_mes_length ;
    int num_noise_mes ;
    int num_noise_proc ;
    int num_repeates ;
    double data(n, x, y) ;
//data { matrices containing test results
}
```

Now, let's concentrate on testing modes which *network_test* application provides.

2.2 Modes of Communications Testing Provided by network_test

There are nine different modes of cluster's communication testing available in the present *network_test* application. All of them are aimed either for finding out properties of behaviour of communications with different load levels (including peak levels), or examination of effectiveness of implementation of certain MPI-function. Load to the cluster's communications is created artificially due to a special way of organizing call to the certain MPI-functions. Special situation in communications of a computational cluster also can be modelled by special order of calling MPI-function. By comparing results from different testing modes, it is possible to make certain conclusions, for example: what kind of MPI-functions in each situation are advantageous.

Let's give short descriptions of these modes.

- **one_to_one mode.** A pair of MPI-processes (i, j) is chosen here. Message passing is initiated between them using blocking call MPI_Send from process i and receiving of the message using blocking call MPI_Recv from process j. It is guaranteed that other MPI-processes are "silent" at this moment. This process is repeated for all pairs of MPI-processing according to the method described above. This mode shows maximum bandwidth of connection channels in communication environment. In case of pairs of the form (i, i) size of delay is assumed to be 0.
- **async_one_to_one mode.** Basically, this mode is analogous to one_to_one, but processes use MPI_Isend and MPI_Irecv in such a way, that two opposite directed streams are organized in communications. If significant difference in time compared to one_to_one, then it is possible that channel between 2 processors, on which MPI-processes work, isn't full duplex. This may be

on evidence of both incorrect design communication environment and some problems related to malfunction of hardware.

- **all_to_all mode.** Unlike described in the previous modes, when all processes except the fixed chosen pair were "silent", this mode all processes start to communicate simultaneously. This is achieved in following way. Every MPI-process runs non-blocking MPI_Isend to all other MPI-processes (including itself), and then it runs MPI_Irecv from all MPI-processes, including itself. After that MPI_Waitany is run in the loop over the number of MPI processes, MPI_Waitany signalizes about completion of (i, j) interchange pair. Delay size is determined by time between initialization of first MPI_Isend and exit from MPI_Waitany after finishing corresponding to it MPI_Irecv. This mode may be used for determining of communications behaviour during "stress" (see figure 5), because large amount of data appears in communications transferred in all directions simultaneously.

- **test_noise** and **test_noise_blocking** modes. These modes are combinations of all_to_all and one_to_one modes. MPI-process with ordinal number 0 on communication MPI_COMM_WORLD splits all processes into three non-intersection groups: "target processes", "silent processes" and "noisy processes". All MPI-processes receive information about their roles with the MPI_Bcast function usage. For each pair of "target processes" delay for call MPI_Recv is computed (for test_noise_blocking), similarly to one_to_one mode. "Noisy processes" are chosen randomly. They imitate background load of network, sending "noise" messages using interaction scheme similar to all_to_all.

- **put_one_to_one** and **get_one_to_one** modes. These modes test ability to directly access memory of a remote process, which was included in MPI-2 standard. Process can put its data using function MPI_put in defined place in memory of remote MPI-process in such a way, that another MPI-process produces no actions, and even don't know how many such operations were performed. Pair operation for MPI_Put is operation MPI_Get, which makes able to "see" memory of another MPI-process. Both discussed modes are organized in similar to one_to_one mode way.

- **bcast mode.** This mode is intended for determining efficiency of MPI collective operations. All MPI-processes call MPI_Bcast. Position (i, j) of result matrix contains delay value counted after MPI-process return from MPI_Bcast. There i process is stated as root in the MPI communicator in such way it becomes source for transmitting data and j process becomes the receiver of data. The work time of function MPI_Bcast for root process in communicator is written into position (i, i) of matrix.

After finishing network_test program a data appears and then we need to analyze this data. Let's discuss the visualisation system.

2.3   Description of Testing Results Visualisation System

Visualization tool displays delays in communications as grayscale images, where growth in the delay of message transfer implies the growth in the intensity of black. When GUI calculates a correspondence between maximal black and white color to the delay values in results of testing, user could select one of the normalization modes: either to minimal and maximal values with fixed message length, or to global minimum and maximum for the whole data in window of messages length. Also, user can change levels of black and white colors on user's own. Among other things user can change the range of messages length which is in correspondence to this window.

There are design of view where user could see simultaneously two values for one (i, j) position from different matrices. The intensity of red shows the ratio between the deviation and the test result. The closer the dot to the red, the closer the measured value to the deviation.

User can optionally in the form shades of red superpose the values of deviations to "signal" (graduation in gray) that were obtained during the process of communications testing. The more dot is red, the close measured value to mean value.

Visualisation tools provide several modes for displaying data.

- Mode of displaying matrix of delays with fixed message length. (Fig. 2)
- Mode of displaying column or row for each message length. This mode can be used for analysis of how one varies the time of transfer from one MPI-process to the other processes. (Fig. 4)
- Mode of displaying (i, j) position from all communicative matrices. Each matrix is corresponding to one of messages length. (Fig. 1)

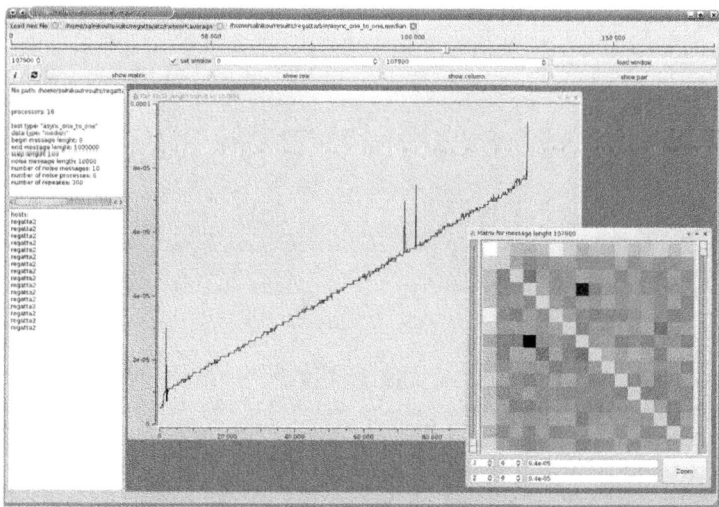

Fig. 1. An example of displaying of data by visualisation system written using QT4.5 framework. Testing mode is async_one_to_one for IBM pSeries 690 computer with 16 processors.

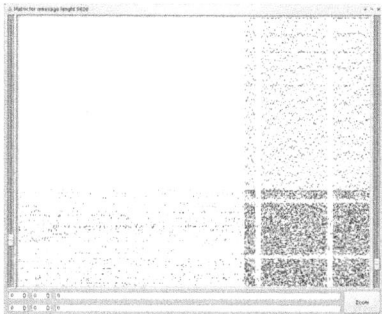

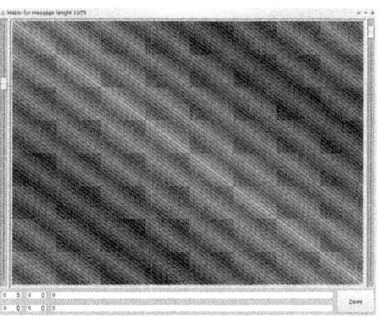

Fig. 2. Displayed variation value for one_to_one testing mode on Lomonosov MSU supercomputer. Image is created basing on message size equaling to 9600 bytes.

Fig. 3. Communication is BlueGene/P supercomputer. Data was acquired in one_to_one mode, Matrix is given for message size equaling 1075 bytes.

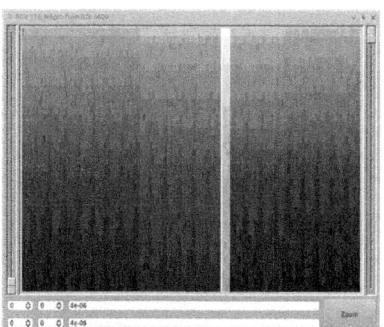

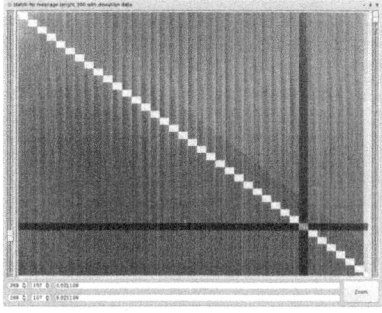

Fig. 4. Row number 176 is chosen and showed for each length of message. Message length is increased from top to bottom. Heterogeneity of cluster is increased with message length growth. Data is give for Lomonosov MSU supercomputer. Range of length is from 0 to 3600.

Fig. 5. In all_to_all mode "stress" for communication environment is organized. Figure shows significant difference between sending and receiving for one of the nodes. Data is acquired for MVS-100K computer when message length is 300 bytes.

2.4   Description of Testing Results Clustering Method

Developed clustering tool is aimed to following: automatically determine regularities in testing results and data compression with preserving fast access to individual values. During visual analysis of testing results it was noticed that sizes of delays are grouped into some, usually rectangular, regions. And these regions are stable with the change of message length in test. This fact leaded to development of described clustering algorithm.

Algorithm itself is quite complicated, and its description is somewhat bulky, so the description won't be discussed in this article, but you can find it out from the article [8].

Automatic selection of threshold values for determining closeness of values falling into one cluster, and algorithm of choosing minimal size of cluster is the subject for further investigation and research.

3   Results and Conclusion

The network_test2 has ran on computational clusters such as MVS-100K, Chebyshev-MSU, Lomonosov-MSU, BlueGene/P (BGP). Due to the inability to present all results in this article, we present only typical moments or most interesting features in communications of such systems.

First we'll discuss topological features. Figure 3 shows topological structure of BGP supercomputer. Such type of architecture has likenesses to other systems as well as differences. For example "cellular" structure is universal, and defined by number of neighbours of cluster node and by number of transits while passing. On the other hand, size and mutual arrangement of this cells are unique. Unique feature of BGP, comparing to other systems, is presence of strips, parallel to main diagonal. These strips are defined by 3D-torus topology, which helps avoiding degradation of transmission rate for distantly located processors. For another popular topology "fat tree" (Lomonosov-MSU, Chebyshev-MSU, MVS-100K are based on it), there is a trend of increase of delay size while moving off the main diagonal.

It seems that topological structure is well-known for the public, however there are some nuances, which are illustrated on the figure 4. This figure shows threshold changes of delay size depending on message length. These data are often provided by developers of testing systems, but usually only for individual pairs of MPI-processes, as it showed in the figures 4, 6, 7, 8, 9. However, the fact of increase of heterogeneity in communications and characteristics of this heterogeneity isn't given. As it follows from the figure 4, we have different topologies of computational system for different message lengths. This can greatly affect computational speed. Figures 2 and 5 show some topological features, that doesn't directly follow from the way of processors connected to each other. Figure 5 illustrates presence of "fail node" in the system. The figure shows cluster node, which transfers messages significantly slower than others. There are several possible explanations for this: temporary state like running service daemon or as permanent state like wire defect.

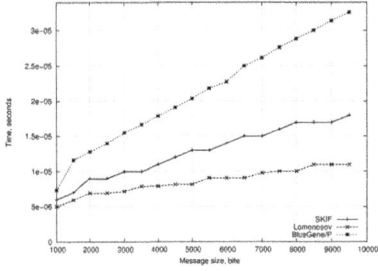

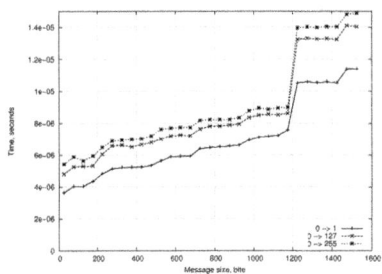

Fig. 6. Delay values obtained from several HPC clusters during the transmission of messages between neighboring nodes

Fig. 7. Data transmission between process pairs on BlueGene/P system. 256 processors in mode 1 process on 4 cores.

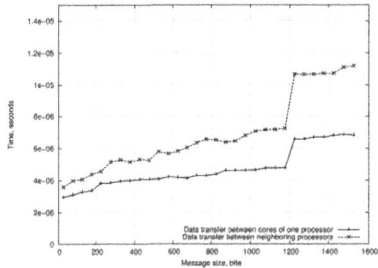

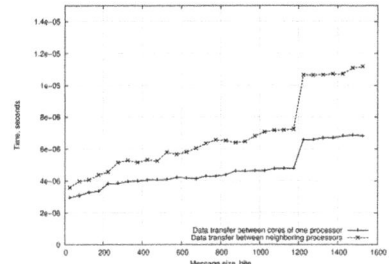

Fig. 8. Data transmission between process pairs on BlueGene/P system. 256 processors in virtual computational node mode. (One process on one core).

Fig. 9. Data transmission between pairs of processors on BlueGene/P system. 512 processors in mode 1 process on 4 cores.

A comparison of transmission delay was done for BGP system. The BGP installed in CMC MSU faculty has minimum partition size is 128 processors. The type of topology that will be available for a MPI-program on BGP extremely depends on the number of requested processors for user's task. Chart on the fig. 9 shows, that delays of data transmission between far and near processors have similar values, however figure 7 shows that transmission time is depending on closeness of processors. 512 processors were used in the first case, and 256 processors in the second case. The difference between the delay values are explained by the fact of usage 3D-tor topology, unlike mesh topology which is used for 128 and 256 processors. Now let's discuss features, related to threshold value. Staircase structure on the figures: 7 and 9 are connected with change of routing algorithm in MPI implementation with change of message length. Chart 8 demonstrates, that routing algorithms affect message passing speed even between cores of one processors.

Developed tools are usable for comparison of different HPC clusters. For example, the figure 6 shows values of delays between neighbour nodes. Maximum bandwidth is reached for Lomonosov-MSU, however the least variation in delays is observed for the BGP system. Developed tools can be useful for system administration to provide stability in communication environment. Programmers can estimate total parallel runtime using information about delay values for data transmission through communication environment. Unfortunately, exhaustive tests of large systems are quite difficult. It is due to a whole set of problems. E.g., poor MPI implementations behaviour on more than 4000 processors, as it is shown in the article [1]. Another problem is large testing time and huge amount of data acquired from systems with such amount of processors. (Lomonosov-MSU has 8892 processors today).

This article results show importance of testing and studying features of communications in HPC clusters by means of the messages passings and visualising of delays. It should allow to extract and understand information about interconnect in clusters that consist of more than 20 thousands of processors.

References

1. Balaji, P., Buntinas, D., Goodell, D., Gropp, W., Kumar, S., Lusk, E., Thakur, R., Träff, J.L.: MPI on a million processors. In: Ropo, M., Westerholm, J., Dongarra, J. (eds.) PVM/MPI. LNCS, vol. 5759, pp. 20–30. Springer, Heidelberg (2009)
2. Dave Turner, X.C.: Protocol-dependent message-passing performance on linux clusters. In: IEEE International Conference on Cluster Computing (CLUSTER 2002), pp. 187–194 (2002)
3. M. P. I. Forum. MPI: A Message-Passing Interface Standard, Version 2.2. High Performance Computing Center Stuttgart (HLRS) (September 2009)
4. Lastovetsky, A., Rychkov, V., O'Flynn, M.: MPIBlib: Benchmarking MPI communications for parallel computing on homogeneous and heterogeneous clusters. In: Lastovetsky, A., Kechadi, T., Dongarra, J. (eds.) EuroPVM/MPI 2008. LNCS, vol. 5205, pp. 227–238. Springer, Heidelberg (2008)
5. Majumder, S., Rixner, S.: Comparing ethernet and myrinet for mpi communication. In: Proceedings of the 7th Workshop on Languages, Compilers, and Run-time Support for Scalable Systems, LCR 2004, pp. 1–7. ACM, New York (2004)
6. Reussner, R., Hunzelmann, G.: Achieving performance portability with sKaMPI for high-performance MPI programs. In: Alexandrov, V.N., Dongarra, J., Juliano, B.A., Renner, R.S., Tan, C.J.K. (eds.) ICCS-ComputSci 2001. LNCS, vol. 2074, pp. 841–850. Springer, Heidelberg (2001)
7. Salnikov, A.N.: PARUS: A parallel programming framework for heterogeneous multiprocessor systems. In: Mohr, B., Träff, J.L., Worringen, J., Dongarra, J. (eds.) PVM/MPI 2006. LNCS, vol. 4192, pp. 408–409. Springer, Heidelberg (2006)
8. Salnikov, A.N., Andreev, D.Y.: Develop tools for monitoring communications environment of computing clusters with a large number of processor elements. In: Proceedings of the Fifth International Conference Parallel Computing and Control Problems, pp. 1187–1208. Russian Academy of Sciences, Moscow (2010)

Parallel Sorting with Minimal Data

Christian Siebert[1,2] and Felix Wolf[1,2,3]

[1] German Research School for Simulation Sciences, 52062 Aachen, Germany
[2] RWTH Aachen University, Computer Science Department, 52056 Aachen, Germany
[3] Forschungszentrum Jülich, Jülich Supercomputing Centre, 52425 Jülich, Germany
{c.siebert,f.wolf}@grs-sim.de

Abstract. For reasons of efficiency, parallel methods are normally used to work with as many elements as possible. Contrary to this preferred situation, some applications need the opposite. This paper presents three parallel sorting algorithms suited for the extreme case where every process contributes only a single element. Scalable solutions for this case are needed for the communicator constructor MPI_Comm_split. Compared to previous approaches requiring $O(p)$ memory, we introduce two new parallel sorting algorithms working with a minimum of $O(1)$ memory. One method is simple to implement and achieves a running time of $O(p)$. Our scalable algorithm solves this sorting problem in $O(\log^2 p)$ time.

Keywords: MPI, Scalability, Sorting, Algorithms, Limited memory.

1 Introduction

Sorting is often considered to be the most fundamental problem in computer science. Since the 1960s, computer manufacturers estimate that more than 25 percent of the processor time is spent on sorting [6, p. 3]. Many applications use sorting algorithms as a key subroutine either because they inherently need to sort some information, or because sorting is a prerequisite for efficiently solving other problems such as searching or matching. Formally, the *sequential sorting problem* can be defined as follows:[1]

Input: A sequence of n items $(x_1, x_2, \ldots, x_n)$, and a relational operator $\leq$ that specifies an order on these items.

Output: A permutation (reordering) $(y_1, y_2, \ldots, y_n)$ of the input sequence such that $y_1 \leq y_2 \leq \cdots \leq y_n$.

This problem has been studied extensively in the literature for more than sixty years. As a result, many practical solutions exist, including sorting algorithms such as *Merge sort* (1945), *Quicksort* (1960), *Smoothsort* (1981), and *Introsort* (1997). Since there can be $n!$ different input permutations, a correct sorting algorithm requires $\Omega(n \log n)$ comparisons. Some of the previously mentioned solutions achieve a worst-case running time of $O(n \log n)$, which makes them therefore asymptotically optimal.

---

[1] To avoid any restrictions, this paper focuses on comparison-based sorting algorithms.

Y. Cotronis et al. (Eds.): EuroMPI 2011, LNCS 6960, pp. 170–177, 2011.

Single-core performance has been stagnant since 2002 and with the trend to have exponentially growing parallelism in hardware due to Moore's law, applications naturally demand a parallel sorting solution, involving p processes. We assume that each process can be identified by a unique rank number between 0 and $p-1$. A necessary condition for an optimal parallel solution is that the n data items are fully distributed over all processes. This means that process i holds a distinct subset of n_i data items, so that $n = \sum_{i=0}^{p-1} n_i$. Usually neglected, our paper investigates the extreme case where every process holds exactly one data item, thus n_i is always 1 and $n = p$. This *parallel sorting problem with minimal data* can be formulated as an extension to the sequential sorting problem:

Input: A sequence of items distributed over p processes $(x_0, x_1, \ldots, x_{p-1})$ so that process i holds item x_i, and a relational operator $\leq$.

Output: A distributed permutation $(y_0, y_1, \ldots, y_{p-1})$ of the input sequence such that process i holds item y_i and $y_0 \leq y_1 \leq \cdots \leq y_{p-1}$.

The communicator creator MPI_Comm_split in the Message Passing Interface requires an efficient solution for the parallel sorting problem with minimal data. Existing implementations as in *MPICH* [5] and *Open MPI* [4] need $O(p)$ memory and $O(p \log p)$ time just to accomplish this sorting task. This paper offers three novel parallel sorting algorithms as suitable alternatives:

1. An algorithm similar to Sack and Gropp's approach [7] in terms of linear resource complexity. Its advantage is simplicity, making it an ideal candidate to implement MPI_Comm_split efficiently for up to 100,000 processes.
2. A modification of the first algorithm to reduce its $O(p)$ memory complexity down to $O(1)$, eliminating this bottleneck at the expense of running time.
3. A scalable algorithm which also achieves this minimal memory complexity, and additionally reduces the time complexity to $O(\log^2 p)$. Experiments prove this method to be the fastest known beyond 100,000 processes.

These algorithms represent self-contained parallel sorting solutions for our case. In combination, they resolve all scalability problems for MPI_Comm_split.

2   Communicator Construction

MPI is an established standard for programming parallel applications, and is especially suited for distributed-memory supercomputers at large scale. Every communication in MPI is associated with a *communicator*. This is a special context where a group of processes belonging to this communicator can exchange messages separated from communication in other contexts. MPI provides two predefined communicators: MPI_COMM_WORLD and MPI_COMM_SELF. Further communicators can be created from a group of MPI processes, which itself can be extracted from existing communicators and modified by set operations such as inclusion, union, and intersection. All participating processes must perform this procedure and provide the same full (i.e., global) information, independently of whether a process is finally included in the new context or not. This way of creating a single communicator is not suited for parallel applications targeting large supercomputers because the resource consumption (i.e., memory and computation) scales linearly with the number of processes in the original group.

The convenience function MPI_Comm_split (MPI-2.2 p. 205f) provides an alternative way to create new communicators and circumvents the MPI group concept. It is based on a *color* and *key* input and enables the simultaneous creation of multiple communicators, precisely one new communicator per color. The key argument is used to influence the order of the new ranks. However, such an abundant functionality comes at a cost: any correct implementation of MPI_Comm_split must sort these <color, key> pairs in a distributed fashion. Every process provides both integers separately. Internally, they can however be combined into a single double-wide value, such as in value=(color<<32)+key for architectures with 32 bit integers, before executing the parallel sorting kernel. The output value together with the original rank number is sufficient for subsequent processing in MPI_Comm_Split, as segmented prefix sums (e.g., exemplified in MPI-2.2 p. 182f) can efficiently compute an identifier offset for the new communicator and the new rank number in $O(\log p)$ time and $O(1)$ space.

The open question is: Can MPI_Comm_split be implemented in a scalable way, in particular with a memory complexity significantly smaller than $O(p)$?

3   Related Work

Parallelizing sorting algorithms has shown to be nontrivial. Although a lot of sequential approaches look promising, turning them into scalable parallel solutions is often complex and typically only feasible on shared-memory architectures [2].

Many popular parallel sorting approaches for distributed memory are based on *Samplesort* (1970) [3]. This algorithm selects (for example at random) a subset of $O(p^2)$ input items called "samples", which need to be sorted for further processing. Unfortunately, methods using this approach do not work for $n < p^2$, and offer as such no solution for our special case with 1 item per process. Even today these samples are still sorted sequentially [8] causing this to be the main bottleneck at large scale. In fact, a scalable solution to the sorting problem with minimal data might even help to eliminate this bottleneck in Samplesort.

Current implementations of MPI_Comm_split based on *Open MPI* as well as *MPICH* do not sort the <color, key> arguments in parallel. Instead, they simply collect all arguments on all processes using MPI_Allgather, then apply a sequential sorting algorithm, and finally pick the resulting value that belongs to the corresponding process. The ANSI C standard library function qsort is used if available, which has an average running time of $O(n \log n)$ but can exhibit the $O(n^2)$ worst case for unfavorable inputs. Both implementations fall back to a slow $O(n^2)$ *Bubblesort* algorithm if *Quicksort* is not provided by the system. This naive approach results in a memory consumption that scales poorly with $O(p)$ and a running time of $O(p \log p)$ or even $O(p^2)$, which is to be avoided.

Sack and Gropp (2010) identified and analyzed this scalability problem for MPI_Comm_split in foresight of the exascale era [7]. Asking for a parallel sorting solution, they proposed to utilize the exact splitting method of Cheng and colleagues (2007) [1] to improve upon the scalability of MPI_Comm_split. Instead of a single sorting process, they propose to partially gather the input items on multiple sorting roots. The exact splitting method is used to partition the gathered data equally for these roots, which then sort the resulting smaller data sets sequentially. The authors evaluated this intricate approach for up to 64 sorting processes and projected an expected speedup of up to 16, representing communicators for 128 million MPI processes. This limited scaling in the order of $\log n$ already reduces the complexities of MPI_Comm_split by a factor of $\log p$ down to $O(p)$ in terms of time and $O(p/\log p)$ in terms of memory. We propose further solutions to this problem in Section 4 to improve upon both complexity terms.

4   Algorithm Designs

All algorithms discussed in this section can be used for the implementation of MPI_Comm_split. They expect one input value per process, sort all values in parallel, and return one output value per process, according to the definition of the parallel sorting problem with minimal data in Section 1.

4.1   Sequential Algorithm

Existing implementations of MPI_Comm_split simply collect all input values on all processes and do the actual sorting work in a redundant sequential fashion.

```
MPI_Comm_rank (comm, &rank);
tmparray = malloc(sizeof(type)*p);
MPI_Allgather (input, 1, type, tmparray, 1, type, comm);
qsort(tmparray, p, sizeof(type), cmpfunc);
output = tmparray[rank];
free(tmparray);
return output;
```

Listing 1.1. Sequential implementation

The MPI_Allgather operation has a time complexity of $O(p)$, but the sequential sorting functionality encapsulated in qsort uses $O(p \log p)$ comparisons on average. Therefore the latter becomes the dominating factor in Listing 1.1, leading to an overall time complexity of $O(p \log p)$. A temporary array capable of holding all p input values is needed, resulting in a memory complexity of $O(p)$.

4.2   Counting Algorithm

An interesting observation helps us to remove the redundant executions of qsort: It is sufficient to count how many values are smaller or equal than a process' own value as the destination for the input value arises directly from this information.

```
tmparray = malloc(sizeof(type)*p);
MPI_Allgather(input, 1, type, tmparray, 1, type, comm);
dest = -1;
for (i = 0; i < p; i++) { if (tmparray[i] <= input) dest++; }
free(tmparray);
MPI_Sendrecv(input, 1, type, dest, tag, output, 1, type,
             MPI_ANY_SOURCE, tag, comm, status);
return output;
```

Listing 1.2. Counting implementation

This parallel sorting algorithm can be made stable by splitting the loop into two parts and using different comparators. To ignore a process' own value, Listing 1.2 initializes dest to -1 instead of 0. After counting, each process knows the corresponding process it has to send its value to. The for loop over p values reduces the total time complexity by a factor of $\log p$ down to $O(p)$. Since this not only makes MPI_Comm_split much faster but also simplifies its implementation by removing the dependencies to external functions such as qsort and own *Bubblesort* implementations, we recommend immediate integration into MPI libraries. The memory requirements do not change and therefore stay $O(p)$.

4.3   Ring Algorithm with $O(1)$ Memory

When memory requirements become a concern (e.g., with huge number of cores), our counting algorithm can be adapted to avoid additional memory. The idea is to mix the gathering and visiting of all values, so that this can be done in smaller chunks—in the extreme case with only a single value. We created a virtual ring of processes by using MPI_Cart_create to embed a one-dimensional and periodic Cartesian topology into the underlying network topology. The convenience function MPI_Cart_shift identifies the left and right neighbor in the ring.

```
dest = 0;
prev = input;
for (i = 1; i < p; i++) {
MPI_Sendrecv(prev, 1, type, left, tag, next, 1, type,
right, tag, ring_comm, status);
if (next <= input) dest++;
prev = next;
}
MPI_Sendrecv(input, 1, type, dest, tag, output, 1, type,
MPI_ANY_SOURCE, tag, comm, status);
return output;
```

Listing 1.3. Ring implementation with O(1) memory

We utilize $p-1$ iterations in Listing 1.3 to ignore a process' own value. The time complexity remains $O(p)$, although the hidden constant[2] is potentially much higher than in Listing 1.2. Fortunately, only a fixed number of variables are needed, reducing the memory complexity down to the minimum of $O(1)$.

4.4   Scalable Algorithm

While the previous algorithms are simple to understand, we sketch now a more sophisticated approach to solve the parallel sorting problem with minimal data. It is based on the divide-and-conquer concept underlying *Quicksort*:

1. globally select a *pivot* value (preferably close to the median of all elements)
2. *divide*: partition all distributed values into the three sets consisting of (i) values that are less than *pivot*, (ii) values that are equal to *pivot* (important for duplicates and stability), and (iii) values that are greater than *pivot*
3. *conquer*: recursively proceed with the set the process belongs to

Assume an $O(\log p)$ time collective communication operation that returns an element close to the median of all provided values. Each process invokes this functionality with its own input value to get a suitable pivot value in return.

$$\vec{v_i} = \begin{cases} (1\ 0\ 0) & \text{if } x_i < \text{pivot,} \\ (0\ 1\ 0) & \text{if } x_i = \text{pivot,} \\ (0\ 0\ 1) & \text{if } x_i > \text{pivot.} \end{cases} \quad \begin{aligned} \vec{y_i} &= \text{PrefixSum}(\vec{v_i}) \\ \vec{w} &= \text{GlobalSum}(\vec{v_i}) \end{aligned} \quad d_i = \vec{v_i} \cdot \left(\begin{pmatrix} 0\ 0\ 0 \\ 1\ 0\ 0 \\ 1\ 1\ 0 \end{pmatrix} \cdot \vec{w}^T + \vec{y_i}^T \right)$$

Fig. 1. Calculating the new location in the divide step

The partitioning can be accomplished by utilizing parallel reduction operations. Each process compares its own value x_i with the ascertained pivot value. Depending on the outcome, it will initialize an array $\vec{v_i}$ as specified in Figure 1. This information is then processed in a prefix summation (cf. MPI_Exscan) and a global summation (cf. MPI_Allreduce) to enable a calculation of the new location d_i (i.e., recipient) of each process' value. A data shuffle via MPI_Sendrecv concludes a single partitioning round with an overall time complexity of $O(\log p)$.

In the conquer step, a process compares its received value against the pivot to decide where to proceed. This will divide the number of values in roughly two halves, causing $O(\log p)$ divide-and-conquer rounds. To avoid the use of $O(\log p)$ stack space, we implemented this tail-recursive conquer step iteratively. Altogether, the memory complexity is $O(1)$ and the running time becomes $O(\log^2 p)$.

Implementation Details. Partitioning leads to subgroups of processes continuing independently in subsequent rounds. Since the algorithm uses collective operations, we could create new communicators. However, existing communicator creation is, with a complexity of $\Omega(p)$, too expensive. Instead, we designed special collective implementations that work on a sub-range of all processes in

---

[2] The counting solution employs only one Allgather which can be implemented to induce $O(\log p)$ network latencies as opposed to $O(p)$ for individual communications.

MPI_COMM_WORLD. In contrast to the hardware-tuned Blue Gene/P collectives, these range collectives slow down our scalable sorting method by a factor of roughly 28, but in exchange achieve the required time complexity of $O(\log p)$.

We use an efficient median-of-3 reduction scheme within a complete ternary tree topology to find an approximate median of all values. Each process provides its input value as one of the leaves. Inner nodes receive three values, determine their median, and forward the result to the next level. This procedure is repeatedly applied in $O(\log p)$ levels until the root gets the result. This single value delivers a good approximation of the median because the $2^{\log_3 p} - 1$ smallest as well as $2^{\log_3 p} - 1$ largest values out of $p = 3^k$ values will never be selected. Analysis reveals that a value close to the median is picked with very high probability.

5   Experimental Evaluation

All measurements were carried out on the full *Jugene* system located at the Jülich Supercomputing Centre in Germany. It consists of $73,728$ compute nodes, each equipped with 2 GiB of memory and a 4-way SMP PowerPC processor running at 850 MHz. Executables were linked against the BG/P MPI library 1.4.2.

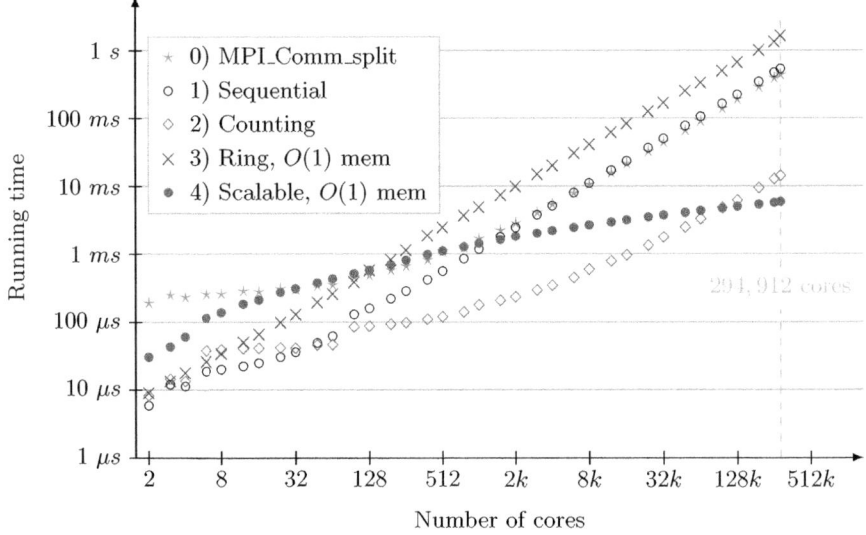

Fig. 2. Performance comparison of the presented algorithms

Figure 2 depicts the runtime of all presented methods for a varying number of cores. Except for the MPI_Comm_split operation which used color=1 and key=rank as input, all sorting algorithms started with a randomly chosen 64-bit value per process. Compared to the extracted sorting kernel, the MPI_Comm_split operation shows some overhead up to 2048 processes, after which both performance curves converge. Our counting solution is up to 687% faster than all other methods to the point of $98,304$ cores. As expected, the ring algorithm is the slowest candidate for larger communicators, but will never run out of memory.

Contrary to the other approaches where the running time increases proportionally to the number of cores, the curve of our scalable algorithm flattens. This makes it the fastest method beyond $100k$ cores, outperforming the current implementation by a factor of 92.2 at full scale. Its performance can be modelled by $t(p) = 17.5 \cdot (\log_2 p)^2$ μs, giving a predicted running time of 12.7 ms for 128 million processes. As such it is a factor of 29.2 faster than Sack and Gropp's best proposed solution while requiring a million times less memory.

6   Conclusion

This paper approaches the problem of parallel sorting with minimal data. Being able to handle a single element per process in a scalable way is crucial for an efficient implementation of the MPI communicator creator MPI_Comm_split. Extending the work of Sack and Gropp, we introduced three novel algorithms to solve this problem. Our first approach is similar to their proposed method in terms of resource complexity, but is much simpler to implement and more efficient in practice, making it an ideal candidate for MPI libraries. In prospect to future systems, we reduced the $O(p)$ memory complexity down to the minimum of $O(1)$ at the expense of performance in our second algorithm. Finally, we sketched a scalable algorithm that solves the parallel sorting problem with minimal data. Measurements on the largest Blue Gene/P installation today showed that this method eventually outperforms all other methods, making it 92.2 times faster than current implementations and a hundred thousand times more memory efficient on $294,912$ cores. Since the algorithm's time complexity of $O(\log^2 p)$ yields excellent scalability without any additional memory, it provides a suitable solution to the tackled problem, at and beyond exascale—closing the open question of a scalable MPI_Comm_split implementation with a positive answer.

References

1. Cheng, D.R., Shah, V., Gilert, J.R., Edelman, A.: A Novel Parallel Sorting Algorithm for Contemporary Architectures. Tech. rep., University of California (2007)
2. Cole, R.: Parallel Merge Sort. SIAM Journal on Computing 17, 770–785 (1988)
3. Frazer, W.D., McKellar, A.C.: Samplesort: A Sampling Approach to Minimal Storage Tree Sorting. Journal of the ACM 17, 496–507 (1970)
4. Gabriel, E., Fagg, G.E., Bosilca, G., Angskun, T., Dongarra, J., Squyres, J.M., Sahay, V., Kambadur, P., Barrett, B.W., Lumsdaine, A., Castain, R.H., Daniel, D.J., Graham, R.L., Woodall, T.S.: Open MPI: Goals, Concept, and Design of a Next Generation MPI Implementation. In: Kranzlmüller, D., Kacsuk, P., Dongarra, J. (eds.) EuroPVM/MPI 2004. LNCS, vol. 3241, pp. 97–104. Springer, Heidelberg (2004)
5. Gropp, W., Lusk, E., Doss, N., Skjellum, A.: A High-Performance, Portable Implementation of the Message Passing Interface Standard. Par. Comp. 22, 789 (1996)
6. Knuth, D.E.: The Art of Computer Programming, Sorting and Searching, 2nd edn., vol. 3. Addison Wesley Longman Publishing, Redwood City (1998)
7. Sack, P., Gropp, W.: A Scalable MPI_Comm_split Algorithm for Exascale Computing. In: Keller, R., Gabriel, E., Resch, M., Dongarra, J. (eds.) EuroMPI 2010. LNCS, vol. 6305, pp. 1–10. Springer, Heidelberg (2010)
8. Shi, H., Schaeffer, J.: Parallel Sorting by Regular Sampling. Journal of Parallel and Distributed Computing 14, 361–372 (1992)

Scaling Performance Tool
MPI Communicator Management

Markus Geimer[1], Marc-André Hermanns[2], Christan Siebert[2],
Felix Wolf[1,2,3], and Brian J.N. Wylie[1]

[1] Jülich Supercomputing Centre, Forschungszentrum Jülich, Germany
{m.geimer,b.wylie}@fz-juelich.de
[2] German Research School for Simulation Sciences, Aachen, Germany
{m.a.hermanns,c.siebert,f.wolf}@grs-sim.de
[3] RWTH Aachen University, Aachen, Germany

Abstract. The Scalasca toolset has successfully demonstrated measurement and analysis scalability on the largest computer systems, however, applications have growing complexity and increasing demands on performance tools. One such application is the *PFLOTRAN* code for simulating multiphase subsurface flow and reactive transport. While *PFLOTRAN* itself and Scalasca runtime summarization both scale well, MPI communicator management becomes critical for trace collection with tens of thousands of processes. Re-design and re-engineering of key components of the Scalasca measurement system are presented which encompass the representation of communicators, communicator definition tracking and unification, and translation of ranks recorded in event traces.

Keywords: MPI communicators, performance measurement tools, scalability.

1 Introduction

Scalasca is an open-source toolset for analyzing the execution behavior of applications based on the MPI and/or OpenMP parallel programming interfaces supporting a wide range of current HPC platforms [7,9]. It combines compact runtime summaries, that are particularly suited for obtaining an overview of execution performance, with in-depth analysis of concurrency inefficiencies via event tracing and parallel replay. With its highly scalable design, Scalasca has facilitated performance analysis and tuning of a range of applications and consisting of unprecedented numbers of processes [17].

Experience with a growing number of HPC applications on leadership IBM Blue Gene and Cray XT systems has shown that they often scale surprisingly well to effectively exploit hundreds of thousands of processor cores [11]. Many codes explicitly use MPI for communication and synchronization, whereas others make extensive use of libraries that encapsulate MPI usage. An example of the latter, the *PFLOTRAN* three-dimensional reservoir simulator [2] has featured prominently in the US Department of Energy SciDAC program, where it

Y. Cotronis et al. (Eds.): EuroMPI 2011, LNCS 6960, pp. 178–187, 2011.

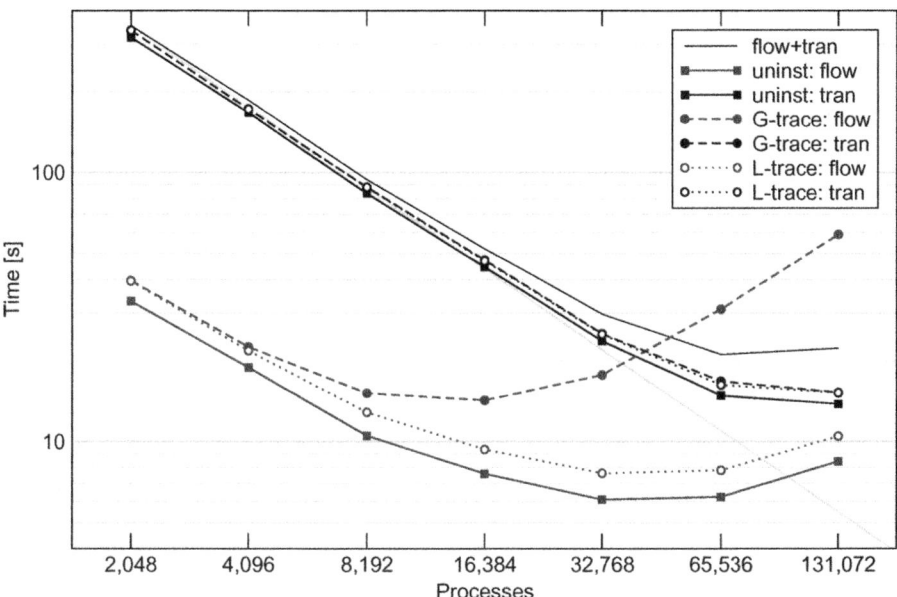

Fig. 1. Average simulation timestep durations reported by *PFLOTRAN* for Scalasca trace experiments on IBM BG/P, with breakdown into 'flow' and 'tran'(sport) phases, compared to reference uninstrumented executions. Distortion in the trace with event communication ranks translated to global ranks (G-trace) is avoided in the trace that records the local ranks (L-trace).

has been used to simulate geologic CO_2 sequestration and migration of radionucleide contaminants in groundwater [8]. Recent measurement and analysis of *PFLOTRAN* execution performance with a 'petascale' dataset on IBM BG/P and Cray XT5 systems with Scalasca [16] identified significant performance opportunities in the application, but also several serious scalability issues with the Scalasca measurement approach that needed to be resolved to produce viable performance analyses. Figure 1 shows the strong scaling of *PFLOTRAN* simulation timesteps on the Jugene BG/P at Jülich Supercomputing Centre along with corresponding time for Scalasca summary and trace experiments, including breakdown of 'flow' and 'tran'(sport) execution phases.

With the provided '2B' test case, *PFLOTRAN* (via the HDF5 and PETSc libraries) was found to create 18 copies of the MPI_COMM_WORLD global communicator and 4 copies of MPI_COMM_SELF on each process. For Scalasca runtime summarization experiments, MPI communicators are ignored, however, for parallel trace analysis it is necessary to record communicator definitions and their usage in MPI communication and synchronization event records (to allow communicators to be reconstructed and used in replaying trace events). The prior implementation of communicator management proved to be inadequate, requiring storage space and processing time that grew linearly or worse with the number of processes, such that collection and analysis of large-scale *PFLOTRAN* traces was not possible.[1] Furthermore, dilation of application execution time during trace collection was found to be severe for the 'flow' phase (as shown in Fig. 1) due to the cost of translating local to global ranks.

To address these issues, we re-designed and re-engineered communicator management and representation for the Scalasca measurement system as described in the remainder of this paper. We start our discussion in Sect. 2 with a review of related work, followed by a description and analysis of the original communicator handling in Sect. 3. Section 4 then discusses the improved data layout and algorithms in detail. Next, in Sect. 5, we show an experimental evaluation of our approach with respect to various key metrics, before concluding the paper in Sect. 6.

2   Related Work

The data that a measurement tool needs to collect and store depends on the analyses that are intended. Even for tools serving similar purposes, communicator management and rank translation can be done very differently, as demonstrated by a brief survey of current open-source software releases. mpiP-3.3 [10] doesn't use communicator recording or rank translation since it doesn't distinguish these in its profile analysis. Periscope-1.3.2 [13] similarly doesn't need to store communicators or translate ranks for its on-line communication analysis. FPMPI-2.1g [3] profiles do provide a matrix of point-to-point communication sources and destinations, however, only in terms of local ranks without distinguishing communicators. For its communication matrix TAU-2.20.2 [15] translates point-to-point source and destination ranks to global ranks during measurement, and it can also distinguish by communicator. Translated ranks also appear in TAU traces of point-to-point communications, but not for the roots of collective communications, which is also the approach adopted by Extrae-2.1.1 [5]. While communicators are distinguished, the communicator composition is neither recorded nor part of their analysis. VampirTrace-5.11 [12], like the Scalasca predecessor from which it derives, translates ranks of both point-to-point and collective communication events. These tools convert local to global ranks using the standard provided `MPI_Group_translate_ranks` routine as communication events are handled during measurement, with shortcuts to avoid unnecessary rank translation for communicators that are identical or congruent to `MPI_COMM_WORLD`. In comparison, the MPE logger provided with MPICH2-1.4 [4] writes traces entirely with local (untranslated) ranks, which are translated when traces are read using communicator rank mappings recorded separately for each communicator and rank.

3   Original Scalasca Scheme

In the original scheme used by Scalasca, each MPI group and communicator was represented by a bitstring where bit i indicates whether the global rank i

---

[1] Notably the amount of trace event data collected, which is often an impediment, was not a limitation for Scalasca in this case.

is part of the group or communicator (=1) or not (=0). Additional fields in the record distinguished between the two types (i.e., group or communicator) and assigned a process-local numerical identifier used by communication events to refer to this definition. As such, multiple distinct communicators required the storage of the full bitstring, even if they comprise the same group of processes. Each PMPI wrapper function creating a new group or communicator determined this bitstring by calling `MPI_Group_translate_ranks` to map the group or the group of the newly created communicator, respectively, onto the group of `MPI_COMM_WORLD` and then setting the corresponding bits. Since in this scheme communicators are defined in terms of global ranks, all events generated for MPI communication operations need to use global rank information as well to allow for a proper determination of sender, receiver or collective root processes. This required another call to `MPI_Group_translate_ranks` in the PMPI wrapper of each communication function to convert the local rank in the communicator provided as arguments into the corresponding global rank, unless the communicator is `MPI_COMM_WORLD` (i.e., the ranks are already global).

To establish a global view, these per-process communicator definitions were "unified" at the end of measurement. That is, communicator definitions from different processes were merged to create a unique set of global communicator definitions, requiring some complicated logic to correctly distinguish between multiple copies of a communicator. Moreover, a per-process mapping from local to global communicator identifiers was created, which could be applied to the corresponding identifiers stored in the communication events while reading the trace data.

Although this solution works reasonably well for small scale measurements, its drawbacks became evident at scale. The $\mathcal{O}(p)$ storage requirements for each local definition mean that a significant amount of memory is already required at measurement time. In particular, the bitstring representation is extremely bad for `MPI_COMM_SELF` and duplicates since only a single bit is set. Moreover, the amount of data to be processed during unification is $\mathcal{O}(p^2)$. While algorithmic improvements in the unification process using a hierarchical scheme [6] successfully parallelized the work, the reduction of the overall workload needed further attention. Since the bitstring for `MPI_COMM_SELF` is different on every process, no merging is possible during unification, leading to $\mathcal{O}(p^2)$ storage requirements for their global definitions. And finally, the bitstring records are also created for every duplicate of a communicator, leading to a lot of redundancy for application codes such as *PFLOTRAN*, quickly resulting in gigabytes of communicator definition records, such that trace analysis was not possible for more than 48k processes. Along with the quadratic growth in size, unification times of the original implementation were also unacceptable as seen in Fig. 2.

Times reported by *PFLOTRAN* for summary and trace collections employing runtime filters on Jugene IBM BG/P compared with reference times from the uninstrumented executions in Fig. 1 show that measurement dilation is generally acceptably small, apart from trace collection with larger configurations of processes. Whereas on Jaguar Cray XT5 the dilation is only significant for 'flow'

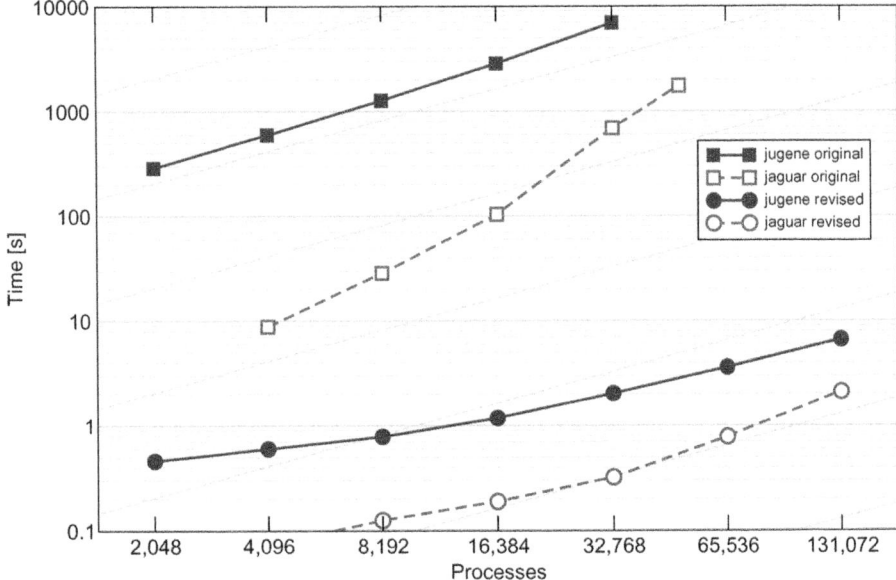

Fig. 2. Time to unify *PFLOTRAN* identifier definitions (and write them to disk with associated mappings for each process) on Jugene IBM BG/P and Jaguar Cray XT5, comparing original and revised Scalasca implementations

at 128k processes (not shown), it is much more pronounced in measurements on Jugene IBM BG/P for even 16k processes and for 128k processes grows to a factor of seven! This difference can be attributed to translations of communication partner/root process ranks (from the local rank in the MPI communicator to the global rank in `MPI_COMM_WORLD`) in every communication operation event recorded, and the relative speeds of computation and communication on both systems.

The cost of the MPI standard routine `MPI_Group_translate_ranks` provided for this conversion increases with the size of the communicator, however, it also depends on the rank(s) being translated with the worst-case cost that for the largest rank. On Jugene BG/P with 128k processes, translation of the largest rank in `MPI_COMM_WORLD` takes 3.0 ms on average.[2] While this is small compared to the time for collective operations like `MPI_Comm_dup` (84 ms for `MPI_COMM_WORLD` at this scale), it is much larger than typical point-to-point communication operations. Even more insidious, the variable cost according to the partner rank results in severe distortion of the measurement.

4   Communicator Management During Trace Collection

To address the scalability limitations described in the previous section, communicator management in Scalasca was completely re-designed. In the following,

---

[2] On Jaguar Cray XT5 with 128k processes the average translation time is 1.0 ms.

we present the solutions we have implemented with respect to scalable communicator tracking, unification and representation.

4.1   Distributed Communicator Tracking

Instead of determining and storing the whole group information on communicator creation on each process at measurement time, we developed a distributed communicator tracking scheme requiring very little memory and allowing the efficient reconstruction of the global communicator information at the end of measurement.

In the distributed tracking scheme, each process stores a record with a globally unique pair of integers as a key, the process-local identifier of this communicator used by the event records referring to it, the process' local rank within the communicator, and its size. The 2-tuple used as key is the foundation for the efficient reconstruction of the global communicator structure. It needs to be globally unique to detect which distributed partial definition records build a global communicator record. The local identifier alone cannot provide this, as it merely represents the information that this record belongs to the ith locally defined communicator. However, different processes can define different communicators, giving this identifier a purely local meaning.

To build these unique keys, each process keeps a state variable during measurement to count the number of communicators where this process was rank 0. For improved readability, we will henceforth write that a process p *defines* a communicator record when it is the process with rank 0 in the corresponding communicator. As the value of this counter is strictly increasing on each rank, and the global rank of the defining process is unique, the combination of those two values forms a unique key for each communicator. Also, both values can be determined during measurement at very low cost.

In principle, each process participating in a newly created communicator can determine the global rank of the defining process by mapping local rank zero onto the group of MPI_COMM_WORLD using MPI_Group_translate_ranks. However, the local state variable of this process is unknown to all but the defining process and needs to be distributed. For simplicity, we avoid the call to MPI_Group_translate_ranks and use a broadcast on the new communicator with the defining process as the root, sending its global rank as well as the aforementioned count. The defining process increments its counter after the broadcast, as its counter value has now been used for the new entry. Since communicator creation is a collective operation – and we are not aware of any MPI implementation not synchronizing all of the participating processes – the additional overhead for this communication operation is negligible.

4.2   Unification of Definition Identifiers

As mentioned before, each process assigns a local numerical identifier to each communicator it is part of. This identifier is used in event records referencing communicators (such as sending or receiving a message), later being translated into a global identifier using a per-process mapping table during analysis.

In the final communicator definition record stored with the trace, the distributed entries created during measurement have to be combined. During this stage the unique 2-tuple key needs to be transformed into the global identifier of the communicator. Here, we need special handling for `MPI_COMM_SELF`-like (i.e., single-process) communicators, which get added to the global list of communicators after applying the unification algorithm presented below. In the remainder of this section, we therefore only refer to multi-process communicators.

For those, we assign strictly increasing values to the communicator records, starting from 0 with the first communicator defined by rank 0, which in any case will be that of `MPI_COMM_WORLD`. All communicators defined by rank 0 will get assigned to the next available identifiers, until the same process is performed with all other communicators and ranks. To facilitate the unique numbering, we use a single exclusive prefix reduction where each process provides the number of communicators it defined. The resulting value on any process k then denotes the number of communicators defined by processes with a rank lower than k. This information is then distributed to every process using `MPI_Allgather`. With this knowledge, local counter values initially used in the tuple can be shifted by the offset of the corresponding defining process, making them globally unique. The resulting record therefore already enables the mapping of local to global communicator identifiers.

The next step assembles the list of global ranks for each process participating in a communicator. First, the total number of multi-process communicators c is broadcast to every process. This value is a by-product of the earlier prefix sum, requiring only one addition on the process with the highest rank number. Finally, we perform c gather operations, where each process provides either its local rank, if it was part of the specific communicator, or -1 to denote that it was not. The root can then assemble the list of processes by extracting them from the gathered values.

In total, our new distributed communicator tracking scheme has a local memory requirement of $\mathcal{O}(1)$ per communicator per process during measurement, and can be unified and consolidated with $\mathcal{O}(c \cdot \log p)$ communications.

4.3   Representation of Communicators

To eliminate the inherent redundancy of the original communicator storage scheme for duplicates, we adopted the approach taken by the MPI standard of separating groups and communicators. In the revised scheme each group is therefore stored only once, potentially being referenced by multiple communicator definition records. These now only consist of two integers, a global communicator identifier and the global identifier of the associated group definition.

Moreover, we no longer represent groups as bitstrings, but rather as an ordered list of integers where the entry at position i stores the rank in the group of

MPI_COMM_WORLD of rank i in the local group. The global rank of a process can then always be reconstructed by a simple table lookup at the corresponding entry in the communicator's group. Also, the memory representation of this rank list is much more compact than the bitstring for sparsely populated communicators.

Special flags are also included in the group record for groups corresponding to the standard MPI communicators, MPI_COMM_SELF and MPI_COMM_WORLD. This provides an obvious space saving for the ubiquitous world-group record, but more importantly the generic self-group record avoids proliferation of distinct records for each rank. Compact representations for other MPI groups have been investigated by others (e.g., [14]) and may be considered in future work.

4.4   Rank Translation

Since *PFLOTRAN* only uses duplicates of the MPI standard communicators, for which rank determination is trivial, it would be straightforward to incorporate special handling for this case. Unfortunately, applications using general MPI communicators would not benefit. However, the new storage scheme of groups and communicators allowed us to use local ranks in communication events (as the global rank can always be reconstructed, if necessary). Although it required changing the trace file format, this unnecessary translation overhead during measurement has therefore been eliminated. Trace reading also needed to be adapted, however, parallel event replay required untranslated ranks in communicators in any case, so analysis performance is not degraded.

5   Evaluation

The effectiveness of avoiding rank translation for every communication event during Scalasca trace measurement is evident in Fig. 1, which compares the *PFLOTRAN* 'flow' and 'tran' phase execution times on BG/P when ranks are globalized (G-trace) versus when they remain local ranks (L-trace). With the new communicator management traces are now collected with minimal dilation as formerly only possible for runtime summarization experiments.

Although communicator definitions are now much more compact, total trace sizes, and the associated storage for buffering event records during measurement, remain essentially unchanged with the new scheme. For the 4.0 TB event trace from 128k processes, unification now takes only 6.7 seconds to produce 10.4 MB of global definitions and 242.9 MB of mappings. Focusing on communicator definition records only, the records for 48k processes executing *PFLOTRAN* originally exceeded 1.4 GB, and were consequently too large for the Scalasca trace analyzer to handle, whereas the new records for 128k processes present no such problem. Figure 3 shows the trace analysis for an execution with 64k processes revealing the distribution of MPI communication and synchronization waiting times that complement the application's inherent computational imbalance [16].

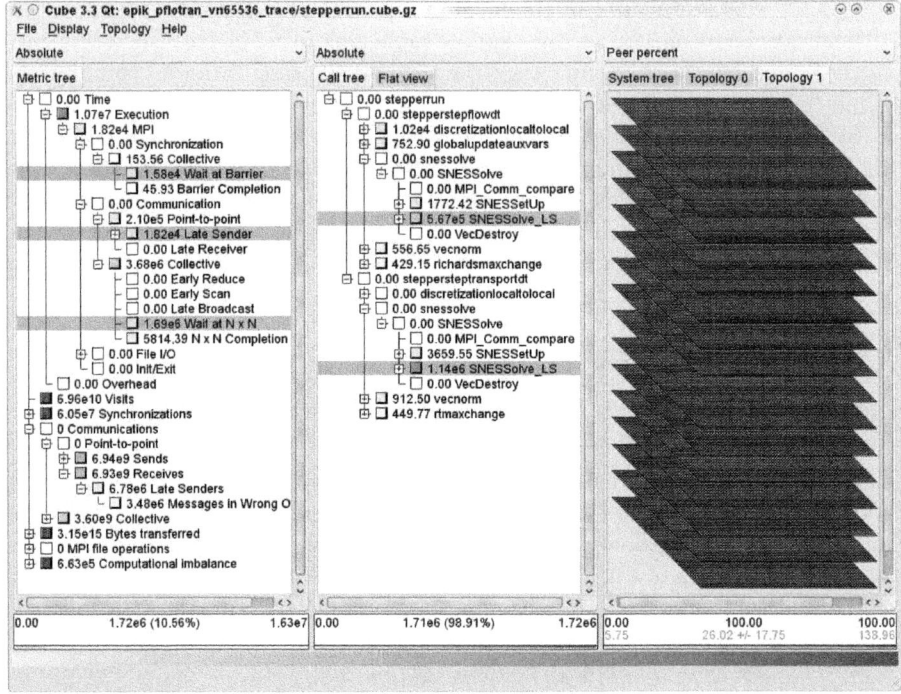

Fig. 3. Scalasca analysis report explorer showing timestep loop extract of *PFLOTRAN* trace experiment with 64k processes on BG/P. MPI communication and synchronization waiting time metrics selected in the left pane correspond to over 10% of the total time. The central pane shows PETSc SNESSolve_LS line search solver calls employed in the flow and transport phases are responsible for 99% of this, whereas the distribution of waiting times for the 64k processes in the right pane reveals that it complements the application's inherent computational load imbalance.

6   Conclusion

For trace collection and analysis of the *PFLOTRAN* application at large scale, Scalasca management of MPI communicators needed to be comprehensively re-engineered. Eliminating the translation to global ranks of communicator ranks of partner and root processes in communication operations to avoid associated measurement dilation also motivated more efficient tracking and storage of communicator specifications required for message replay during analysis. With the revised implementation, formerly impossible trace analysis with 128k and more processes has now been achieved. Small extensions are under investigation for the rare applications using MPI inter-communicators. The new communicator management scheme has also been contributed to the open-source Score-P measurement system [1] being developed for the next generation of the Scalasca, Periscope, TAU and Vampir performance tools.

Acknowledgments. Financial support from the Helmholtz Association of German Research Centers through Grant VH-NG-118 is gratefully acknowledged.

References

1. an Mey, D., Biersdorff, S., Bischof, C., Diethelm, K., Eschweiler, D., Gerndt, M., Knüpfer, A., Lorenz, D., Malony, A.D., Nagel, W.E., Oleynik, Y., Rössel, C., Saviankou, P., Schmidl, D., Shende, S.S., Wagner, M., Wesarg, B., Wolf, F.: Score-P–A unified performance measurement system for petascale applications. In: Proc. Competence in High Performance Computing, HPC Status Konferenz der Gauß-Allianz e.V., CiHPC, Schwetzingen, Germany. Springer, Heidelberg (2010) (to appear)
2. ANL/LANL/ORNL/PNNL/UIUC: PFLOTRAN, http://ees.lanl.gov/pflotran/
3. Argonne National Laboratory, USA: FPMPI-2.1g (August 2010), http://www.mcs.anl.gov/research/projects/fpmpi/
4. Argonne National Laboratory, USA: MPICH2-1.4 MPE (June 2011), http://www.mcs.anl.gov/research/projects/mpich2/
5. Barcelona Supercomputing Centre, Spain: Extrae-2.1.1 (March 2011), http://www.bsc.es/ssl/apps/performanceTools/
6. Geimer, M., Saviankou, P., Strube, A., Szebenyi, Z., Wolf, F., Wylie, B.J.N.: Further improving the scalability of the Scalasca toolset. In: Proc. PARA 2010, Reykjavík, Iceland. LNCS. Springer, Heidelberg (2010)
7. Geimer, M., Wolf, F., Wylie, B.J.N., Ábrahám, E., Becker, D., Mohr, B.: The Scalasca performance toolset architecture. Concurrency and Computation: Practice and Experience 22(6), 702–719 (2010)
8. Hammond, G.E., Lichtner, P.C.: Cleaning up the Cold War: Simulating uranium migration at the Hanford 300 Area. In: Proc. Scientific Discovery through Advanced Computing, SciDAC, Chattanooga, TN, USA. Journal of Physics: Conference Series. IOP Publishing (July 2010)
9. Jülich Supercomputing Centre, Germany: Scalasca toolset for scalable performance analysis of large-scale parallel applications, http://www.scalasca.org/
10. Lawrence Livermore National Laboratory, USA: mpiP-3.3 (June 2011), http://mpip.sourceforge.net/
11. Mohr, B., Frings, W. (eds.): Jülich Blue Gene/P Extreme Scaling Workshop. FZJ-JSC-IB reports 2010-02, 2010-03 & 2011-02, Jülich Supercomputing Centre (2009, 2010 & 2011), http://www2.fz-juelich.de/jsc/bg-ws11/
12. Technische Universität Dresden, Germany: VampirTrace-5.11 (June 2011), http://www.tu-dresden.de/zih/vampirtrace/
13. Technische Universität München, Germany: Periscope-1.3.2 (February 2011), http://www.lrr.in.tum.de/periscope/
14. Träff, J.L.: Compact and efficient implementation of the MPI group operations. In: Keller, R., Gabriel, E., Resch, M., Dongarra, J. (eds.) EuroMPI 2010. LNCS, vol. 6305, pp. 170–178. Springer, Heidelberg (2010)
15. University of Oregon, Eugene, USA: TAU-2.20.2 (May 2011), http://tau.uoregon.edu/tau/
16. Wylie, B.J.N., Geimer, M.: Large-scale performance analysis of PFLOTRAN with Scalasca. In: Proc. 53rd CUG Meeting, Fairbanks, AK, USA. Cray User Group, Inc. (May 2011)
17. Wylie, B.J.N., Geimer, M., Mohr, B., Böhme, D., Szebenyi, Z., Wolf, F.: Large-scale performance analysis of Sweep3D with the Scalasca toolset. Parallel Processing Letters 20(4), 397–414 (2010)

Per-call Energy Saving Strategies in All-to-All Communications*

Vaibhav Sundriyal and Masha Sosonkina

Department of Electrical and Computer Engineering
Ames Laboratory
Iowa State University
Ames, IA 50011 USA
{vaibhavs,masha}@scl.ameslab.gov

Abstract. With the increase in the peak performance of modern computing platforms, their energy consumption grows as well, which may lead to overwhelming operating costs and failure rates. Techniques, such as Dynamic Voltage and Frequency Scaling (called DVFS) and CPU Clock Modulation (called throttling) are often used to reduce the power consumption of the compute nodes. However, these techniques should be used judiciously during the application execution to avoid significant performance losses. In this work, two implementations of the all-to-all collective operations are studied as to their augmentation with energy saving strategies on the *per-call* basis. Experiments were performed on the OSU MPI benchmarks as well as on a few real-world problems from the CPMD and NAS suits, in which energy consumption was reduced by up to 10% and 15.7%, respectively, with little performance degradation.

Keywords: Collective Communications, MPI, DVFS, CPU Throttling.

1 Introduction

Power consumption is rapidly becoming one of the critical design constraints in modern high-end computing systems. While the focus of the high-performance computing (HPC) community has been to maximize the performance, the system operating costs and failure rates can reach a prohibitive level.

The Message Passing Interface[1] has become a *de facto* standard for the design of parallel applications. It defines both point-to-point and collective communication primitives widely used in parallel applications. This work examines the nature of all-to-all communications because they are among the most intensive

---

* This work was supported in part by Iowa State University under the contract DE-AC02-07CH11358 with the U.S. Department of Energy, by the Director, Office of Science, Division of Mathematical, Information, and Computational Sciences of the U.S. Department of Energy under contract number DE-AC02-05CH11231, and by the National Science Foundation grants NSF/OCI – 0749156, 0941434, 1047772.
[1] MPI Forum: http://www.mpi-forum.org

Y. Cotronis et al. (Eds.): EuroMPI 2011, LNCS 6960, pp. 188–197, 2011.
© Springer-Verlag Berlin Heidelberg 2011

and time consuming collective operations while being wide-spread in parallel applications. By definition, a collective operation requires the participation of all the processes in a given communicator. Hence, such operations incur a significant amount of the network phase during which there exist excellent opportunities for applying energy saving techniques, such as DVFS and CPU throttling. As a rule of thumb, the latter complements well the DVFS although larger energy savings are generally obtained with DVFS than with throttling when each is used separately. The experiments presented in this work emphasize the beneficial effect of throttling to augment the savings provided by the DVFS and are in line with the earlier experiments [11] — done by the authors with realistic electronic structure calculations in the GAMESS [9] package — that judged the DVFS gains by the resulting power consumption of the entire node as opposed to processor-only energy savings. The all-to-all operation is studied here on the *per-call* (fine-grain) basis as opposed to the "black-box" approach that treats communication phase as indivisible operation contributing to the parallel overhead. In this work, the energy saving strategies are incorporated within the existing all-to-all algorithms.

CPU Throttling and DVFS in Intel Architectures. The current generation of Intel processors provides various P-states for DVFS and T-states for throttling. In particular, the Intel "Core" microarchitecture, which provides four P-states and eight T-states from T_0 to T_7, where state T_j refers to introducing j idle cycles per eight cycles in CPU execution. The delay of switching from one P-state to another can depend on the current and desired P-state and is discussed in [8]. The user may write a specific value to Model Specific Registers (MSR) to change the P- and T-states of the system.

Infiniband has become one of most popular interconnect standard marking its presence in more that 43% of the systems in the TOP 500[2] list. Several network protocols are offloaded to the Host Channel Adapters (HCA) in an Infiniband[3] network. Here, MVAPICH[4] implementation of MPI, which is designed for Infiniband networks, is considered. MVAPICH2 uses "polling" communication mode by default since a lower communication overhead is incurred with polling when an MPI process constantly samples for the arrival of a new message rather than the with "blocking", which causes CPU to wait for an incoming message.

1.1 Effect of CPU Throttling on Communication

Since synchronous point-to-point communication operations underlie collectives, it is reasonable to analyze the CPU throttling effects on them first. Fig. 1(a) shows the point-to-point internode communication times for the communicating processes at T-states T_0 and T_5. Similarly, Fig. 1(b) depicts the change in intranode communication time for the states T_0 and T_1. It can be observed that

---

[2] http://www.top500.org/
[3] http://www.infinibandta.org/
[4] http://mvapich.cse.ohio-state.edu/

the effect of throttling on internode communication is minimal. In fact, the average performance loss was just 5% at state T_5 for various message sizes. Also, the performance loss observed was inversely proportional to the message size. However, introducing just one idle cycle per eight cycles degrades the intranode communication considerably (about 25%). This is expected since intranode communication uses more CPU cycles for a message transfer whereas in internode transfers RDMA offloads a large part of the communication processing to the NICs [6].

The difference between the intra- and inter-node message transfer types with respect to CPU throttling becomes the basis for the energy saving strategy proposed in this work. The Intel Xeon processor, which is used in this work, supports CPU throttling on the core level of granularity. Hence, the appropriate T-state for a core is selected depending on the communication type it is involved in. The lowest T-state T_0 is chosen when a core communicates intranode. Conversely, a higher throttling state T_5 is selected when the core performs internode communication. In the experiments, a throttling state higher than T_5 resulted in a significant performance loss, which is not desirable since the aim is to minimize the energy consumption without sacrificing the performance. However, if a core is idle during the collective operation, then it can be throttled at the highest state T_7. To summarize,

○ Core communicates intranode → T_0;
○ Core communicates internode → T_5;
○ Core does not participate in communication → T_7.

The rest of the paper is organized as follows. Section 2 describes the proposed energy savings in the all-to-all operation. Section 3 shows experimental results while Sections 4 and 5 provide related work and conclusions, respectively.

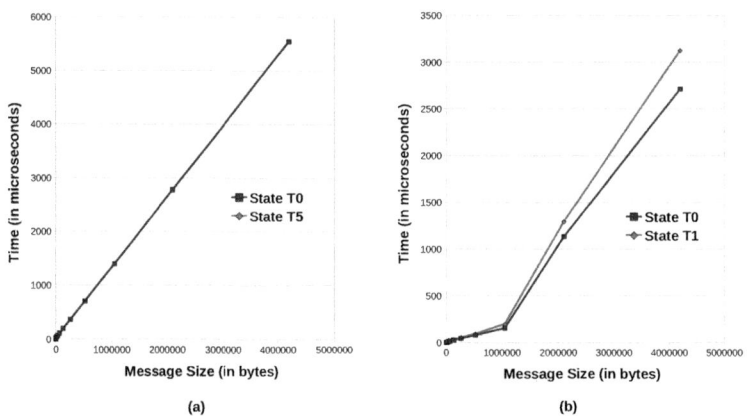

(a)   (b)

Fig. 1. MPI ping-pong test to determine the effect of CPU throttling on (a) internode communication (b) intranode communication

2   All-to-All Energy Aware Algorithm

MVAPICH2 implementations of all-to-all are considered in this work. They are based on three algorithms: 1) Bruck Index, used for small — less than 8KB — messages with at least eight participating processes; 2) Pairwise Exchange, used for large messages and when the number of processes is a power of two; 3) Send To rank $i + k$ and Receive From rank $i - k$, used for all the other processor numbers and large messages. These algorithms are referred further in text as BIA, PEA, and STRF, respectively. In this work, they are implemented using all the available cores on a node, while having the number of MPI ranks smaller than the number of cores may be considered in the future.

Bruck Index first does a local copy with the upward shift of the data blocks from the input to output buffer. Specifically, a process with the rank i rotates its data up by i blocks. The communication starts such that, for all the p communicating processes in each communication step k ($0 \leq k < \lceil \log_2 p \rceil$), process i, ($i = 0, \ldots, p - 1$), sends to $(i + 2^k) \bmod p$ (with wrap-around) all those data blocks whose kth bit is 1 and who receive from $(i - 2^k) \bmod p$. The incoming data is stored into the blocks whose kth bit is 1. Finally, the local data blocks are shifted downward to place them in the right order. Fig. 2 shows $N = 3$ nodes with $c = 8$ cores — placed on two sockets — each and the total number of $p = 8N$ processors performing the first four steps of the BIA. The rank placement is performed in block manner using consecutive core ordering. Note that, until the kth step where $2^k < c$, the communication is still *intranode* for any core in the cluster. However, after the kth step, the communication becomes purely *internode* for all the participating cores. Thus, from this step on, the throttling level T_5 may be applied to all the cores without incurring a significant performance loss.

"Send-To Receive-From" and Pairwise Exchange. For the block placement of ranks, in each step k ($1 \leq k < p$) of STRF, a process with rank i sends data to $(i + k) \bmod p$ and receives from $(i - k + p) \bmod p$. Therefore, for the initial and the final $c - 1$ steps, the communications are not purely internode. The PEA uses *exclusive-or* operation to determine the rank of processes for data exchange. It is similar to the BIA in terms of communication phase since after step k where $k = c$, the communication operation remains internode until the end.

Energy Saving Strategy. Because all three algorithms exhibit purely internode communications at a certain step k, the following energy saving strategy may be applied in stages to each of them.

Stage 1   At the start of all-to-all, scale down the frequency of all the cores involved in the communication to the minimum.

Stage 2   During the communication phase, throttle all the cores to the state T_5 in step k if
  - BIA: $2^k \geq c$,
  - STRF: $c \leq k < p - c$.
  - PEA: $k > c$.

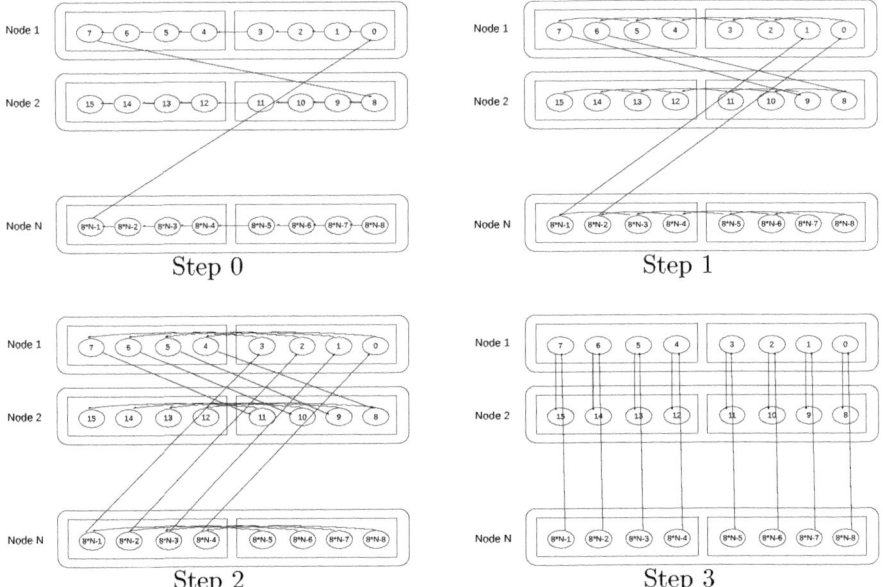

Fig. 2. The first four communication steps of the Bruck Index all-to-all algorithm on three nodes with two sockets (shown as rectangles) and eight cores (ovals) each. Internode communications are shown as straight lines across the node boundaries.

Stage 3 For STRF: throttle to state T_0 at the communication step $k = p - c$.
Stage 4 At the end of all-to-all, throttle all the cores to state T_0 (if needed) and restore their operating frequency to the maximum.

Following [7], the DVFS and CPU throttling policies are defined by the specific points within an algorithm where these techniques are applied and by a set of conditions indicating when to apply them. In the proposed energy saving strategy, Stages 1 and 4 are the points of the DVFS unconditional application while Stages 2, 3, and 4 state the conditions on throttling along with specifying its application points within all-to-all.

Rank Placement Consideration. MVAPICH2 provides two formats of rank placements on multicores, namely *block* and *cyclic*. In the block strategy, ranks are placed such that any node j ($j = 0, 1, \ldots, N - 1$) contains ranks from $c \times j$ to $c \times (j+1) - 1$. In the cyclic strategy, all the ranks i belong to j if ($i \bmod N$) equals j. The block rank placement calls for only two DVFS and throttling switches in the proposed energy saving strategy, and thus minimizes the switching overhead. In the cyclic rank placement, however, after a fixed number of steps the communication would oscillate between intra- and inter-node, requiring a throttling switch at every such step. Therefore, the block rank placement has been considered for the energy savings application.

2.1   Power Consumption Estimates

Let a multicore compute node has frequencies f_i, $(i = 1, \ldots, m)$, such that $f_1 < \ldots < f_m$, and throttling states T_j, $(j = 0, 1 \ldots, n)$. When all the c cores of the node execute an application at frequency f_i, each core consumes the dynamic power P_i proportional to f_i^3. Let P_{ij} be the power consumed by the entire node at the frequency f_i and throttling state T_j, P_s be the total static power consumption, and P_d be the dynamic power consumption of the compute node components, such as memory, disk, and NIC, which are different from the processor. Then, the power consumption with no idle cycles (at T_0) may be assumed as $P_{i0} = c \times P_i + P_s + P_d$, so, at T_j, it is

$$P_{ij} = \frac{j \times (P_s + P_d) + (n - j)(P_{i0})}{n} . \tag{1}$$

The P_{i0} expression serves just to give an idea of the effect of frequency scaling on power consumption. It may vary with the application characteristics since each application may have a different power consumption pattern depending its utilization of the compute node components.

The BIA power consumption for the first $2^k < c$ steps is P_{10} in each node, since the execution is at the minimum frequency and no throttling is applied. After this step, the power consumption is equal to P_{15} since T_5 is applied. As the number of nodes increase, the internode communication becomes dominant and hence, the power consumption starts to approximate P_{15}. Similar to the BIA in the STRF, the prevailing portion of the execution falls on the intermediate $c \leq k < p - c$ steps with the increase in node numbers. Thus, the all-to-all power consumption nears P_{15}, on average. For more details on the modeling and verification of the same, see [10].

3   Experimental Results

The experiments were performed on the computing platform Dynamo[5], which comprises ten Infiniband DDR-connected compute nodes, each of which has 16 GB of main memory and two Intel Xeon E5450 Quad core processors arranged as two sockets with the operating frequency ranging from 2 GHz to 3 GHz and the eight levels of throttling from T_0 to T_7. For measuring the node power and energy consumption, a Wattsup[6] power meter is used with a sampling rate of 1 Hz. Due to such a low measuring resolution, a large number of all-to-all operations have to be performed. For determining the average power consumption at a particular message size, 100 samples are taken for that message size followed by averaging. Specifically, at first the time spent in the all-to-all operation is measured for a given message size then the number of iterations is determined, so that the all-to-all executes for 100 seconds on Dynamo. A higher resolution meter will be considered in the future.

---

[5] funded and operated jointly by Iowa State University and Ames Laboratory.
[6] https://www.wattsupmeters.com

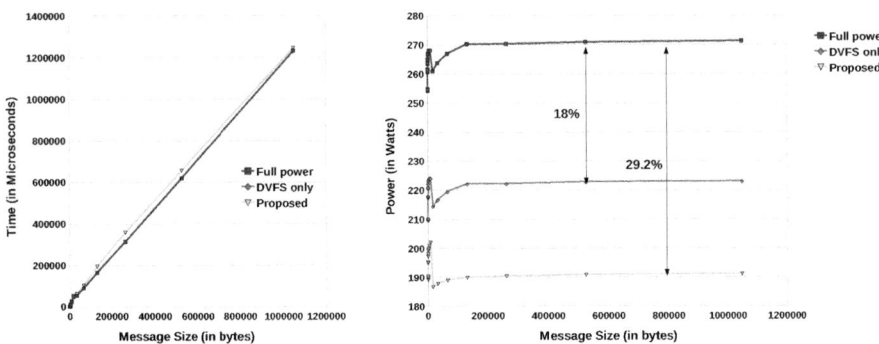

Fig. 3. The all-to-all execution time on 80 processes (left) and the power consumption across a compute node (right) for the three cases: Executing at the highest frequency and no throttling (`Full power`); only frequency scaling without throttling (`DVFS only`); and using the proposed energy saving strategies (`Proposed`)

OSU MPI Benchmarks. This set of benchmarks[7] are used here to determine the change in execution time and power consumption of "stand alone" all-to-all operations. From Fig. 3(right), it can be observed that the execution time for all-to-all has very low performance penalty when the proposed energy savings are used. The average performance loss observed for various message sizes was just 0.97% of that for the *Full power* case. While somewhat higher than in the *DVFS only* case, which was 0.5%, it is quite acceptable taking into the consideration large reductions in the power consumption achieved (Fig. 3(left)) with the *Proposed* strategy. Note, however that, in all the cases, the power consumption increases with the message size since the memory dynamic power consumption increases because of message copying [6]. Similar power reductions have been obtained for the all-to-all vector operation.

Application Testing CPMD (CarParrinello Molecular Dynamics)[8] is a ab-initio quantum mechanical molecular dynamics real-world application using pseudopotentials and a plane wave basis set. Eleven input sets from the CPMD application are used here. MPI_Alltoall is the key collective operation in CPMD. Since most messages have the sizes in the range of 128 B to 8 KB, the BIA is used. From the NAS benchmarks [1], FT and IS Class C benchmarks are chosen because they use the all-to-all operation. Fig. 4 shows the execution time and energy consumption of CPMD inputs and NAS benchmarks on 80 and 64 processes, respectively, normalized to the `Full power` case. For the CPMD with the `Proposed` strategies, the performance loss ranges from 0.4% to 4.3% averaging 2.78% leading to the energy savings in the range of 9.8% to 15.7% (13.4% on average). For the NAS, the performance loss ranges from 1.1% to 4.5% and the average energy savings

---

[7] OSU MPI Benchmarks: http://mvapich.cse.ohio-state.edu

[8] CPMD Consortium: http://www.cpmd.org

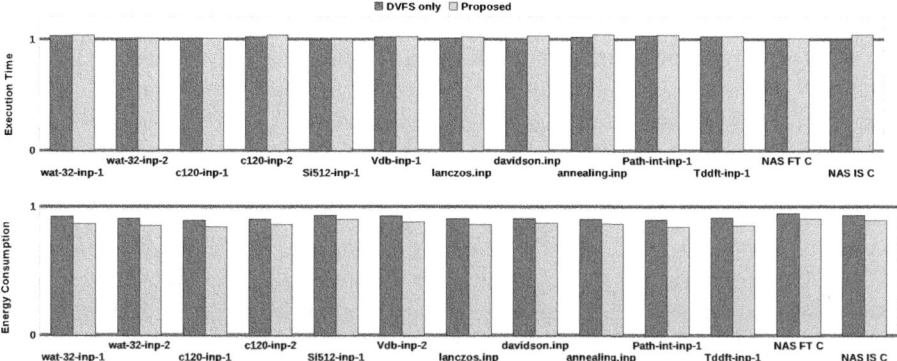

Fig. 4. Execution time (top) and energy consumption (bottom) of 11 CPMD inputs on 80 processors and of NAS benchmarks on 64 processes for the DVFS only and Proposed cases normalized to the Full power

are about 10%. Hence, the applications tested suffer from little performance loss and have significant energy savings.

4 Related Work

The energy efficiency delivered by the modern interconnects in high performance clusters is discussed in [12]. The communication phase characterization to obtain energy savings by using DVFS is studied in, e.g., in [5] and [2].In [3], authors have developed a tool which estimates power consumption characteristics of a parallel application in terms of various CPU components. In [4], algorithms to save energy in the collectives, such MPI_Alltoall and MPI_Bcast, are proposed. They differ significantly with the approach presented in this paper. Specifically, [4] assumes that throttling has a negative effect on the internode communication and thus, redesigns the all-to-all operation, such that a certain set of sockets does not participate in the communication at some point of time in order to be throttled. However, since the number of cores within a node continues to increase, forcing the sockets to remain idle during the communication, can introduce significant performance overheads. The power savings achieved in [4] are equivalent to operating two sockets at the minimum frequency and throttling state T_4, whereas the approach proposed here achieves power saving by keeping both sockets at the minimum frequency while throttling them to a higher state T_5. An experimental comparison of the two algorithms is left as future work.

5 Conclusions and Future Work

Energy-saving strategies are proposed for the all-to-all operation and implemented as the MPI_Alltoall collective in MVAPICH2 *without* modifying existing algorithms. The sensitivity of inter- and intra-node message transfers to CPU throttling has been assessed. It was observed that throttling has almost no negative effect on the performance of internode communications. Thus, both DVFS and CPU throttling were applied in the *purely internode* communication steps within three different all-to-all implementations in MVAPICH2. The experiments demonstrate that the proposed strategies can deliver up to 15.7% of energy savings without introducing significant performance penalty for the CPMD application inputs and NAS application benchmarks, which is representative of the potential benefits to scientific applications in general.

Similar energy saving strategies may be extended to other collectives including reduction operations. Furthermore, as the number of cores within a node keeps increasing, the opportunity of applying throttling in intranode communication, on a per-core level, must be also explored leading to the consideration of the point-to-point operations. As the interconnect technology becomes more efficient, the DVFS and throttling switching overheads may become significant relatively to the actual communication time. Therefore, the energy saving strategies need to be made aware of the message size transferred in certain types of collective operations.

References

1. Bailey, D.H., Barszcz, E., Barton, J.T., Browning, D.S., Carter, R.L., Dagum, L., Fatoohi, R.A., Frederickson, P.O., Lasinski, T.A., Schreiber, R.S., Simon, H.D., Venkatakrishnan, V., Weeratunga, S.K.: The NAS parallel benchmarks–summary and preliminary results. In: Proceedings of the 1991 ACM/IEEE Conference on Supercomputing, pp. 158–165 (1991)
2. Freeh, V.W., Lowenthal, D.K.: Using multiple energy gears in MPI programs on a power-scalable cluster. In: Proceedings of the Tenth ACM SIGPLAN Symposium on Principles and Practice of Parallel Programming, pp. 164–173 (2005)
3. Ge, R., Feng, X., Song, S., Chang, H.C., Li, D., Cameron, K.W.: PowerPack: Energy profiling and analysis of high-performance systems and applications. IEEE Transactions on Parallel and Distributed Systems 21, 658–671 (2010)
4. Kandalla, K., Mancini, E.P., Sur, S., Panda, D.K.: Designing power-aware collective communication algorithms for InfiniBand clusters. In: 2010 39th International Conference on Parallel Processing, ICPP 2010, pp. 218–227 (2010)
5. Lim, M.Y., Freeh, V.W., Lowenthal, D.K.: Adaptive, transparent frequency and voltage scaling of communication phases in MPI programs. In: Proceedings of the 2006 ACM/IEEE conference on Supercomputing (2006)
6. Liu, J., Poff, D., Abali, B.: Evaluating high performance communication: a power perspective. In: Proceedings of the 23rd International Conference on Supercomputing, pp. 326–337 (2009)
7. Martonosi, M., Malik, S., Xie, F.: Efficient behavior-driven runtime dynamic voltage scaling policies. In: Third IEEE/ACM/IFIP International Conference on Hardware/Software Codesign and System Synthesis, CODES+ISSS 2005, September 2005, pp. 105–110 (2005)

8. Park, J., Shin, D., Chang, N., Pedram, M.: Accurate modeling and calculation of delay and energy overheads of dynamic voltage scaling in modern high-performance microprocessors. In: 2010 International Symposium on Low-Power Electronics and Design (ISLPED), pp. 419–424 (2010)
9. Schmidt, M.W., Baldridge, K.K., Boatz, J.A., Elbert, S.T., Gordon, M.S., Jensen, J.H., Koseki, S., Matsunaga, N., Nguyen, K.A., Su, S., Windus, T.L., Dupuis, M., Montgomery Jr., J.A.: General atomic and molecular electronic structure system. J. Comput. Chem. 14, 1347–1363 (1993)
10. Sundriyal, V., Sosonkina, M.: Percall energy saving strategies in all-to-all communications. Technical Report 11-05, Computer Science Department, Iowa State University, Ames, IA, 5011 (May 2011)
11. Sundriyal, V., Sosonkina, M., Liu, F., Schmidt, M.: Dynamic frequency scaling and energy saving in quantum chemistry applications. In: Proceedings of the International Parallel and Distributed Processing Symposium (IPDPS 2011), May 16-20 (2011)
12. Zamani, R., Afsahi, A., Qian, Y., Hamacher, C.: A feasibility analysis of power-awareness and energy minimization in modern interconnects for high-performance computing. In: Proceedings of the 2007 International Conference on Cluster Computing, pp. 118–128 (2007)

Data Redistribution Using One-sided Transfers to In-Memory HDF5 Files

Jerome Soumagne[1,2], John Biddiscombe[1], and Aurélien Esnard[2]

[1] Swiss National Supercomputing Centre
Galleria 2, Via Cantonale, 6928 Manno, Switzerland
[2] INRIA Bordeaux Sud-Ouest
351 cours de la Liberation, 33405 Talence, France

Abstract. Outputs of simulation codes making use of the HDF5 file format are usually and mainly composed of several different attributes and datasets, storing either lightweight pieces of information or containing heavy parts of data. These objects, when written or read through the HDF5 layer, create metadata and data IO operations of different block sizes, which depend on the precision and dimension of the arrays that are being manipulated. By making use of simple block redistribution strategies, we present in this paper a case study showing HDF5 IO performance improvements for "in-memory" files stored in a distributed shared memory buffer using one-sided communications through the HDF5 API.

Keywords: Data Redistribution, Distributed Shared Memory, HDF5, One-sided Communication.

1 Introduction

HDF5 [11], the Hierarchical Data Format, allows users to write data output in a very flexible manner. One file can be composed of different datasets, usually containing a large amount of data, and of attributes, storing small pieces of information. Datasets can be simple scalars or N-dimensional vectors written in parallel using hyperslab selections – these selections depend entirely on the code implementation. Parallel writes or reads can be issued in a uniform manner or can follow a totally random pattern. Concurrent with these data IO operations, HDF5 metadata is written and can be accessed several times if objects are opened, created or closed or if the metadata has not been previously cached. Therefore a complete HDF5 file write or read in parallel may consist of a large number of accesses in a complex pattern.

The HDF5 architecture allows the creation of customized IO methods called drivers, one well-known parallel driver is the MPI-IO driver, discussed in section 2. Disk IO being a significant and now commonplace bottleneck in simulations, we developed a parallel virtual file driver called the DSM driver which allows one to redirect HDF5 IO operations in parallel to a distributed shared memory (DSM) buffer (the reader is referred to [9] and [8] for a more complete introduction to the DSM driver and communicators). Simulation processes may write

Y. Cotronis et al. (Eds.): EuroMPI 2011, LNCS 6960, pp. 198–207, 2011.

in-memory HDF5 files using various types of communication, the principal intended use of these in-memory files being code-coupling of parallel applications. The original implementation made use of two-sided communication only; we recently extended it to make use of one-sided communication – we focus in this paper only on one-sided transfers and consider the case where the nodes hosting the DSM are different from the nodes hosting the simulation processes, i.e. where traffic between them *must* traverse the network. We present in section 3 the MPI one-sided communicator used for this study, along with an additional communicator specially designed for the Cray XE6.

In the original implementation, HDF5 files are written using a linear address space where the file grows in size by extending upwardly the address range used. The addresses are spread (evenly by default) across a series of DSM host processes so that as the file grows in size and data is written into higher addresses more network links are utilized and the higher the transfer bandwidth *should* be. In practice, most data reads/writes for datasets or hyperslabs are significantly smaller than the entire file and thus use only a small number of memory partitions – and hence network links, at any given time, which limits the bandwidth reached. We extend this strategy in section 4 by remapping the address space nonlinearly using varying block sizes among DSM host processes (thereby distributing traffic more evenly). We present a case study showing the performance obtained in section 5 and compare it to related studies in section 6.

2   HDF5 File IOs

HDF5 IOs can be produced in very different ways. As mentioned above, *drivers* allow users to select a suitable IO mechanism for the system. One frequently used driver is the MPI-IO driver, best suited for parallel file systems, since it uses MPI-IO underneath. Whilst MPI-IO and implementations such as ROMIO [10] have been optimized for various types of accesses depending on the file system used, HDF5 also provides its own ways of tuning and writing data in parallel. For instance, the chunking mechanism allows files and particularly datasets to be stored in a non-contiguous form, i.e. in equally sized chunks, which can be helpful for parallel file systems, over which datasets can therefore be striped. Additional optimizations have also been made in the MPI-IO driver and HDF5 library itself for specific file systems such as the Lustre file system [6].

These enhancements are particularly useful in a traditional pipeline model where data is archived and post-processed from file systems, however bandwidth offered by file systems is limited. Introducing the DSM driver in the pipeline allows us to couple two different applications in parallel through the network by using the HDF5 interface. This offers an additional exchange method before saving post-processed data to disk for archiving purposes. Parallel optimizations implemented in the HDF5 library can be re-used by the DSM driver, such as the chunking mechanism, but other types of accesses specific to file systems need to be adapted and re-optimized within the driver itself.

3   DSM Driver and Communicators

As opposed to the MPI-IO driver, where the application is effectively *coupled* to the file-system, when using the DSM driver, two applications – parallel simulation and DSM host (integrating post processing code) – are coupled together through a communication layer, referred to as an inter-communicator. The DSM architecture being modular, permits different inter-communicator types to be implemented, which can follow one-sided or two-sided communication patterns. For this case study, two different one-sided inter-communicators are considered: one based on MPI RMA and one specific to Cray systems, based on an API called DMAPP.

MPI RMA Inter-communicator. The MPI RMA communicator makes use of the passive MPI RMA communication mechanism [5]. When the DSM is allocated, `MPI_Alloc_mem` is called and the window is defined as the size of the requested HDF5 file. `MPI_Put` can then be issued in a one sided manner using `MPI_Win_lock` and `MPI_Win_unlock` between transactions.

The communicator can be dynamically created (using the dynamic process management set of functions) but due to the numerous restrictions imposed by MPI implementations, on large systems (e.g. on Cray systems), the communicator has to be defined using an `MPI_Intercomm_create` call within an MPMD job (where the global communicator has been previously split between applications).

DMAPP Inter-communicator. The DMAPP communicator is derived from the aforementioned MPI RMA communicator. On Cray machines that support the latest generation of interconnect, Gemini [3], Cray defines the Distributed Memory Application API, referred as DMAPP [4]. This API is used on these systems to implement one-sided libraries such as Cray SHMEM and is also used by PGAS compilers (Co-array Fortran and UPC). We have implemented a communicator taking direct advantage of this lower level one-sided communication library. On the simulation side, to avoid memory overheads created by symmetric memory usage, we make use of non-symmetric memory, allocated and registered to the DMAPP API on the DSM hosts only. This registration step provides memory segment information which is then exchanged with the simulation (only once at initialization time, assuming that the DSM size is fixed between time steps). `dmapp_put` calls can then be issued to transfer data into the DSM.

4   Redistribution Strategies

In our implementation the DSM is distributed among p processes, each process allocating l bytes of data, which gives a total DSM length of $L = l \times p$. Using linear addressing, the DSM is contiguously filled from process rank 0 to process rank $(p-1)$. If a simulation writes a file of size S, the actual number of processes used to receive data will thus be $\left\lceil \frac{S}{l} \right\rceil$ with $S \leq L$. Whilst this method can provide relatively good performance when $S \simeq L$, if the file written is composed of several different datasets (i.e. each much smaller than L), which are contiguously (and sequentially) mapped onto the DSM, individual simulation processes will waste bandwidth by using only a small partition of the network links available – particularly so when datasets are divided between simulation processes and written using hyperslab selections. We therefore sought better strategies which could be enabled on demand.

4.1   Mask Redistribution

When $S \ll L$, a first simple strategy is to automatically re-size the DSM window to the requested file size without any concrete memory reallocation. This can effectively improve the overall bandwidth by making $S \simeq L$ but this brings two main drawbacks: the most evident one is that it wastes memory allocated on the DSM, the second one is that it does not solve the multiple dataset problem mentioned above.

4.2   Block Cyclic Redistribution

The second strategy to be considered in this case study is a block cyclic redistribution [13]. It is a simple strategy and it potentially allows a good load balance between DSM processes. A block size s being fixed, the DSM address mapping is decomposed into $\frac{L}{s}$ blocks. For convenience, the DSM length L is adapted so that it becomes a multiple of s. Blocks are distributed in a round-robin fashion, the Bth block is sent to the process rank $(B \bmod P)$ or $(B \bmod B_c)$ (if B_c, the number of blocks in a cycle is not equal to P, the number of processes). Hence every address a is associated to the following triplet (p, o, i) which can be written as:

$$a \mapsto \left(B \bmod P, \left\lfloor \frac{B}{P} \right\rfloor, a \bmod s \right) \tag{1}$$

the first term p being the process index within the DSM, o the local block offset in a process and i the local address offset within a block.

   This method presents two obvious advantages: bandwidth is not wasted even if $S \ll L$; data chunks are load balanced, which is especially beneficial when multiple datasets are written. However this method can potentially create a huge number of data transactions, depending on the block size chosen, which could result in a performance drop.

4.3   Random Block Redistribution

The third strategy tested consists in re-using the algorithm previously described, scattering the DSM address space into pieces of size s. Another step is then added to the redistribution pipeline, shuffling the blocks in a randomized but constant manner (so that blocks can be retrieved).

   This method can present another advantage compared to the previous solution (but keeps the same main drawback), it may avoid a possible network congestion if two simulation processes were sending data to the same DSM process using the block cyclic redistribution algorithm – which may occur with a periodic frequency introduced by certain communication patterns and data distributions in the file.

5   Performance Evaluation

For these tests, we use two systems: an InfiniBand QDR 4X cluster with MVA-PICH2 [2] composed of 15 nodes (180 cores) and a Cray XE6 system composed of two racks, i.e. 176 compute nodes (4224 cores), with Cray MPT (derived from MPICH2 [1] [7]). To be able to evaluate the performance obtained using the previously defined strategies, we first run micro-benchmarks on these two machines.

5.1   Internode Micro-Benchmark

The micro-benchmarks are derived from the OSU test suite [2] and identify the bandwidth performance on the different systems for different sizes of packets between two different nodes. Only put operations are tested here. Results are shown below in figure 1.

A careful examination of these charts shows a performance drop point with MVAPICH2 for packets of 16KB, though the overall bandwidth reflects Infini-Band QDR 4X performance. For the XE6, theoretical unidirectional performance is estimated at 5GB/s. Here the DMAPP interface performs better than the MPI one-sided interface. Two main drop points can however be noticed, 4KB for DMAPP and 1KB for MPI – these points correspond to the standard offload thresholds, making use of the RDMA engine for large messages.

5.2   Single Dataset Benchmark

For the following benchmarks, write bandwidth tests can be seen as basic client-server tests: a first set of processes (servers) hosts the DSM and waits for in-coming data, a second set of processes (clients) writes HDF5 data in parallel to the DSM using the HDF5 DSM driver. The measured bandwidth corresponds to the average time of a complete file write (HDF5 create, write and close opera-tions). The first benchmark writes one file composed of one single dataset using hyperslab selections.

Contiguous/Linear Distribution. The DSM is distributed among 8 nodes (32 processes, 4 per node) on the InfiniBand cluster and among 88 nodes (176 processes, 2 per node) on the XE6. To keep a certain consistency between the systems, the local buffer size allocated per node is kept to 512MB, which creates a DSM of $8 \times 512 = 4$GB on the InfiniBand cluster and a DSM of $88 \times 512 = 44$GB on the XE6. Given this fixed DSM (file) size, a single dataset of the matching size is written from the combined send nodes (smaller pieces per process as number of processes increases).

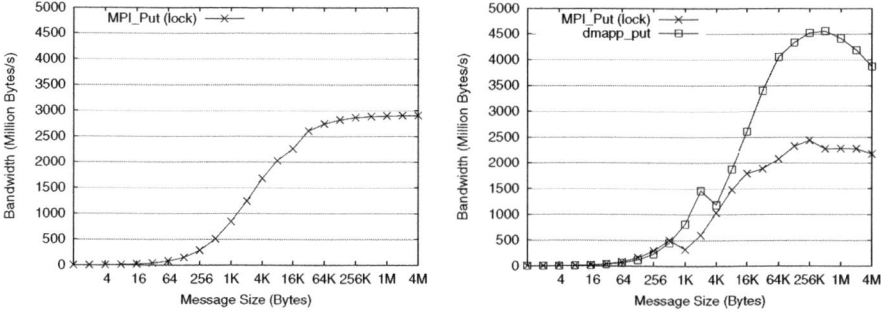

Fig. 1. Internode bandwidth micro-benchmark – (Left) InfiniBand QDR 4X cluster with MVAPICH2 – (Right) Cray XE6 with Cray MPT and DMAPP

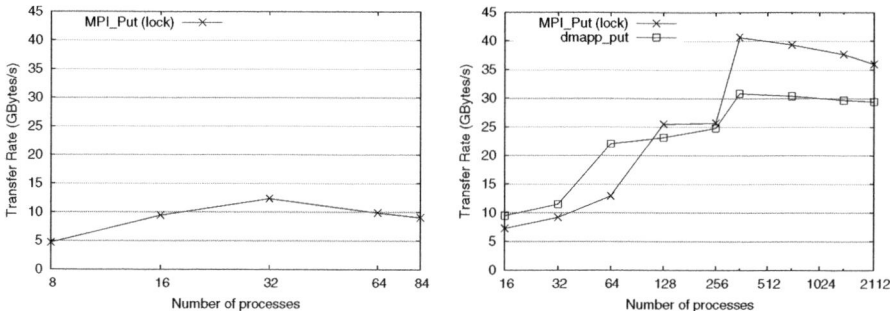

Fig. 2. Write transfer rate of an (in-memory) HDF5 file composed of one single dataset using contiguous distribution – (Left) InfiniBand QDR 4X cluster – (Right) Cray XE6

For writing, on the XE6, the number of processes is 4 per send node until 88 nodes are reached (352 processes) at which point, processes per send node are increased up to 24 – giving 2112 processes writing data in total. On the Infiniband cluster, 7 send nodes are available and 4 processes per node are used initially and then incremented to 12 per send node giving a maximum of 84 send processes. (Note that 4 processes per send node were selected as the starting point, because with fewer processes injecting data, we are unable to fully utilize the individual network links). Therefore, as shown in figure 2: on the InfiniBand cluster, a peak bandwidth is observed at 12.5GB/s with 32 processes (8 receive and 7 send links active); on the XE6, at 40.5GB/s with 352 processes (88 send and receive links active). Note that the XE6 system used for the tests has a 2D torus ($1 \times 6 \times 16$), and the resulting bandwidth is lower than that achievable using a 3D torus.

Block Cyclic and Random Block Redistributions. For different block sizes, we run the same benchmark as above, using a single dataset. This test allows us to evaluate block redistribution advantages as opposed to a simple contiguous distribution. Results are presented in figures 3 and 4.

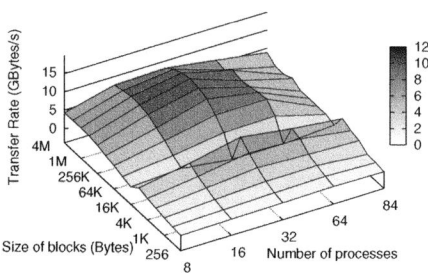

Fig. 3. Write transfer rate on InfiniBand QDR 4X cluster of an (in-memory) HDF5 file composed of one single dataset using block cyclic redistribution

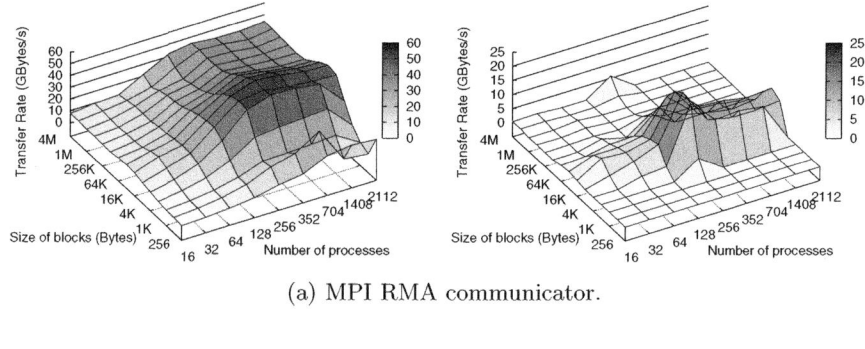

(a) MPI RMA communicator.

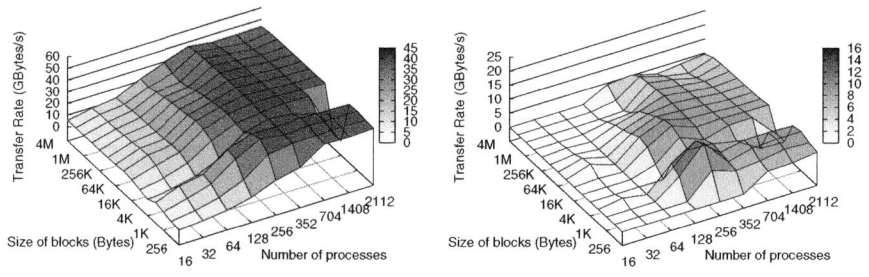

(b) DMAPP communicator.

Fig. 4. Write transfer rate on Cray XE6 of an (in-memory) HDF5 file composed of one single dataset – (Left) Block cyclic redistribution – (Right) Performance comparison (difference) between block cyclic and contiguous distributions

On both systems, the bandwidth drop points of section 5.1 can be observed. While these drop points had a small effect on the micro-benchmark, they lead to a significant slow-down in the HDF5 write operations when those block sizes are used repeatedly. On the XE6 system, a significant improvement compared to the contiguous write is evident for block sizes belonging to the [16KB; 64KB] interval with the MPI RMA communicator and for block sizes below 4KB for the

DMAPP communicator. Since metadata operations are usually very small transfers, being able to use this communicator in combination with the MPI RMA communicator is an advantage for this system. However one can also notice in figure 3 that there is no improvement at all on the InfiniBand cluster when using a block cyclic method if *only a single dataset* is written into the file (compare the peak transfer rate to that of the left of figure 2). The transfer rates for the full DSM sized dataset are in general slower using the block/random redistribution on the InfiniBand system and this is because breaking the data writes into many smaller blocks does not improve performance as can be seen from the micro-benchmark result of figure 1.

Random block results are not shown here – for brevity – but globally increase the bandwidth as one may see in the next benchmark, and avoid possible congestion issues in the DSM.

5.3   Multiple Dataset Benchmark

To reflect the behaviour of a common simulation code, the previous benchmark is reused here, this time creating a file composed of 10 datasets instead of a single one. The same configuration is used as the previous tests, each of the datasets has the same fixed size and their sum is the size of the allocated DSM, i.e. 4GB for the InfiniBand cluster and 44GB for the XE6. Results are shown in figure 5.

It is evident from this figure that writing using block redistribution is much more efficient than linear mapping. Since each dataset *in the linear HDF5 memory space* is contiguous, writing 10 datasets in parallel but sequentially in time, causes only one tenth of the links to become active for each individual dataset. By redistributing blocks for each of the much smaller datasets across all processes, we make use of all of the links for all of the transfers. For block cyclic redistribution, providing each dataset is at least $s \times P$ in size, the data will be well distributed.

It is perhaps surprising that the graph of figure 5 (left) shows a significant drop in transfer rate as the number of send processes increases. The drop is smaller for random distribution than for cyclic and this can be explained by

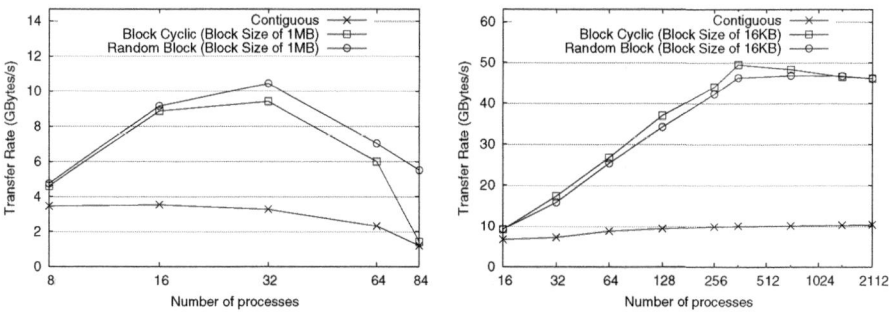

Fig. 5. Write transfer rate using MPI RMA communicator of an (in-memory) HDF5 file composed of 10 datasets – (Left) InfiniBand QDR 4X cluster – (Right) Cray XE6

noting that we have used 4 processes per listening node, so in fact the cyclic redistribution hits the same link 4 times in succession, which will not generally happen for the random distribution. We therefore see a more gradual fall off in line with figure 2 (left) for the random mode, the overall drop being caused by the increase in individual number of transfers as the effect of latency and lower performance for smaller packets dominates.

6   Related Work and Discussion

The results presented here appear to be typical for the kinds of system tested, but can however be affected by the network topology and capabilities, system configuration, number of nodes used, number of processes per node, and so on. The space of potential combinations of parameters for plots is beyond what can be presented in a short paper so certain decisions as the number of processes to use per node were made to try to maximize the data injection and network saturation to give representative results. Note that the implementation of MPI RMA for the XE6 is not yet optimized and the measured performance for large messages (above 1KB) should be improved in the future [7]. Absolute bandwidths may not therefore be indicative of all installations – though this does not affect our results.

The improvements in transfer rates found when using redistribution are broadly in line with expectations. In fact the advantages of data redistribution are well known and date back to the origins of message passing [13]. Many projects have made use of block cyclic distribution as a means of improving performance for scattered data, in particular PGAS languages (such as UPC [12]) provide options for shared array allocation using block cyclic layouts, which can improve algorithmic performance. In fact our flexible communicator design opens up the possibility that a PGAS based layer could be used directly instead of MPI or DMAPP as we have presented here and we shall pursue this in future work.

The observation that certain packet sizes are handled better by different APIs also allows the possibility of further fine tuning transfers. IO operations from HDF5 applications typically consist of many small metadata and larger heavy data requests and these different needs can be served by switching communicators on the fly to use DMAPP for metadata and MPI RMA for heavy data decomposed into blocks.

7   Conclusion

We presented in this paper a case study where HDF5 files are sent to a DSM using one-sided transfers and found that implementing redistribution strategies significantly improves the performance of data writes for typical use cases where multiple datasets are written into a much larger file. By choosing block sizes that are optimal for the underlying hardware and matching the number of send/receive nodes, we are able to improve the data bandwidth. Codes coupled using the DSM driver are now able to communicate at speeds approaching the maximum possible on the systems tested.

Acknowledgements. The authors would like to thank Nina Suvanphim from Cray for her support in addressing the various issues encountered on the Cray XE6 used in this paper and installed at the Swiss National Supercomputing Centre. The work presented in this paper is supported by the "NextMuSE" project receiving funding from the European Community's Seventh Framework Programme (FP7/2007-2013) under grant agreement 225967.

References

1. MPICH2, `http://www.mcs.anl.gov/research/projects/mpich2`
2. MVAPICH: MPI over InfiniBand, 10GigE/iWARP and RoCE, `http://mvapich.cse.ohio-state.edu/index.shtml`
3. Alverson, R., Roweth, D., Kaplan, L.: The Gemini System Interconnect. In: Symposium on High-Performance Interconnects, pp. 83–87 (2010)
4. Bruggencate, M., Roweth, D.: DMAPP – An API for One-sided Program Models on Baker Systems. In: Proceedings of Cray User Group (2010)
5. Gropp, W., Lusk, E., Thakur, R.: Using MPI-2: Advanced Features of the Message-Passing Interface. MIT Press, Cambridge (1999)
6. Howison, M., Koziol, Q., Knaak, D., Mainzer, J., Shalf, J.: Tuning HDF5 for Lustre File Systems. In: Proceedings of Workshop on Interfaces and Abstractions for Scientific Data Storage (2010), LBNL-4803E
7. Pritchard, H., Gorodetsky, I.: A uGNI-Based MPICH2 Nemesis Network Module for Cray XE Computer Systems. In: Proceedings of Cray User Group (2011)
8. Soumagne, J., Biddiscombe, J.: Computational Steering and Parallel Online Monitoring Using RMA through the HDF5 DSM Virtual File Driver. Procedia Computer Science 4, 479–488 (2011), ICCS 2011
9. Soumagne, J., Biddiscombe, J., Clarke, J.: An HDF5 MPI Virtual File Driver for Parallel In-situ Post-processing. In: Keller, R., Gabriel, E., Resch, M., Dongarra, J. (eds.) EuroMPI 2010. LNCS, vol. 6305, pp. 62–71. Springer, Heidelberg (2010)
10. Thakur, R., Gropp, W., Lusk, E.: Optimizing noncontiguous accesses in MPI-IO. Parallel Computing 28(1), 83–105 (2002)
11. The HDF Group: Hierarchical Data Format Version 5 (2000–2011), `http://www.hdfgroup.org/HDF5`
12. UPC Consortium: UPC Language Specifications, v1.2. Tech report (2005), `http://upc.gwu.edu/docs/upc_specs_1.2.pdf`
13. Walker, D.W., Otto, S.W.: Redistribution of Block-Cyclic Data Distributions Using MPI. Concurrency - Practice and Experience 8(9), 707–728 (1996)

RCKMPI – Lightweight MPI Implementation for Intel's Single-chip Cloud Computer (SCC)

Isaías A. Comprés Ureña, Michael Riepen, and Michael Konow

Microprocessor and Programming Research Labs (MPR)
Theodor-Heuss-Straße 7, 38122 Braunschweig, Germany
{isaias.a.compres.urena,michael.riepen,michael.konow}@intel.com

Abstract. The Single-chip Cloud Computer (SCC) is an experimental processor created by Intel Labs. It is a distributed memory architecture that provides shared memory possibilities and an on die Message Passing Buffer (MPB). This paper presents an MPI implementation (RCKMPI) that uses an efficient mix of MPB and DDR3 shared memory for low level communication. The on die buffer found in the SCC provides higher bandwidth and lower latency than the available shared memory. In spite of this, message passing can be faster through DDR3, due to protocol overheads related to the small size of the MPB and the necessity to split and reassemble large packages, together with the possibility that the data is not available in the cache. These overheads take over after certain message sizes, requiring run time decisions with regards to which type of buffers to use, in order to achieve higher performance. In the current implementation, the decision is based on remaining bytes to transfer from in transit packets. MPI benchmarks are shown to demonstrate that the use of both types of buffers results in equal or lower transmission times than when communicating through the on die buffer alone.

Keywords: Many-Core Processors, Message Passing, MPI, RCKMPI.

1 Introduction

The Single-chip Cloud Computer (SCC) experimental processor [4] is a 48-core 'concept vehicle' created by Intel Labs as a platform for many-core software research. Its memory organization is distributed, in the sense that the computing elements have private memory. Certain amounts of the DDR3 memory pool can be configured as shared, although lacking cache coherency. To allow for low latency communication, on chip buffers are also present. These are called Message Passing Buffers (MPBs), and amount to 8KB per core. They differ from caches in that they are directly controlled by software.

For distributed memory systems, message passing is the most common parallel programming model. The Message Passing Interface (MPI) is the de facto standard for message passing implementations. MPI applications can run on the SCC [3], with unmodified MPI libraries configured to use TCP/IP sockets. A network driver, called rckmb, is available for SCC Linux. The driver provides a network device that can be used for communication between the cores, as if they were nodes in a cluster. Other

Y. Cotronis et al. (Eds.): EuroMPI 2011, LNCS 6960, pp. 208–217, 2011.

message passing libraries are available for the SCC. The RCCE [1] library and community supplied extensions for it, support collective operations as well as non-blocking communication [2]. RCCE, and similar solutions, have been shown to clearly outperform MPI libraries configured to use the rckmb network driver [3]. However, the available software and tools for these programming models are limited when compared to MPI.

It is clear that an MPI compliant library, with performance similar to the low weight message passing solutions, is desirable. RCKMPI is such a library, and in contrast to other solutions, it exploits both possible forms of communication: Shared Memory and MPB. It works with both types of buffers at the same time and makes run time decisions with regards to how to reach the target process more efficiently.

This paper introduces the memory organization of the SCC, explains how RCKMPI makes use of it, and finally presents performance results. In section 2, the SCC's general architecture, details about its memory organization and the Message Passing Buffer (MPB) are introduced. Section 3 describes the MPI library's design as well as the multi buffer type operation. Performance results are shown in section 4, followed by a summary.

2   SCC Hardware Architecture

The Single-chip Cloud Computer is a distributed memory architecture. It has the possibility of setting up 16MB chunks of shared memory, although without cache coherency. The chip contains dedicated message passing hardware in the form of on die memory that can be used for synchronization. This opens up a window for research on programming models that scale on non-coherent hardware. Furthermore, the system offers fine grain power management features for DVFS research.

2.1   General Overview

The IA core is based on the P54C Pentium® processor and therefore operates in-order, and is two-way superscalar. Each core has 16KB of separate code and instruction level 1 cache, as well as 256KB of level 2 cache. The core has moderate performance and its small footprint allows the integration of a relatively high core count, which is desirable to support research on many core programming models. Two cores form a tile and the chip is organized in a 6×4 2D mesh of such tiles. Mesh connections are established through high speed routers.

Each of the routers is connected to a tile as well as the neighboring mesh nodes. These neighbor nodes may be routers or periphery devices (e.g. the memory controller on the west port of the lower left router in figure 1). Each tile also contains the Message Passing Buffer which is on die memory that, as it name suggests, can be used to perform message passing. The system I/F connects to the System Interface FPGA (SIF-FPGA) which works as a reconfigurable chipset. This chipset allows the connection to external devices through different channels, including (but not limited to) a 4× PCIe connection, Gigabit Ethernet and a SATA interface. The PCIe port can be connected to a Management Console PC (MCPC) and exposes the mesh protocol to it, so that it is possible to access all registers and memory locations (DDR3 as well as MPBs) from the PC. This provides powerful debugging capabilities to the developer.

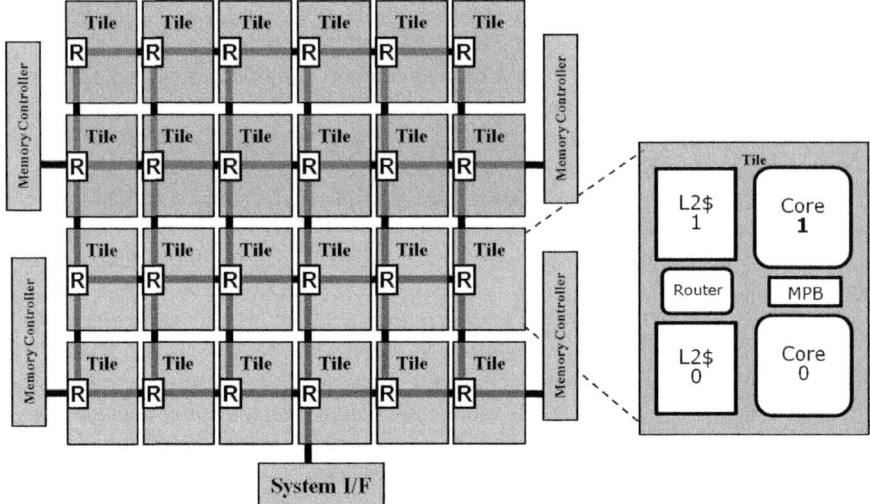

Fig. 1. SCC Architecture Overview

2.2 SCC Memory Architecture

SCC has access to a maximum of 64 GB external DDR3 memory through four on-die memory controllers, each addressing up to 16GB of external memory. This results in a 34 bit global address space for the memory controllers while each of the cores only has a 32 bit physical address space. In order to address all available memory, a new address translation layer has been introduced: the Look Up Tables (LUTs). There is one LUT per core, each with 256 entries.

The LUTs allow mapping 16MB chunks of the core's address space (physical address) to any SCC system address. System addresses can be memory locations on the DDR3 memory as well as memory mapped registers in the tiles themselves. Memory mapped registers are also used to modify the state of each core (e.g. for resets and IRQs) which give developers great flexibility to influence system behavior. It is possible for each core to modify the registers of any other core (including its own), in order to modify voltages and frequencies or to map the SCC system memory to their liking by modifying the LUTs themselves. The latter possibility opens the door to dynamic mappings of the DDR3 memory. This flexibility makes SCC an interesting research vehicle for distributed shared memory applications. Applications that are running on the cores can simply map selected windows of the shared memory to their physical address space. This method is also used by the shared memory implementation that is discussed in section 3.2. LUT consistency is the responsibility of the programmer.

2.3 Hardware Support for Message Passing

Each tile contains 16KB (8KB per core) of on die SRAM memory that is dedicated to message passing and is called the Message Passing Buffer (MPB). As each MPB can also be addressed via the LUT address translation, each core has access to all 24 MPBs. Messages passed through the MPBs see a large performance improvement over messages sent through DDR3. The latency to the MPBs is 15 times lower when compared to DDR3 [4].

When cores access the MPB, locality matters latency wise. As the P54C core has an in-order architecture, it means that memory read transactions are blocking while write access is not because of the superscalar implementation. So programming models should make use of local read and remote write to transfer messages from one core to another.

3   RCKMPI Architecture

The RCKMPI library consists of the addition of SCC specific channels to MPICH2 [6]. There are three new channels: one MPB based, one shared memory based and one that operates with both types of memory to handle MPI traffic. The general architecture of MPICH2 [5] and its CH3 device are left intact. All other channels were removed, due to incompatibilities with the SCC. The only exception is the TCP/IP sockets channel, since it works flawlessly with the rckmb network driver.

In this section, the general design of the channels, the shared memory strategy developed for message passing and the multiple buffer type operation present in one of the channels, are presented.

3.1   RCKMPI Channels Design

All new channels in RCKMPI share the same core design: exclusive write address ranges per remote process (receive buffers) and a progress engine based on busy polling. The latter is a design characteristic that is shared with other channel implementations in the MPICH2 library (e.g. Nemesis [5]). The address ranges are called Exclusive Write Sections (EWS) in the library, and only the owner remote process is allowed to write in them.

The use of EWSs is the main characteristic of the RCKMPI channel implementations. This approach allows for writes to happen concurrently without a global blocking mechanism, since the consistency of a buffer depends only on the sender and receiver process pair. The EWSs can be of MPB or Shared Memory type, and a combination of both is possible.

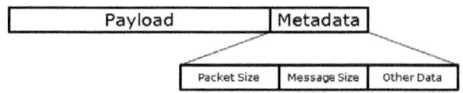

Fig. 2. Exclusive Write Section containing payload and metadata

To ensure consistency, a channel needs to control message sequence, multiple message packets, message size, etc. This requirement is enforced by the communication protocol, which is based on polling of metadata contained in the receive buffers. The polling is done in round robin fashion.

As per MPICH2's CH3 device design, send operations can be done entirely in a send function, if the channel can handle the full transfer immediately after the call. Otherwise, a send request is queued to be handled by the progress engine later. Message reception is done exclusively by the progress engine (i.e. there are no receive operations in the channels).

3.2 Shared Memory Communication

As mentioned in section 2, the SCC lacks any form of cache coherency. There are no DMA controllers available, so data movement has to be done by the cores. This means that zero copy memory transfers and overlapping of memory movement and computation are not possible. Fortunately, due to the existence of the LUTs, and the possibility that each core has to modify them, single copy transfers are possible.

RCKMPI is designed to work with SCC Linux. In this mode of operation, one Linux image per SCC core is loaded at initialization. Shared memory was previously available by means of 4 predefined LUT entries (64MB total). This shared memory is used by existing tools and is relatively small and fragmented. To achieve more flexibility and a large (not fragmented) shared memory pool, extra functionality was developed in the form of a Linux kernel patch and a user space library.

The Linux kernel patch allows the cores to allocate and pin contiguous chunks of 16MB of private RAM (16MB aligned, therefore corresponding to LUT entries). The location of the first LUT and the total number of chunks per core are then written to a predetermined location in the fixed shared memory area.

The library reads the first LUT per SCC core, and the total number of pages each core provides. With this information, it builds a page list and provides a memcpy like interface. The library modifies a single LUT entry at the local core, allowing it to have a 16MB memory window to the desired location. The corresponding 16MBs of that LUT entry are mapped into the address space of the MPI process.

With this functionality, when an MPI process wants to send a message through DDR3, it copies the contents to the remote shared memory EWS, and then signals the remote core. The remote core then handles the MPICH2 packet in place, and when done, signals the availability of the buffer for the next message.

The default size for the shared memory areas is 16MBs per core (for a total of 768MB visible to all cores). This can be adjusted to 80MBs per core (3.84GB total). The library will detect the number of pages at initialization, and adjust the size of the DDR3 EWSs accordingly (from 341KB to 1.66MB in the 48 process case).

3.3 Multiple Buffer Type Operation

While low overhead communication was already achieved with the use of the message passing buffer by RCKMPI, extra efforts were made to exploit the flexibility of the SCC memory organization. The MPB size of 8KB per core (which is divided in EWSs of equal size) results in the necessity to split and rebuild, relatively small, MPICH2 packets when they do not fit in a target receive buffer.

In spite of the lower latency and higher bandwidth, depending of the message size, the protocol overhead of using the MPB can make operations longer than using the larger receive buffers located in shared memory. This means that for RCKMPI, the decision of how to send a packet needs to be done at runtime, since it is not always clear which medium is faster. Fig. 3 shows a concrete scenario where this happens, with core clocks set at 533MHz.

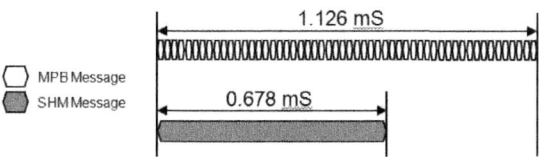

Fig. 3. Transmission time for 8KB MPICH2 packets (48 processes, 160 byte MPB EWSs)

Switching from MPB to DDR3 based messaging is done strictly based on pending bytes to send of in transit MPICH2 packets. The byte value where the switch is done is set statically at MPI job startup, and depends on the MPI job size (i.e. number of processes) and the size of the level 2 cache. The number of processes correlates directly to the size of the EWSs (from 8KB for 2 processes, to 160 bytes for 48 processes). The time to transfer a message is deterministic and depends on the number of round trips the protocol needs to complete the packet transfer. Taking into account the size of the L2 cache is based on the assumption that communication with each additional remote process increases the probability of a cache miss when the sender starts communication. Strategies such as this one, that depend on a single computation at startup, are effective on the SCC due to the limited performance of the P54C cores in relation to the bandwidth available on the on chip mesh and the DDR3 memory (i.e. the architecture is compute bound). Other strategies were evaluated as part of exploratory research, which proved detrimental to overall performance due to their added computational overhead. For the specific 48 process case, the switch to DDR3 is done at 5.6KB (the effectiveness of this value can be observed in the next section).

4 Performance Results

To explore the performance characteristics of RCKMPI, tests with the three different channels are done: MPB, shared memory and dual buffer type operation. Benchmarks are done on a Rocky Lake SCC system running at 800MHz for the cores, 800MHz for the routers and 800MHz for the DDR3 RAM. An experimental 2.6.38.2 SCC Linux image is loaded in each core. RCKMPI was configured with the MPD process manager and all optimizations enabled. The GCC compiler version 4.5.2 was used to compile the library, the applications and the Linux kernel. All benchmarks are run with 48 processes and specified settings, unless stated.

The following figure illustrates the effectiveness of the switching criteria for a simple case of ping-pong with MPI_Send and MPI_Recv. The operations are run 1000 times per each process pair (47 pairs in this case) and then averaged. The results show that for these point to point operations, the switch is done optimally. This is also the case for MPI_Sendrecv and other point to point operations, as well as for one sided communication, with MPI jobs from 2 to 48 processes.

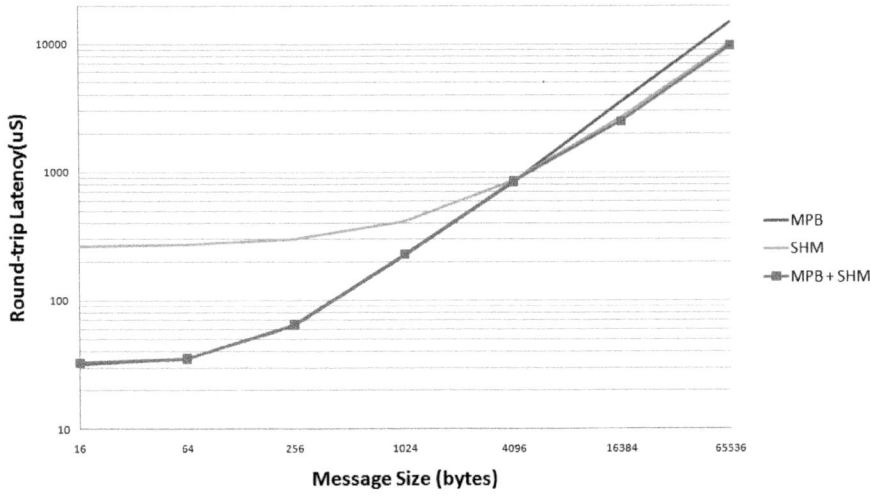

Fig. 4. Roundtrip latency for RCKMPI channels (48 process case)

The collective operations on RCKMPI are not SCC optimized. They are the default algorithms found in the 1.2.1p1 release of MPICH2, which rely on the non-blocking point to point functionality provided by the channel implementations. MPICH2 uses several algorithms for each collective operation [7,8]. These are selected based on MPI job size and the particular buffer size. No attempt was made to optimize the selection criteria for the SCC. A short description of the algorithms and how they are selected is found in the MPICH2 source code, at the beginning of each implementation file under `src/mpi/coll/`.

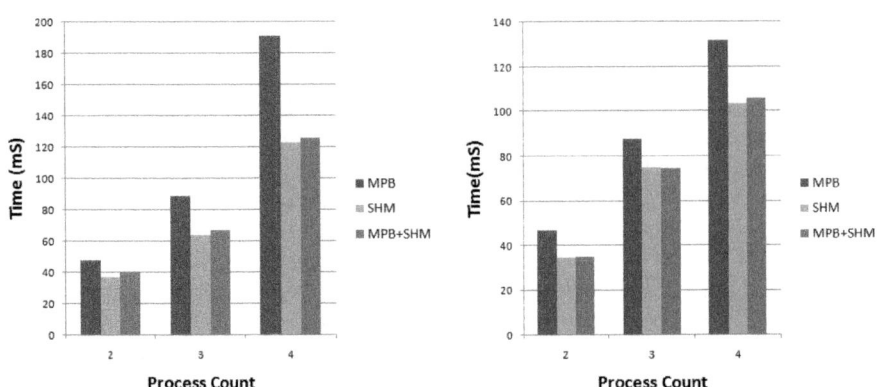

Fig. 5. MPI_Alltoall (left) and MPI_Allgather (right) performance for 256KB buffers

Figure 5 shows the results for MPI_Allgather and MPI_Alltoall with 256KB buffers, when going from 2 to 4 processes. Similarly to point to point operations, for smaller buffers, both the MPB and MPB + DDR3 channels perform nearly identical, while the DDR3 only channel is significantly slower. In general for all collective operations, as message sizes and process counts are increased, the MPB only channel falls behind both the DDR3 only and DDR3 + MPB channels. In the 256KB case, MPICH2 selects a ring based algorithm with wrap-around for MPI_Allgather, and a pair-wise exchange algorithm for MPI_Alltoall.

Barrier operations involve minimal size transfers when compared to other collectives, and benefit from the lower latency of the MPB [4]. The default implementation of MPI_Barrier uses a dissemination algorithm [9]. In figure 6, results for MPI_Barrier operations are shown for different MPI process counts.

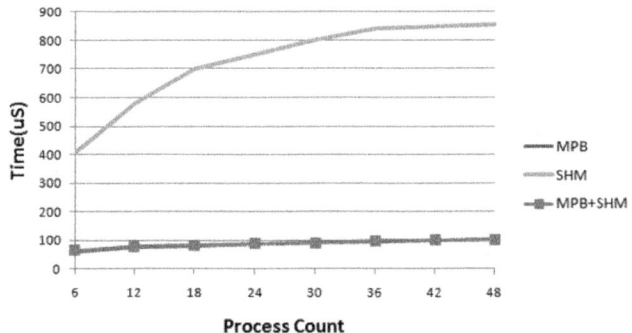

Fig. 6. MPI_Barrier scaling with the number of participating processes

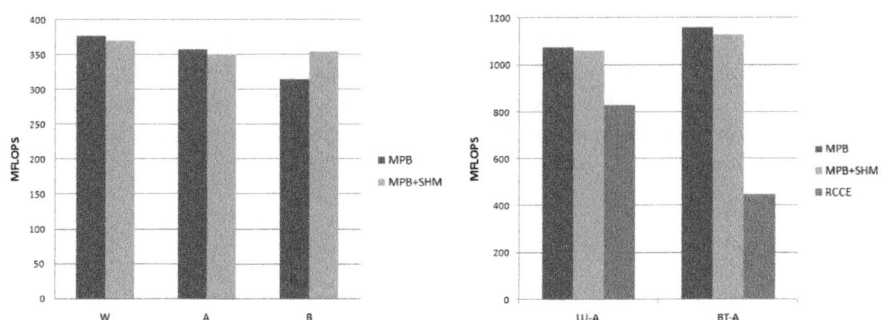

Fig. 7. NAS 3.3 benchmarks: FT size W, A and B (left), BT and LU size A (right)

Comparisons with the RCCE and derived solutions are limited to applications that have been ported to these tools. The NAS 3.3[10] BT and LU benchmarks have been ported in the case of problem size A. Figure 8 presents 2 bar charts. On the right, it is shown how the C ports running on RCCE compare to the unmodified FORTRAN originals from NASA. On the left, a performance comparison of the MPB and the

MPB + DDR3 channels is presented, when running the FT (1D fast Fourier transform) benchmark. In general for the NAS suite (BT, LU, CG, FT, MG and SP were tested), the MPB only operation slightly outperforms the MPB+DDR3 scheme until problems of size B or greater are run. Further examination of the source code for these benchmarks, shows that larger message sizes are used when the problem size is increased. For these benchmarks, the stable Linux 2.6.16 image was used, as well as a core clock of 533MHz, to ensure compatibility and stability with the latest RCCE snapshot. The BT benchmarks are run with 36 processes, while the FT and LU are run with 32, since these are the largest possible below 48.

5   Concluding Remarks

RCKMPI reached comparable performance, with respect to the specialized but non standard message passing solutions available for the SCC, through the introduction of an SCC specific MPICH2 channel. This channel did away with the TCP/IP overhead introduced by the use of a network device, and took advantage of the higher bandwidth and lower latency of the on die buffer of the architecture. Efforts were made to further exploit the possibilities found in the SCC's unique architecture. These further improvements provided significantly lower latencies for larger MPI messages.

The combination of on die buffers and shared memory present in the SCC, results in unique opportunities for message passing algorithms. It is safe to assume that on die buffers will be relatively small in relationship to last level caches and especially off chip RAM, in future architectures. The superiority in latency and bandwidth of on die buffers provide excellent message passing performance for small messages. After certain message sizes, shared memory strategies can prove more effective, due to the reduced protocol overhead and higher probabilities of data not being present in the caches. For optimal performance in MPI applications, effective use of these resources is necessary, and involves run time decisions depending on the kind of operations and the amount of bytes to transfer. In this work, we showed that a strategy based on remaining bytes to transfer of in transit packets was effective for the SCC's architecture. Performance results show the benefit of on chip and addressable SRAM, in combination with traditional external memory, particularly for improving message passing performance.

References

1. Mattson, T.G., Van der Wijngaart, R.F., Riepen, M., et al.: The 48-core SCC processor: the programmer's view. In: Proceedings of the 2010 ACM/IEEE Conference on Supercomputing , SC 2010, New Orleans, Louisiana (November 2010)
2. Clauss, C., Lankes, S., Galowicz, J. Bemmerl, T.: iRCCE: A Non-blocking Communication Extension to the RCCE Communication Library for the Intel Single-Chip Cloud Computer, Chair for Operating Systems, RWTH Aachen University, December 17 (2010)
3. van der Wijngaart, R.F., Mattson, T.G., Haas, W.: Light-weight Communications on Intel's Single-Chip Cloud Computer Processor

4. Howard, J., Dighe, S., Hoskote, Y., et al.: A 48-Core IA-32 Message-Passing Processor with DVFS in 45nm CMOS. In: Proceedings of the International Solid-State Circuits Conference (February 2010)
5. Buntinas, D., Mercier, G., Gropp, W.: Implementation and Shared-Memory Evaluation of MPICH2 over the Nemesis Communication Subsystem. Mathematics and Computer Science Division, Argonne National Laboratory
6. Argonne National Laboratory: MPICH2, http://www.mcs.anl.gov/mpi/mpich2
7. Thakur, R., Rabenseifner, R., Gropp, W.: Optimization of Collective Communication Operations in MPICH. International Journal of High Performance Computing Applications (Spring 2005)
8. Thakur, R., Gropp, W.D.: Improving the performance of collective operations in MPICH. In: Dongarra, J., Laforenza, D., Orlando, S. (eds.) EuroPVM/MPI 2003. LNCS, vol. 2840, pp. 257–267. Springer, Heidelberg (2003)
9. Hensgen, D., Finkel, R., Manber, U.: Two algorithms for barrier synchronization. International Journal of Parallel Programming (1988)
10. NASA Advanced Supercomputing Division Parallel Benchmarks, http://www.nas.nasa.gov/Resources/Software/npb.html

Hybrid OpenMP-MPI Turbulent Boundary Layer Code Over 32k Cores

Juan Sillero[1,*], Guillem Borrell[1], Javier Jiménez[1], and Robert D. Moser[2]

[1] School of Aeronautics, Universidad Politécnica de Madrid, 280040 Madrid, Spain
[2] Department of Mechanical Engineering and Institute for Computational
Engineering and Sciences, University of Texas at Austin, Austin, TX 78735, USA
sillero@torroja.dmt.upm.es

Abstract. A hybrid OpenMP-MPI code has been developed and optimized for Blue Gene/P in order to perform a direct numerical simulation of a zero-pressure-gradient turbulent boundary layer at high Reynolds numbers. OpenMP is becoming the standard application programming interface for shared memory platforms, offering simplicity and portability. For architectures with limiting memory as Blue Gene/P, the use of OpenMP is especially well suited. MPI communications overhead are also improved due to the decreasing number of processes involved. Two boundary layers are simultaneously run due to physical considerations, represented by two different MPI groups. Different node mappings layouts have been investigated reducing communication times in a factor of two. The present hybrid code shows approximately linear weak scaling up to 32k cores.

Keywords: OpenMP, MPI, data locality, blocking, node topology.

1 Introduction

Modern parallel programming paradigms are now often used in clusters, combining Message Passing Interface (MPI) paradigm [2] for across the nodes with Open Multi-Processing (OpenMP) [1] within the nodes, known as hybrid OpenMP-MPI. The use of a hybrid methodology has some important advantages with respect to the traditional use of MPI: it is easy to implement through the use of directives, has low latency, high bandwidth, fine granularity, implicit communications versus explicit communications at node level, etc.

Previous TBL codes by our group were developed using MPI [6]. This choice was justified because of the computer architecture and the relatively low number of cores used. Nevertheless, using tens of thousands of cores with only MPI may degrade the code scalability and thus, its performance. This is one of the reasons to modify the original TBL code to a new hybrid OpenMP-MPI. Despite that, the main reason to port the code is the available memory per core. In order to simulate smooth $Re_\theta \approx 6650$ and rough $Re_\theta \approx 4200$ TBLs, allocation

---

* Corresponding author.

Y. Cotronis et al. (Eds.): EuroMPI 2011, LNCS 6960, pp. 218–227, 2011.

time in Intrepid at Argonne National Laboratory (USA) and Jugene at Jülich Forschungszentrum (Germany) have been granted through an INCITE award and a PRACE project respectively. Both codes are similar, and, from now on, we will just describe the smooth-wall one. The available memory per core is in both cases 512 Mb, instead of 2 GB as is the case of Mare Nostrum (MN, Barcelona). The previous $Re_\theta \approx 2000$ TBL was run on MN facility under the RES (Red Española de Supercomputación) project. With this available memory and the current TBL problem size, the use of OpenMP has naturally arisen as the simpler solution to overcome this issue. With the usage of OpenMP, some of the extra communication overhead associated with the use of MPI within the node is avoided as well. Nevertheless, other problems such as locality, false sharing, data placement [4] can arise from its usage.

2   The Numerical Code

The boundary layer is simulated in a parallelepiped over a smooth wall, spatially periodic spanwise, but with nonperiodic inflow and outflow in the streamwise direction. The code uses a relatively classical fractional-step method [7,8] to solve the incompressible Navier-Stokes equations expressed in primitive variables, using spectral expansions in the spanwise direction, and compact finite differences in the other two. A three sub-step, semi-implicit low storage Runge-Kutta scheme is used to evolve the equations in time.

For the problem here considered, both spectral methods and compact finite differences are tightly coupled operations. Our code is constructed in such way that only single data lines, along one of the coordinate directions, have to be accessed globally. However, the three directions have to be treated in every substep.

The code uses single precision in the I/O operations and communications and double precision in the differentiation and interpolation operations where the implicit part of the compact finite differences can cause loss of significance.

Compared to other highly scalable DNS/LES codes like FrontTier, Nek5000 or PHASTA, this code is specifically designed an tuned for a single purpose: to solve a zero-pressure-gradient turbulent boundary layer over a flat plate.

2.1   Computational Setup

The simulation is split in two concatenated domains with different boundary conditions as showed in figure 1. The planes π_i and π_i' are given inflow boundary conditions, and outflow boundary conditions are assigned to π_e and π_e'. The boundary conditions in π_t and π_t' impose a zero pressure gradient on the domain. Finally, the spanwise direction is considered periodic. The mission of the first boundary layer (BL_1) is to provide accurate inflow boundary conditions to the second one (BL_2). The inflow of BL_1 is obtained from its own plane π_1 that is rescaled using a method based on the one proposed by Lund, Wu and Squires[5]. The physical length of BL_1 is chosen to be long enough to let the large scales recover from an unrealistic initial condition and, once this asymptotic state has been reached, the plane π_2 is used to give BL_2 its inflow boundary condition. As a consequence, a small portion of the BL_1 simulation is thrown away.

Given that the goal of BL_1 is to allow the large scales to reach their asymptotic state and, given that the smaller scales take much shorter to reach a similar condition, BL_1 is run at a coarser resolution than BL_2. This setup permits computing a single boundary layer with significantly less computational work.

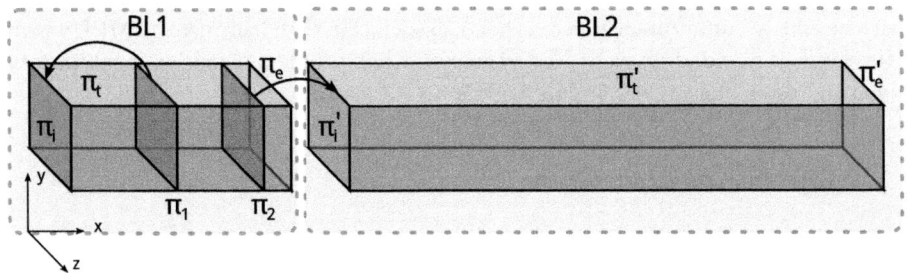

Fig. 1. Scheme of the computational domain and boundary conditions

Each of these two boundary layers is mapped to an MPI group. The first group runs the auxiliary simulation at coarse resolution and it consists of 512 nodes while the second MPI group comprises 7680 nodes and runs the main one in high resolution. The first MPI group is only about 8.5% of the total computational cost. This information is shown in table 1.

Table 1. Computational setup for the auxiliary BL_1 and main BL_2 boundary layers: N_t is the total number of degree of freedoms in giga points; Time/DoF is the amount of total CPU (core) time spent to compute a degree of freedom for every step

Case	Re_θ	Nodes	$N_x \times N_y \times N_z$	N_t (Gp)	Time/DoF
BL_1	1100-3000	512	$3585 \times 315 \times 2560$	2.89	13.98 μs
BL_2	2800-6650	7680	$15361 \times 535 \times 4096$	33.66	18.01 μs

MPI groups communicate each other only twice per sub-step by means of the MPI_COMM_WORLD communicator, while communications within each group occur via a local communicator defined at the beginning of the program. The first global operation is a SEND/RECEIVE of the π_2 plane, from BL_1 to BL_2. The second global operation is an MPI_ALL_REDUCE to set the time step for the temporal Runge-Kutta integrator, thus synchronizing both groups. The work done by each group must be balanced since each MPI group must wait for the other one in global operations, otherwise one group will slow down the second one that must remain idle during that time. The worst case scenario is when the auxiliary simulation slows down the main one. The time employed in communications for the auxiliary simulation has been improved using a customized node topology described in section 3.

2.2   Domain Decomposition

The parallelization distributes the simulation space over the different nodes, and to avoid global operations across nodes, it does a global transpose of the whole flow field twice every time sub-step (back and forth). The domain decomposition is sketched in figure 3 and can be classified as a *plane to pencil* domain decomposition. This strategy is motivated by the limited amount of memory in the Blue Gene/P nodes. Only transverse planes Π_{ZY} can fit in a node, and longitudinal planes Π_{XY} must be decomposed in X lines, i.e, pencils $\mathcal{P}_X$. According with the values presented in table 1, transverse planes are 25 Mb, longitudinal planes 94 Mb and pencils 120 Kbytes. Sixteen double precision buffers are need, and $3\Pi_{ZY}$ planes per node are used in the main simulation. Hence, the memory node occupation is close to 60%.

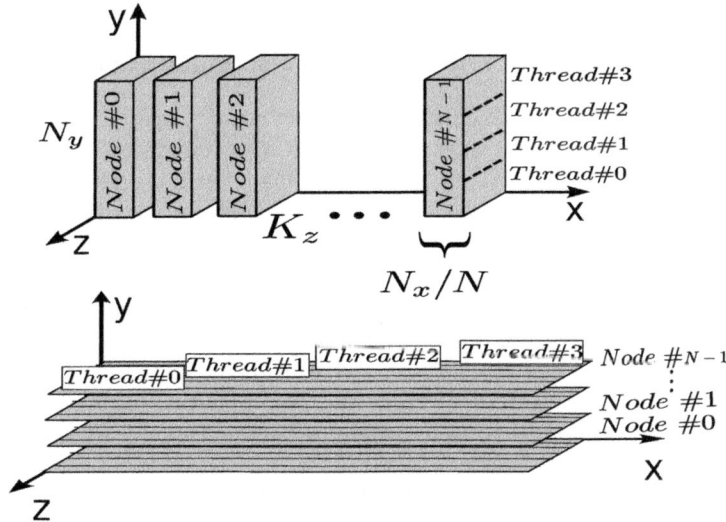

Fig. 2. Partition of the computational domain for OpenMP-MPI for $\mathcal{N}$ nodes and four threads. Top, Π_{ZY} planes; bottom, $\mathcal{P}_X$ pencil.

Each node contains $N_x/\mathcal{N}$ cross-flow planes, where N_x is the number of grid points in streamwise direction and $\mathcal{N}$ the total number of nodes. Each node is an MPI process, and OpenMP is applied within the node, splitting the sub-domain in a number of pieces equal to the available number of threads, four in Blue Gene/P.

The variables are allocated in memory as $\psi(K_z, N_y, N_x/\mathcal{N})$, where K_z is the number of modes in spanwise direction $(2/3N_z)$ and N_y the number of Y grid points. Each thread works in the same memory region of the shared variables using a first-touch data placement policy [3], maximizing data locality and diminishing cache missed [4], thus improving performance. The most common configuration that a team of threads can find is presented in figure 3, in which each thread works over a portion of N_y with static scheduling. This scheduling is defined manually through *thread_private* indexes, which maximizes memory reuse. In that way, each thread always works in the same portion of the array. Nevertheless, when loop dependencies in Y direction are found (i.e, LU decomposition) threads work over portions of K_z. For such loops, blocking techniques are used, putting the innermost loop index to the outermost part, thus maximizing data locality since strips of the arrays fit into the cache at the same time that threads can efficiently share the work load. The block size has been tuned for Blue Gene/P architecture comparing the performance of several runs.

In this configuration, operations in Y and Z are then performed. For operations in X direction global transposes are used to change variables memory layout to $\psi(N_x, K_z N_y / \mathcal{N})$. Now, each node contains a number of $K_z N_y / \mathcal{N}$ pencils. Each OpenMP thread works over a packet of $(K_z N_y / \mathcal{N}) / N_{thread}$, where N_{thread} is the total number of threads. As in the previous configuration, workload is statically distributed among threads using *thread_private* indexes.

2.3   Global Transposes and Collective Communications

Roughly 45% of the overall execution time is spent transposing the variables from planes to pencils and back, therefore it was mandatory to optimize the global transpose as much as possible. Preliminary tests revealed that the most suitable communication strategy was to use the alltoallv routine and the BG/P torus network, twice as fast than our previous custom transpose routine based on point to point communication over the same network.

The global transpose is split into three sub-steps. The first one changes the alignment of the buffer containing a variable and casts the data from double to single precision to reduce the amount of data to be communicated. If more than one Π_{ZY} plane is stored in every node then, the buffer comprises the portion of contiguous data belonging to that node in order to keep message sizes as big as possible.

The second sub-step is a call to the MPI_ALLTOALLV routine. It was decided not to use MPI derived types because the transpose operations that change the data alignment and the double to float casting are parallelized with OpenMP.

The third and last sub-step transpose the resulting buffer aligning the data $\mathcal{P}_X$-wise. This last transpose has been optimized using a blocking strategy because the array to be transposed has many times more rows than columns. The whole array is split into smaller and squarer arrays that are transposed separately. The aspect ratio of those smaller arrays is optimized for cache performance using collected data from a series of tests. Finally the data is cast to double precision again.

The procedure to transpose from $\mathcal{P}_X$ pencils to Π_{ZY} planes is similar and is split in three sub-steps too.

3   Blue Gene/P Mapping

Mapping virtual processes onto physical processors is one of the essential issues in parallel computing, being a field of intense study in the last decade. Proper mapping is critical to achieve sustainable and scalable performance in modern supercomputing systems.

Blue Gene/P has a torus network topology, except for allocations smaller than 512 nodes, in which the torus degenerates to a mesh. Therefore, each node is connected to six nodes by a direct link. The location of a node within the torus can be described by three coordinates $[X, Y, Z]$.

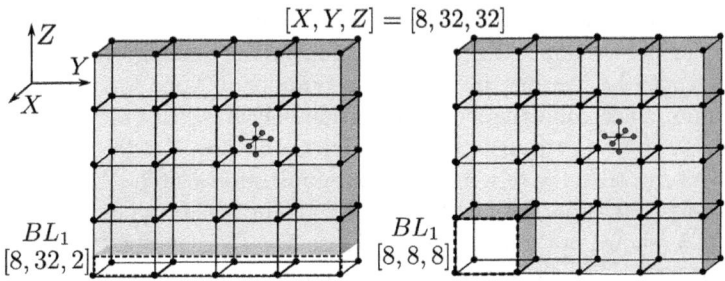

Fig. 3. Predefined (left) and custom (right) node mapping for a 8192 node partition in a $[8, 32, 32]$ topology. The predefined mapping assigns to BL_1 the nodes in a $[8, 32, 2]$ sub-domain. Custom mapping assigns the nodes to a $[8, 8, 8]$ sub-domain. BL_2 is mapped to the rest of the domain till complete the partition.

Different physical layouts of MPI tasks onto physical processors are predefined depending of the number of nodes to be allocated. The predefined mapping for a 512 node partition is a $[8, 8, 8]$ topology, while for 8192 nodes it is $[8, 32, 32]$ as it is shown in figure 3. Users can specify their desired node topology by using the environment variable BG_MAPPING and specifying the topology in a text file.

Changing the node topology completely changes the graph embedding problem and the path in which the MPI message travels. This can increase or decrease the number of hops needed to connect one node to another, and as a result, alter the communication time to send a message. Fine tuning for specific problems can considerably improve the time spent in communications. Table 2 shows different mappings that have been evaluated for our specific problem size. The custom mapping reduces the communication time for BL_1 by a factor of two. The work load for BL_1 is projected using this new communication time while the load for BL_2 is fixed. Balance is achieved minimising the time in which BL_1 or BL_2 are idle in the global communications.

The choice of a user-defined mapping is motivated due to the particular distribution of nodes and MPI groups. The first boundary layer BL_1 runs in 512 MPI processes mapped onto the first 512 nodes, while BL_2 runs in 7680 MPI processes mapped on the nodes ranging form 513 to 8192. Note that at

Table 2. Time spent in communication during global transposes. Different node topologies are presented for 10 time steps and for each boundary layer. Times are given in seconds.

Topology	Nodes	Comm BL_1	Comm BL_2
Predefined $[8, 8, 8]$	512	27.77	—
Custom $[32, 32, 8]$	8192	79.59	86.09
Predefined $[32, 32, 8]$	8192	160.22	85.44

the moment the communicator is split such that $Comm_{BL_1} \cup Comm_{BL_2} = MPI\_COMM\_WORLD$, neither $Comm_{BL_1}$ nor $Comm_{BL_2}$ can be on a 3D torus network. The communications will drop down to a 2D mesh with sub-optimal performance. Therefore, the optimum topology for our particular problem would be the one in which the number of hops for each MPI group is minimum, since collective communications occur locally for each group. For a single 512 node partitions the optimum is the use of a $[8, 8, 8]$ topology, in which messages travel within a single communication switch. We have found the optimum mapping for BL_1 to be a $[8, 8, 8]$ sub-domain within the predefined $[8, 32, 32]$, as shown in the right side of figure 3. BL_2 is mapped to the remaining nodes using the predefined topology and no other mappings have been further investigated. Although a $[8, 8, 8]$ topology is used for BL_1 by analogy with the single 512 node partition, communication time is nevertheless greater. This is due to the sub-optimal performance of using a 2D mesh instead of a 3D torus network, as already discussed. Ultimately, the reason can be found in the new hardware connection, since the 512 nodes and 8192 nodes of the 3-Dimensional torus network are physically connected in a different way. This leads to the increase in the number of hops for BL_1 collective communications, since messages cannot travel within a single communication switch anymore.

The methodology to optimize communications for another size partitions would be similar to the one just described: mapping virtual processes to nodes that are physically as close as possible so the number of hops is minimized.

4   Scalability Results in Blue Gene/P

4.1   OpenMP Scalability

It is important to state that the reason to mix concurrency and parallelism was not driven by the need for more performance but because the small memory capacity of the Blue Gene/P node, which does not allow a physically-significant block of data to be allocated to each core.

Some tests were run in a 512 node configuration after porting the code to OpenMP. The results are shown in table 3. These samples suggest that almost no penalty is paid when the computations are parallelized with OpenMP. In addition, the problem size per node and the MPI message size can be increased by a factor of four while using all the node's resources.

Table 3. OpenMP scalability test performed on 512 nodes. Two efficiencies (η) given: one based on the computation time (*Comp. T*) and one based on the total time (*Total T.*). Times are given in seconds.

$N_{threads}$	Comp. T	η	Total T.	η
1	60.820	1	70.528	1
2	30.895	0.984	38.951	0.905
4	16.470	0.923	24.438	0.721

4.2   MPI Scalability

Extensive data about MPI scalability was collected during the test runs in a BG/P system. The most relevant cases are listed in the table 4.

Table 4. Data collected from the profiled test cases. Time/DoF is the amount of total CPU (core) time spent to compute a degree of freedom for every step; N_t is the size in GiB of a buffer of size $N_x \times N_y \times N_z$; Comm, Transp and Comp are the percentage of the communication, transpose and computation time respect to the total.

Nodes	$N_x \times N_y \times N_z$	N_t	Time/DoF	Comm.	Transp.	Comp.	Symbol
512	$1297 \times 331 \times 768$	0.33	10.6 μs	17.9%	8.29%	73.8%	►
1024	$3457 \times 646 \times 1536$	3.43	17.6 μs	44.7%	7.52%	47.8%	◄
2048	$6145 \times 646 \times 1536$	6.10	17.4 μs	46.0%	5.31%	48.8%	▲
4096	$8193 \times 711 \times 1536$	8.94	17.6 μs	44.6%	5.23%	53.2%	▼
8192	$8193 \times 711 \times 2048$	11.93	19.4 μs	37.4%	8.30%	57.6%	◆
8192	$16385 \times 801 \times 4608$	60.47	19.3 μs	39.7%	8.41%	51.9%	■

All the simulations run show a linear weak scaling up to 8192 nodes (32768 cores). The same code is expected to scale further without modifications, although at this time, higher node partitions have been not tested.

Figure 4(b) shows that the communications time is typically 40% of the total run time, and that both computation and communications are scaling as expected. The global transpose implementation shows an excellent scalability in all the test cases as shown in figure 4(a). It is important to mention that in the BG/P supercomputer architecture, the linear scaling is kept even when the estimated message size is about 1 kB in size. All our previous implementations of the global transpose broke the scalability near the 3 kB estimated message size limit.

5   Parallel I/O

Intermediate stages of the simulation in the form of flow fields (velocities and pressure) are an important result and are saved even more often than what checkpointing would require. Another mandatory feature to maintain the scalability with a high node count is the support for parallel collective I/O operations

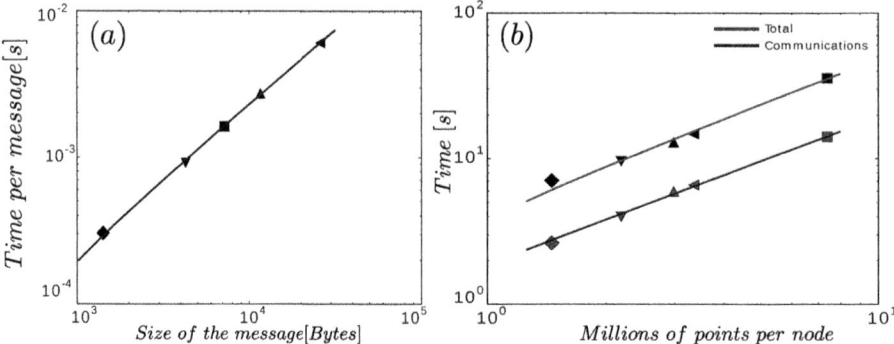

Fig. 4. Latency analysis (a) and scalability of the total and communication time for different test cases (b). Solid lines are linear regressions computed before taking logarithms of both axis.

when a parallel file system is available. A handful of alternatives have been tested, mainly upon GPFS, like raw posix calls enforcing the file system block size, sionlib (developed at JFZ) and parallel hdf5.

Hdf5 is a more convenient choice for storing and distributing scientific data than the alternatives tested because, despite having better performance, they require to translate the resulting files to a more useful format. Unfortunately sufficient performance could not be acheived without tuning the I/O process. Hdf5 performance depends on the availability of a cache in the file system. The observed behaviour in the BG/P systems was that writing was one, and sometimes two, orders of magnitude slower than reading because in the GPFS used the write cache was turned off. To overcome this issue, when the MPI I/O driver for hdf5 is used, the sieve buffer size parameter of hdf5 can be set to the file system block size. As a result, the write bandwidth for 8192 nodes was increased up to 16GiB/s, similar to the read bandwidth 22GiB/s and closer to the estimated maximum.

6   Conclusions

A hybrid OpenMP-MPI code has been developed from its original MPI version to perform direct numerical simulations of smooth and rough turbulent boundary layers at high Reynolds numbers. The code has been tested in a Blue Gene/P computer using up to 8192 nodes for MPI processes, and four threads per process for OpenMP, showing good scalability for both MPI and OpenMP. Two different domain decompositions are used to perform global operations in each of the 3-dimensional directions, employing collective communications to perform global transposes. Customized mappings of processes onto physical processors has been used for each of the two MPI groups, representing the auxiliary low resolution and the main high resolution simulation, speeding communications up by a factor of two.

Acknowledgments. This research used resources of the Argonne Leadership Computing Facility at Argonne National Laboratory, which is supported by the Office of Science of the U.S. Department of Energy under contract DE-AC02-06CH11357. Also funded by the Prace initiative. The work at the UPM was funded by CICYT under grant TRA2009-11498 and Consolider CSD2007-00050. J.A. Sillero was supported by an FPU fellowship from the UPM.

References

1. OpenMP Architecture Review Board OpenMP Specifications, http://www.openmp.org
2. Gropp, W., Lusk, E., Skjellum, A.: Using MPI: portable parallel programming with the message-passing interface, 2nd edn. MIT Press, Cambridge (1999)
3. Marchetti, M., Kontothanassis, L., Bianchini, R., Scott, M.: Using Simple Page Placement Policies to Reduce the Cost of Cache Fills in Coherent Shared-Memory Systems. In: Proceeding of the 9th International Parallel Processing Symposium, Santa Barbara, CA, pp. 480–485 (April 1995)
4. Nikolopoulos, D.S., Papatheodorou, t.S., Polychronopoulos, c.D., Labarta, J., Ayguade, E.: Is Data Distribution Necessary in OpenMP? IEEE (2000)
5. Lund, T.S., Wu, X., Squires, K.D.: Generation or turbulent inflow data for spatially-developing boundary layer simulations. J. Comput. Phys. 140, 233–258 (1998)
6. Simens, M.P., Jiménez, J., Hoyas, S., Mizuno, Y.: A high-resolution code for turbulent boundary layers. J. Comput. Phys. 228, 4218–4231 (2009)
7. Kim, J., Moin, P.: Application of a fractional-step method to incompressible Navier-Stokes equations. J. Computat. Phys. 59, 308–323 (1985)
8. Perot, J.B.: An analysis of the fractional step method. J. Computat. Phys. 108, 51–58 (1993)

CAF versus MPI - Applicability of Coarray Fortran to a Flow Solver

Manuel Hasert[1], Harald Klimach[1], and Sabine Roller[1,2]

[1] German Research School for Simulation Sciences GmbH, 52062 Aachen, Germany
[2] RWTH Aachen University, 52062 Aachen, Germany

Abstract. We investigate how to use coarrays in Fortran (CAF) for parallelizing a flow solver and the capabilities of current compilers with coarray support. Usability and performance of CAF in mesh-based applications is examined and compared to traditional MPI strategies. We analyze the influence of the memory layout, the usage of communication buffers against direct access to the data used in the computation and different methods of the communication itself. Our objective is to provide insights on how common communication patterns have to be formulated when using coarrays.

Keywords: PGAS, Coarray Fortran, MPI, Performance Comparison.

1 Introduction

Attempts to exploit parallelism in computing devices automatically have always been made, and it was successfully done by compilers in a restricted form such as vector operations. For more general parallelization concepts with multiple instructions on multiple data (MIMD), the automation was less successful and programmers had to support the compiler by directives such as in OpenMP. The Message Passing Interface (MPI) offers a rich set of functionality for MIMD applications on distributed systems and high-level parallelization is supplied by APIs from libraries. The parallelization however, has to be elaborated in detail by the programmer. Recently, an increasing effort in language-inherent parallelism is made to leave parallel implementation details to the compiler. A concept developed in detail is the partitioned global address space (PGAS), which was brought into the Fortran 2008 standard with the notion of coarrays. Parallel features are turned into intrinsic language properties and allow a high-level formulation of parallelism in the language itself [10]. Coarrays minimally extend Fortran to allow the creation of parallel programs with minor modifications to the sequential code. The optimistic goal is to obtain a language which inherits parallelism and allows the compiler to concurrently consider serial aspects and communication for code optimization.

In CAF, shared data objects are indicated by an additional index in square brackets, for which the remote location in terms of the process number of the shared variable is defined. Contrary to MPI, there are no collective operations defined in the current standard. These have been shown to constitute a large part

Y. Cotronis et al. (Eds.): EuroMPI 2011, LNCS 6960, pp. 228–236, 2011.

of nowadays' total communication on supercomputers [9]. thus posing a severe limitation to a pure coarray parallelization, which is a major point of criticism by Mellor-Crummey et al. [8]. Only few publications actually give advice on how to use coarrays in order to obtain a fast and scalable parallel code. Ashby & Reid [1] ported an MPI flow solver to coarrays and Barrett [2] uses a finite difference scheme to assess the performance of several coarray implementations.

The goal of this paper is to compare several parallelization implementations with coarrays and MPI. We compare these approaches in terms of performance, flexibility and ease of programming. Speedup studies on two machines with a different network interface are performed. The work concentrates on the implementation provided by Cray.

2   Numerical Method

We use a lattice Boltzmann method (LBM) fluid solver as a testbed. This explicit numerical scheme uses direct neighbor stencils and a homogeneous, Cartesian grid. The numerical algorithm consists of streaming and collision performed at each time step (Listing 1.1). The cell-local collision mimics particle interactions and streaming represents the free motion of particles, consisting of pure memory copy operations from nearest neighbor cells. Neighbors are directly accessed on a cubic grid, which is subdivided into rectangular sub-blocks for parallel execution.

```
do k=1,nz; do j=1,ny; do i=1,nx
  ftmp(:)=fIn(:,i-cx(:),j-cy(:),k-cz(:)) ! advection from offsets cx, cy, cz
  ... ! Double buffering: Read values from fIn, work on ftmp and write to fOut
  fOut(:) = ftmp(:) - (1-omega)*(ftmp(:) -feq(:)) ! collide
enddo; enddo; enddo
```

Listing 1.1. Serial stream collide routine

2.1   Alignment in Memory

The Cartesian grid structure naturally maps to a four-dimensional array, where the indices $\{i, j, k\}$ represent the fluid cells' spatial coordinates. Each cell holds $n_{nod}=19$ density values. It is represented by the index l and its position in the array can be chosen. Communication then involves strided data access, where the strides depend on the direction and memory layout. There are two main arrays fIn and fOut which by turns hold the state of the current time step. These arrays are of the size $\{n_{nod}, n_x, n_y, n_z\}$, depending on the position of l. In Fortran the first index is aligned successively in memory yielding a stride one access. With the density-first *lijk*, the smallest data chunk for communication has at least n_{nod} consecutive memory entries. The smallest memory chunks of n_{nod} entries for communication occur in the x-direction. Communication in y-direction involves chunks of $n_{nod} \cdot n_x$ and in z-direction $n_{nod} \cdot n_x \cdot n_y$. When the density is saved last *ijkl*, the x-direction again involves the smallest data chunks, but only of a single memory entry with strides of $n_{stride} = n_x \cdot n_y \cdot n_z$.

2.2   Traditional Parallelization Approach

A time-dependent mesh-based flow solver is usually parallelized following a SPMD (single program multiple data) approach. All p processes execute the same program but work on different data. In LBM, each fluid cell induces the same computing effort per time step. For an ideal load-balancing, the work is equally distributed among the processes and a simple domain decomposition can be performed, splitting the regular Cartesian mesh into p equal sub-domains. An update of a given cell requires the information from the neighboring cells and itself from the previous iteration. At the border between sub-domains it is then necessary to exchange data in each iteration. A common approach is to use halo cells, which are not updated by the scheme itself but provide valid data from the remote processes. This allows the actual update procedure to act just like in the serial algorithm. With MPI, data is exchanged before the update of the local cells at each time step, usually with a non-blocking point-to-point exchange.

2.3   Strategy Following the Coarray Concept

Using coarrays, data can be directly accessed in the neighbor's memory without using halos. Coarray data objects must have the same size and shape on each process. The cells on remote images can then be accessed from within the algorithm like local ones, but with the additional process address (nbp) appended. Synchronization is required between time steps to ensure data consistency across all processes, but no further communication statements are necessary. We present different approaches to coarray implementations.

In the *Naive Coarray Approach* every streaming operation, i.e. copy from neighbor to the local cell, is done by a coarray access, even for values which are found locally. This requires either the calculation of the address for the requested value, or a lookup table for this information. Both approaches result in additional run time costs and the calculation of the neighbor process number and the position there obscure the kernel code.

```
do k=1,nz; do j=1,ny; do i=1,nx
  ! streaming step (get values from neighbors)
  do l=1,nnod        ! loop over densities in each cell
   xpos = mod(crd(1)*bnd(1)+i-cx(1,1)-1,bnd(1))+1
   xp(1)= (crd(1)*bnd(1)+i-cx(1,1)-1)/bnd(1)+1
   ... ! analoguous for the other directions
   if(xp(1) .lt. 1) then ... ! correct physical boundaries
   nbp=image_index( caf_cart_comm,xp(1:3) )      ! get image num
   ftmp( l)=fIn( l,xpos,ypos,zpos )[nbp]  ! coarray get
  enddo
  ... ! collision
enddo; enddo; enddo
```

Listing 1.2. Naive streaming step with coarrays

The copy itself is easy to implement, but the logic for getting the remote data address requires quite some effort (Listing 1.2). Significant overhead is generated by the coarray access itself and repetitive address calculations. If the position of each neighbor is determined in advance and saved, memory demand and, more important, memory access increases. As the LBM is a memory bandwidth-bound algorithm, this puts even more pressure on the memory interface.

With the *Segmented Coarray Approach*, the inner nodes of each partition are treated as in the serial version. Coarray access is only used where necessary, i.e. on the interface cells. Separate subroutines are defined for the coarray and non-coarray streaming access, which are called according to the current position in the fluid domain (Listing 1.3). With this approach, there are two kernels, one with additional coarray access as above and additionally the loop for determining the kernel. This raises the required lines of code again.

```
call stream_collide_caf(fOut,fIn,1,nx,1,ny,1,1)
do k=2,nz-1
   call stream_collide_caf(fOut,fIn,1,nx,1,1,k,k)
   do j=2,ny-1
      call stream_collide_caf(fOut,fIn,1,1,j,j,k,k)
      call stream_collide_sub(fOut,fIn,j,k)
      call stream_collide_caf(fOut,fIn,nx,nx,j,j,k,k)
   end do
   call stream_collide_caf(fOut,fIn,1,nx,ny,ny,k,k)
end do
call stream_collide_caf(fOut,fIn,1,nx,1,ny,nz,nz)
```

Listing 1.3. Segmented stream collide with coarrays

3 Tested Communication Schemes

3.1 Data Structures

Message passing based parallelization requires data structures for collecting, sending and receiving the data. MPI types or regular arrays can be used in MPI, whereas with CAF, regular arrays or derived types with arrays can be employed. With *Regular Global Arrays* and the same-size restriction for coarrays, separate data objects for each neighbor buffer are required. This applies both for send and receive, which increases implementation complexity. The usage of *Derived Types* provides the programmer with flexibility, as the arrays inside the coarray derived types do not have to be of the same size. Before each communication or alternatively at every array (de)allocation, information about the array size of every globally accessible data object has to be made visible to all processes.

```
! Regular global arrays as coarrays
real,dimension(:,:,:,:)[:],allocatable :: caf_snd1,..
! Derived types
type caf_dt    ! Coarray derived type with regular array
   real,dimension(:,:,:,:),allocatable :: send
end type caf_dt
type(caf_dt) :: buffer[*] !< Coarrray definition
type reg_dt    ! Regular derived type with coarray inside
   real,dimension(:,:,:,:)[:],allocatable :: snd1,.. snd19
end type reg_dt
type(reg_dt) :: buffer    !< Regular array definition
```

Listing 1.4. Derived type and regular global coarrays

Table 1. Machine setup

	CPU	Rev	Cores	GHz	L2(KB)	L3(MB)	Memory	ASIC	CCE	MPI
XT5m	Shanghai	23C2	4	2,4	512	6	16GB	Seastar2	7.2.4	MPT 5.0.0
XE6	MagnyCours	6128	8	2,0	512	12	32GB	Gemini	7.3.0	MPT 5.1.1

3.2 Buffered and Direct Communication

The usage of *explicit buffers*, i.e. using halos, requires separate send/receive buffers with potentially different sizes. The communication is done in an *explicit*, dedicated routine, and before its start, the required data is collected and placed into the halo buffer, from where it is put back into the main array after the communication. One-sided communication models allow *direct remote memory access* to locations, which are not available to the local process, but where access is routed through the network. This allows the construction of either *explicit* message passing of buffers or access to remote data within the kernel itself, here referred to as *implicit* access.

4   Experimental Results

We performed investigations on Cray XT and XE systems (see Table 1), as they are among the few supporting PGAS both by hardware and compiler. These architectures mainly differ in the application-specific integrated circuits (ASIC) [4], which connect the processors to the system network and offloads communication functions from the processor. The Cray XT5m nodes use SeaStar ASICs, which contain among other a direct memory access engine to move data on the local memory, a router connecting with the system network and a remote access memory engine [3]. The XE6 nodes are equipped with the Gemini ASIC [5], which supports a global address space and is optimized for efficient one-sided point-to-point communication with a high throughput for small messages. We used the Cray Fortran compiler from the Cray Compiling Environment (CCE). We first evaluate the single core performance for various domain sizes and memory layouts, from which we choose a suited method for parallel scaling. Coarray and MPI are then compared on both machines. We perform a three-dimensional domain decomposition, for $p > 8$ with an equal amount of subdivisions in all three directions. For $p \leq 8$, the domain is split in z-direction only.

4.1 Influence of the Memory Layout in Serial and Parallel

The serial performance is measured with the physics-independent quantity million lattice updates per second (MLUPs), as a function of the total fluid cells n_{tot}. Performance studies of the LBM by Donath [6] have revealed a significant influence of memory layout and domain size. With *density-first* lijk, the performance decreases with increasing domain size. In Fig. 1 (left), the cache hierarchy is clearly visible in terms of different MLUPs levels, especially for *lijk*. For *density-later* iljk,ijlk,ijkl, the performance is relatively constant with cache-thrashing occurring for *ijkl*.

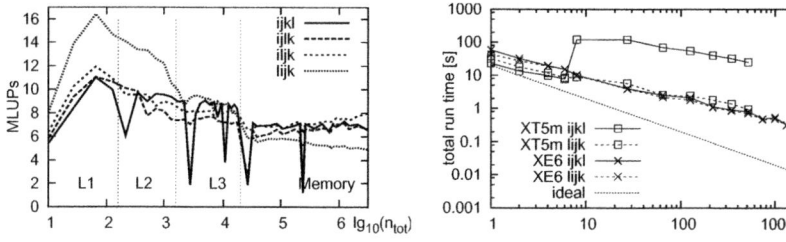

Fig. 1. Impact of memory layout in serial (XT5m) and parallel (direct CAF)

The memory layout not only plays a large role for serial execution, but also when data for communication is collected. Depending on data alignment in memory, the compiler has to collect data with varying strides resulting in a large time discrepancy of communication in different directions.

In figure 1 on the right, all layouts scale linearly for a one-dimensional domain decomposition for $p \leq 8$ and a domain size of $n_{tot} = 200^3$. Invoking more directions leads to a strong degradation of performance on the XT5m for the *ijkl* layout, probably due to the heavily fragmented memory, that needs to be communicated. The smallest chunks in *lijk* remains 19, as all densities are needed for communication and stored consecutively in memory. On the XE6 with the new programming environment, this seems to be resolved.

4.2 Derived Type and Regular Coarray Buffers

On the XT5m there is an obvious difference between the two implementations. The regular array implementation scales in a nearly linear fashion. The derived type version even increases linearly in the run time when using more processes inside a single node. When using the network, it scales nearly linear for $p \geq 16$. This issue also seems to be resolved on the new XE6 architecture. Within a sin-

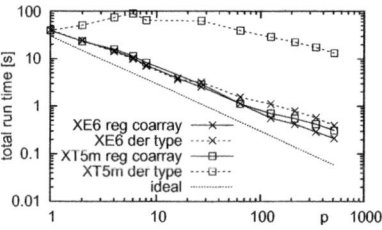

Fig. 2. Derived type coarrays

gle node, there are virtually no differences between the two variants. However, the derived type variant seems to scale a little worse beyond a single node.

4.3 MPI Compared to Coarray Communications

Here we compare various coarray schemes to regular non-blocking, buffered MPI communication. We start with a typical explicit MPI scheme and work our way to an implicit communication scheme with coarrays. The MPI and MPI-style CAF schemes handle the communication in a separate routine. The implicit coarray schemes perform the access to remote data during the streaming step. We use the *lijk* memory layout, where all values of one cell are saved contiguously, which results in a minimal data pack of 19 values. Strong and weak scaling experiments are performed for the following communication schemes:

1. Explicit MPI: buffered isend-irecv
2. Explicit CAF buffered: same as MPI but with coarray GET to fill buffers
3. Explicit CAF direct access: no buffers but direct remote data access
4. Implicit CAF segmented loops: coarray access on border nodes only
5. Implicit CAF naive: coarray access on all nodes

Strong scaling. A fluid domain with a total of 200^3 grid cells is employed for all process counts. In Fig. 3 the total execution time of the main loop is shown as a function of the number of processes. Increasing the number of processes decreases domain sizes per process, by which the execution time decreases. Ideal

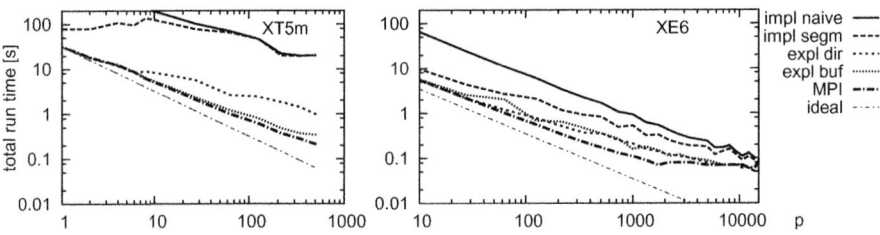

Fig. 3. Strong scaling comparison of MPI and CAF with *lijk* layout and $n_{tot} = 200^3$

scaling is plotted for comparison. The MPI implementation is fastest and scales close to ideal on both machines, but scaling scaling stalls for $p > 2000$. The coarray implementations show, that there was a huge progress made from the XT5m to the XE6. Implicit coarray access schemes shows a much slower serial runtime. The naive coarray implementation is slower than explicit ones even by a factor of 30, due to coarray accesses to local data, although it scales perfectly on the XE6. On the XE6, for $p > 10000$, all schemes tend to converge against the naive implementation, which is expected, as this approach pretends all neighbors to be remote, essentially resulting in 1 cell partitions. However this results in a very low parallel efficiency with respect to the fastest serial implementation. Due to an unexplainable loss of scaling in the MPI implementation beyond 2000 processes, coarrays mimicking the MPI communication pattern get even slightly faster in this range.

Weak scaling. The increasingly high parallelism due to power and clock frequency limitation [7] combined with limited memory resources lead to extremely distributed systems. Small computational domains of $n = 9^3$ fluid cells, which fit completely in cache are used to anticipate this trend in our analysis (Fig. 4). With such domain sizes, latency effects prevail. The MPI parallelization scales similar on both machines, and yields the best run time among the tested schemes. Both explicit buffers in coarray scale nearly perfect for $p > 64$. Whereas implicit coarray addressing within the algorithm clearly looses in all respects.

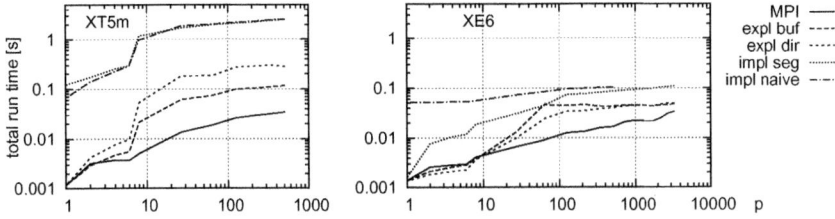

Fig. 4. Weak scaling in three directions with $n = 9^3$ cells per process (latency area)

5   Conclusion

We presented different approaches on how to parallelize a fluid solver using coarrays in Fortran. We compared the ease of programming and performance to traditional parallelization schemes with MPI. The code complexity for simple and slow CAF implementations is low, but quickly increases with the constraint of a high performance. We showed that the achievable performance of applications using coarrays depends on the definition of data structures. The analysis indicates, that it might get beneficial to use coarrays for very large numbers of processes, however on the systems available today, MPI communication provides highest performance and needs to be mimicked in coarray implementations.

Acknowledgments. We thank the DEISA Consortium (www.deisa.eu), funded through the EU FP7 project RI-222919, for support within the DEISA Extreme Computing Initiative. We also want to express our grateful thanks to M. Schliephake from the PDC KTH Stockholm for providing us the scaling runs on the XE6 system Lindgren and U. Küster from HLRS Stuttgart for insightful discussions.

References

1. Ashby, J.V., Reid, J.K.: Migrating a Scientific Application from MPI to Coarrays. Technical report, Comp. Sc. Eng. Dept. STFC Rutherford Appleton Laboratory (2008)
2. Barrett, R.: Co-Array Fortran Experiences with Finite Differencing Methods. Technical Report (2006)
3. Brightwell, R., Pedretti, K., Underwood, K.D.: Initial Performance Evaluation of the Cray SeaStar Interconnect. In: Proc. 13th Symp. on High Performance Interconnects (2005)
4. CRAY. Cray XT System Overview (June 2009)
5. CRAY. Using the GNI and DMAPP APIs. User Manual (2010)
6. Donath, S.: On optimized Implementations of the Lattice Boltzmann Method on contemporary High Performance Architectures. Master's thesis, Friedrich-Alexander-University, Erlangen-Nürnberg (2004)

7. Kogge, P.: Exascale computing study: Technology challenges in achieving exascale systems. Technical report, Information Processing Techniques Office and Air Force Research Lab (2008)
8. Mellor-Crummey, J., Adhianto, L., Scherer, W.: A Critique of Co-array Features in Fortran. Working Draft J3/07-007r3 (2008)
9. Rabenseifner, R.: Optimization of collective reduction operations. In: Bubak, M., van Albada, G.D., Sloot, P.M.A., Dongarra, J. (eds.) ICCS 2004. LNCS, vol. 3036, pp. 1–9. Springer, Heidelberg (2004)
10. Reid, J.: Coarrays in the next Fortran Standard (March 2009)

The Impact of Injection Bandwidth Performance on Application Scalability

Kevin T. Pedretti, Ron Brightwell, Doug Doerfler,
K. Scott Hemmert, and James H. Laros III

Sandia National Laboratories*
Albuquerque, NM, 87185, USA
{ktpedre,rbbrigh,dwdoerf,kshemme,jhlaros}@sandia.gov

Abstract. Future exascale systems are expected to have significantly reduced network bandwidth relative to computational performance than current systems. Clearly, this will impact bandwidth-intensive applications, so it is important to gain insight into the magnitude of the negative impact on performance and scalability to help identify mitigation strategies. In this paper, we show how current systems can be configured to emulate the expected imbalance of future systems. We demonstrate this approach by reducing the network injection bandwidth performance of a 160-node, 1920-core Cray XT5 system and analyze the performance and scalability of a suite of MPI benchmarks and applications.

Keywords: bandwidth configurability, benchmarking, exascale co-design.

1 Introduction

An individual compute node's network injection bandwidth to computational performance ratio is an important metric used when designing and comparing large-scale parallel computers. The ratio must be high enough to support the breadth of applications intended to run on a system, but not so high that bandwidth, which is expensive, both in terms of dollars and power, goes unused and is wasted. Determining the right design point is a significant challenge.

Historically, a rule of thumb has been that a "balanced" system should support one byte per second of network I/O injection bandwidth for each floating point operation per second (FLOPS) computed. Such a system would be generally applicable and support a broad range of applications. Unfortunately, a variety of factors has resulted in system balance ratios decreasing over time and the trend is predicted to accelerate. A recent DARPA-sponsored study [10] concluded that even with optimistic technology scaling assumptions, realizing the next milestone of exa-FLOPS capable parallel computers will require that system

---

* Sandia National Laboratories is a multi-program laboratory managed and operated by Sandia Corporation, a wholly owned subsidiary of Lockheed Martin company, for the U.S. Department of Energy's National Nuclear Security Administration under contract DE-AC04-94AL85000.

Y. Cotronis et al. (Eds.): EuroMPI 2011, LNCS 6960, pp. 237–246, 2011.

balance targets be reduced by more than an order of magnitude from what is considered acceptable today. As a point of comparison, Table 1 lists the network injection to compute balance ratios of two prior-generation systems along side the hypothetical exascale system forecast in the report.

Table 1. Comparison of system balance over time

System	ASCI Red	Jaguar-PF	Exascale Prediction
Year	1997	2009	2015
System Peak TFLOPS	3.15	2,332	1,029,901
Node Inj. Bandwidth (GB/s)	0.400	2.2	3.68
Node Peak GFLOPS	0.666	124.8	4,600
Net/Compute Balance Ratio	**0.6**	**0.018**	**0.0008**

With this hardware prediction in view, it is important to gain more insight into how this change in system balance will affect existing application software, the design of future applications, and, ultimately, provide feedback to influence the design requirements of future exascale computers as part of the co-design process [2]. The approach we are currently pursuing is to leverage the extensive configurability of existing supercomputer platforms, specifically the Cray XT and XE architectures, to emulate the expected behavior of future systems. The resulting experimental platform is then used to run important production applications with realistic input problems to observe how they respond to the change in system balance. Our initial efforts have focused on application sensitivity to network injection bandwidth, since it plays a critical role in determining how the processor and network interface are connected and the level of integration required. It is one of the earliest hardware design decisions that has arisen.

The remainder of this paper is organized as follows. Section 2 describes the relevant aspects of our test platform. Our approach is then described in Sect. 3, followed by results in Sect. 4. Section 5 discusses work that is related to our experimental approach. Finally, conclusions are given in Sect. 6.

2   Test Platform

All of our injection bandwidth experiments to date have been performed on the "XTP" Cray XT5 system at Sandia. This system consists of 160 12-core compute nodes, each with 32 GB of memory and a Cray SeaStar [4] network interface. Each node connects to its SeaStar via a point-to-point HyperTransport link, and each SeaStar then connects to the node's six nearest neighbors via proprietary Cray point-to-point network links. The overall network topology forms a 3-D torus.

The XTP system is considerably smaller than the systems we ultimately seek to target, but is enlightening nonetheless. Our work on XTP has validated the soundness of our approach and produced a set of baseline scaling results for our target applications. Moving on to larger-scale Cray XT and XE systems can leverage the XTP work directly, and is expected to be straightforward.

3   Approach

In order to reduce a compute node's network injection bandwidth, we are leveraging source-level access to the Cray XT and XE *coldstart* boostrap infrastructure, which serves the same purpose as a traditional BIOS. A separate instance of coldstart runs on each compute node, but all instances are the same and configure each of the compute nodes identically.

Very early in the power-on sequence, coldstart initializes the HyperTransport (HT) point-to-point link that connects the Opteron host processor to the SeaStar network interface. The speed of this link determines a compute node's injection bandwidth capability into the network fabric. The Opteron processor supports several HT links, but we only degrade the single link that connects to the SeaStar. All other HT links continue to operate at full speed.

By default, coldstart configures the Opteron-to-SeaStar HT link for the maximum speed supported by both the Opteron and SeaStar. We modify this negotiation process to select one of the alternative speeds listed in Table 2, which is the cross-product of the settings supported by the Opteron and the settings supported by the SeaStar. The basic configuration parameters for the HT link are width, either 8-bits or 16-bits wide, and operating frequency, 200 MHz, 400 MHz, or 800 MHz. This results in the theoretical peak uni-directional HT link bandwidths listed in the table. However, HT and SeaStar protocol overheads limit the achievable injection bandwidth to far less than the theoretical peak.

Table 2. Possible Opteron-to-SeaStar HT link bandwidth configurations

Link Frequency & Width	8-bit	16-bit
200 MHz	400 MB/s	800 MB/s
400 MHz	800 MB/s	1600 MB/s
800 MHz	1600 MB/s	3200 MB/s

3.1   Benchmarks and Applications

Our analysis includes several benchmarks and applications. We briefly describe them here.

Communication Micro-Benchmarks: We use a standard MPI ping-pong micro-benchmark to measure latency and bandwidth between two nodes over the network and an MPI streaming bandwidth benchmark to measure an entire node's injection bandwidth using all cores simultaneously. The ping-pong bandwidth benchmark is part of the Intel MPI Benchmark Suite and the streaming bandwidth test is from Ohio State University. These benchmarks are primarily used to verify that the injection bandwidth degradation performs as expected.

HPC Challenge (HPCC) Benchmarks: The HPC Challenge [12] benchmarks are commonly used to evaluate the performance of supercomputers. Although they are not production applications, they have been constructed to represent a disjoint set of application characteristics common in scientific computing. We selected four of these benchmarks: HPL, RandomAccess, PTRANS, and FFT. The High Performance Linpack (HPL) benchmark measures the floating point performance obtained when solving a dense linear system of equations. RandomAccess measures the rate of random updates to a large pool of 64-byte integer values. The RA_SANDIA_NOPT version of the algorithm was used, which results in communication between processes consisting primarily of 8 KB messages. The PTRANS benchmark performs a parallel matrix transpose and is network bandwidth intensive. FFT measures the floating-point performance obtained for a double-precision, complex, one-dimensional discrete Fourier transform (DFT).

Applications: Four applications were evaluated for their sensitivity to network injection bandwidth: CTH, Sage, xNOBEL, and Charon. These codes were selected due to their importance to the Advanced Simulation and Computing (ASC) program within the U.S. Department of Energy and because of their ability to scale to very large-scale machines. The combination of these particular applications also comprise a disjoint set of communication patterns that are representative of a large majority of other ASC applications. Realistic input problems were used in a weak scaling fashion, resulting in computational work per core being held roughly constant regardless of scale. We briefly describe each of these application below.

CTH [5] is a multi-material, large deformation, strong shock wave, solid mechanics code. The test problem used for this study was a 3D shaped charge simulation discretized to a rectangular mesh. In this configuration, inter-process communication consists primarily of large, multi-megabyte sized messages communicated among neighboring nodes. CTH is therefore limited primarily by point-to-point network bandwidth.

SAGE [14] is a multidimensional, multi-material Eulerian hydrodynamics code. SAGE inter-process communication consists primarily of a bulk-synchronous gather/scatter abstraction, which aggregates messages into large doubly-indexed arrays. Message sizes are generally in the hundreds of kilobytes to one megabyte range.

xNOBEL [7] is a one-,two-,or three-dimensional, multi-material Eulerian hydrodynamics code. It was developed for solving a variety of high-deformation flow of materials problems, with the distinguishing characteristic of being able to model high explosives. The problem used for this study was a shape charge simulation in two dimensions. Network communication consists of relatively small messages in the tens of bytes to hundreds of kilobytes.

Charon [11] is a semiconductor device simulation code. The problem used for this study is a 2D steady-state drift-diffusion simulation for a bipolar junction transistor. Charon is sensitive to small message latency for point-to-point and global reduction operations.

4 Results

Measured network bandwidth between two adjacent nodes is plotted in Fig. 1 for varying levels of injection bandwidth degradation. The labels in this figure and all others in this paper use the following convention: *None* = (800 MHz, 16 bit), *Half* = (400 MHz, 16 bit), *Quarter* = (200 MHz, 16 bit), and *Eighth* = (200 MHz, 8 bit). The bandwidth curves for the streaming test (Fig. 1b) ramp up much more quickly and reach slightly higher asymptotic levels than the ping-pong test (Fig. 1a). Applications that send few message at a time will behave more like what is observed for the ping-pong test, while applications that send many messages at a time will behave more like the streaming test. Small message latency was also measured for each configuration and found to be approximately the same as with no degradation ($< 1.0\ \mu s$).

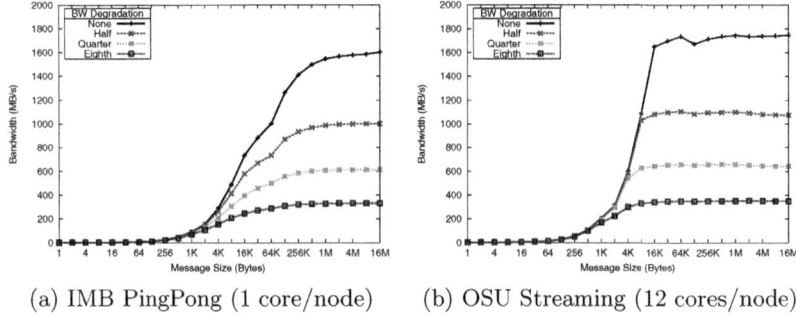

(a) IMB PingPong (1 core/node)   (b) OSU Streaming (12 cores/node)

Fig. 1. MPI micro-benchmark uni-directional bandwidth measurements

The resulting network injection bandwidth to compute balance ratios achieved using our test system are listed in Table 3. These ratios were calculated using the asymptotic maximum node-level bandwidths measured in Fig. 1b (in GB/s) divided by the peak 115.2 GFLOPS capability of each compute node. The lowest achieved balance ratio of 0.0030 is a factor of five worse than the full-speed baseline, but still a factor of 3.75 higher than the exascale system prediction listed in Table 1. Nevertheless, the test system exhibits a significant decrease in network-to-compute balance that can be leveraged to characterize application sensitivity to injection bandwidth.

Table 3. Achieved network injection bandwidth to compute balance ratios using experimental platform

Inj. BW Degradation	Net/Compute Balance Ratio	Factor Worse
None	0.0151	–
Half	0.0094	1.6
Quarter	0.0056	2.7
Eighth	0.0030	5.0

HPCC and application results are presented as performance degradation relative to full injection bandwidth, and are therefore unit-less. For example, a value of 1.5 indicates that the performance measured with the degraded injection bandwidth configuration was 1.5 times slower than with no degradation. In order to obtain the highest level of imbalance possible in practice, we used all twelve processor cores per compute node, running one MPI process per core. Therefore, the number of nodes used for each data-point in Figures 2 and 3 is obtained by dividing the x-axis label by twelve and rounding up to the nearest integer.

Results for the HPCC benchmarks are shown in Fig. 2. While network performance is important, its impact on HPL performance is secondary compared to computational capability, which we do not modify. As shown in Fig. 2a, our experimental results confirm this, showing relatively little performance degradation for HPL (note the reduced y-axis range compared to the other plots).

The results shown in Fig. 2b for RandomAccess were surprising to us. Half and quarter configurations result in almost no performance penalty, while eighth results in much larger penalties that appear to grow with scale. Our current theory is that 8 KB `MPI_Sendrecv` operations are at a point on the bandwidth curve that has roughly the same bandwidth for none, half, and quarter configurations, and much less for eighth. Such an effect can be seen for 4 KB messages in Fig. 1b, and we hypothesize that something similar is happening here for 8 KB messages. Further investigation at larger-scale is required.

As expected for a network bandwidth bound benchmark like PTRANS, the effect of injection bandwidth degradation shown in Fig. 2c is dramatic. The degradation factors relative to full injection bandwidth for half, quarter, and eighth configurations are approximately 1.25, 2, and 3.8, respectively, and the penalties do not appear to grow with scale. These degradation factors are reasonably close to the theoretical maximums listed in Table 3. FFT demonstrates behavior similar to PTRANS, but with lower degradation magnitudes.

Results for the applications evaluated are shown in Fig. 3. CTH is observed to be relatively insensitive to half injection bandwidth, moderately sensitive to quarter bandwidth, and significantly sensitive to eighth bandwidth (Fig. 3a). The performance degradation appears to be increasing with scale, but we believe this is an artifact of the communication to computation ratio increasing logarithmically with scale, as observed for CTH in [6]. Larger-scale experiments are needed to confirm this.

SAGE demonstrates a performance degradation profile similar to CTH, but of a lesser degree (Fig. 3b). SAGE's message sizes are generally not as large as CTH's, and are likely falling at a point on the message size vs. bandwidth curve where the differences between the injection bandwidth configurations are not as pronounced. As with CTH, larger-scale experiments are required to determine if performance degradation continues to increase.

xNOBEL demonstrates different scaling behavior than was observed for CTH and SAGE (Fig. 3c). Half and quarter injection bandwidth configurations have very little impact on performance. However, with one-eighth speed injection

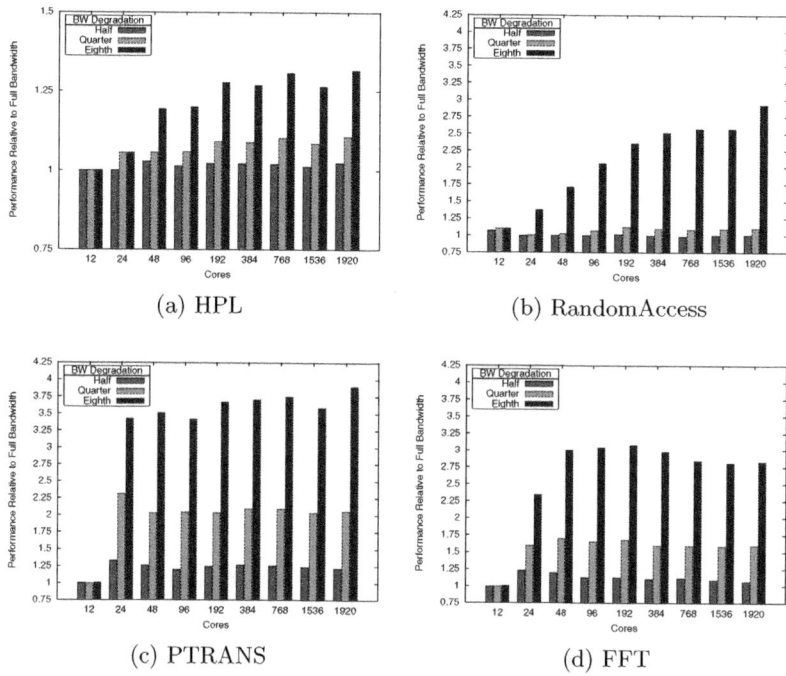

(a) HPL

(b) RandomAccess

(c) PTRANS

(d) FFT

Fig. 2. HPCC Results

bandwidth, significant performance degradation is observed but only at the largest scales tested. We believe this to be caused by xNOBEL suddenly losing the ability to significantly overlap communication and computation due to the severely degraded injection bandwidth. If true, this demonstrates the importance of providing sufficient network performance so that well-designed applications can maximize their ability to overlap communication and computation.

As expected for a latency bound application, Charon is unaffected by injection bandwidth degradation (Fig. 3d). For the test problem used, the average message size is less than 1 kilobyte, which, according to Fig. 1, results in essentially identical performance no matter what injection bandwidth configuration is used. Given Charon's insensitivity to injection bandwidth, it would clearly be advantageous to save power by using a low-bandwidth, low-latency interconnect.

5   Related Work

Parameterized analytic performance models have been developed for a number of large-scale parallel applications [1,9] and been shown to closely track experimental measurements. While clearly worthwhile, a downside to analytic modeling is that it is time consuming, requiring expert-level knowledge of a given application,

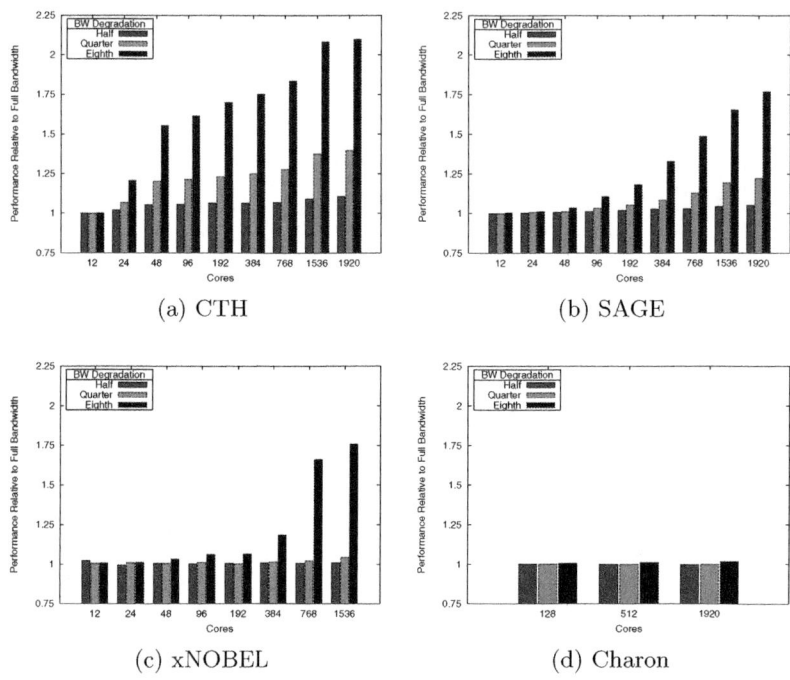

Fig. 3. Application Results

and is necessarily application specific. Our in-situ experimental-based approach can be complementary to analytic modeling by accelerating the application understanding process and by assisting in model validation.

Simulation-based approaches are highly configurable and flexible. The downside of this approach is often performance – slowdowns of 1000x or more are common, and increase with the simulation's fidelity. Model-based simulators can be much faster, but make simplifying assumptions. In contrast, our approach allows full applications to be evaluated in real-time. However, a drawback of our approach is that it can generally only slow-down the performance parameter being studied. The Structural Simulation Toolkit (SST) [13] is an example of a simulation platform targeted at simulating large-scale parallel computers. Our approach could be used to assist in validating SST component models and other simulation tools.

Finally, experimental methods such as our approach use existing systems and real applications to evaluate the impact of the parameter under study. Software-based de-tuning was used in [6] to characterize the performance penalty with scale of various levels of operating system interference. Recently, results from this study were reproduced via simulation [8], demonstrating the important relationship between simulation and experiment. Another software-based experimental approach is described in [3], where a special MPI library was used to degrade network performance. Our approach is similar, but uses direct hardware manipulation instead of software techniques. This eliminates the potential of introducing artificial software-induced overhead.

6   Conclusions and Future Work

Future exascale computing systems are expected to have significantly reduced network and memory bandwidth relative to computational performance than current systems. In this paper, we have demonstrated how existing large-scale parallel computers can be used to more closely emulate the expected imbalance of future systems. The resulting experimental platform can then be used for evaluating application sensitivity to the parameters under study. Our specific focus has been on network injection bandwidth, but many other hardware parameters can be examined as well. Results from our application scaling studies on a 160-node, 1920-core Cray XT5 system indicate that some applications experience sudden drops in performance at certain network injection bandwidth thresholds. Our ongoing work involves performing much larger-scale experiments to determine if the trends observed continue at higher core counts. We hope to leverage the empirical results obtained to assist in the development and validation of application models and simulation tools. Ultimately, we seek to gather information on the system balance requirements of existing highly-scalable parallel applications and leverage this insight to guide the design of future exascale supercomputers.

References

1. Alam, S.R., Vetter, J.S.: An analysis of system balance requirements for scientific applications. In: ICPP 2006: Proceedings of the International Conference on Parallel Processing (2006)
2. Alvin, K., Barrett, B., Brightwell, R., Dosanjh, S., Geist, A., Hemmert, S., Heroux, M., Kothe, D., Murphy, R., Nichols, J., Oldfield, R., Rodrigues, A., Vetter, J.: On the Path to Exascale. International Journal of Distributed Systems and Technologies 1(2), 1–22 (2010)
3. Ang, J.A., Barnette, D., Benner, B., Goudy, S., Malins, B., Rajan, M., Vaughan, C.: Supercomputer and cluster performance modeling and analysis efforts: 2004-2006. Tech. rep., Sandia National Laboratories Technical Report, SAND2007-0601 (2007)
4. Brightwell, R., Hudson, T., Pedretti, K., Underwood, K.D.: SeaStar Interconnect: Balanced bandwidth for Scalable Performance. IEEE Micro 26(3), 41–57 (2006)
5. Hertel Jr., E.S., Bell, R., Elrick, M., Farnsworth, A., Kerley, G., McGlaun, J., Petney, S., Silling, S., Taylor, P., Yarrington, L.: CTH: A Software Family for Multi-Dimensional Shock Physics Analysis. In: Proceedings of the International Symposium on Shock Waves, pp. 377–382 (July 1993)
6. Ferreira, K.B., Bridges, P., Brightwell, R.: Characterizing application sensitivity to OS interference using kernel-level noise injection. In: SC 2008: Proceedings of the International Conference on High-Performance Computing, Networking, Storage, and Analysis (November 2008)

7. Gittings, M., Weaver, R., Clover, M., Betlach, T., Byrne, N., Coker, R., Dendy, E., Hueckstaedt, R., New, K., Oakes, W.R., Ranta, D., Stefan, R.: The rage radiation-hydrodynamic code. Computational Science & Discovery 1(1), 015005 (2008)
8. Hoefler, T., Schneider, T., Lumsdaine, A.: Characterizing the Influence of System Noise on Large-Scale Applications by Simulation. In: SC 2010: Proceedings of the International Conference for High Performance Computing, Networking, Storage, and Analysis (November 2010)
9. Hoisie, A., Johnson, G., Kerbyson, D.J., Lang, M., Pakin, S.: A performance comparison through benchmarking and modeling of three leading supercomputers: Blue Gene/L, Red Storm, and Purple. In: SC 2006: Proceedings of the International Conference on High-Performance Computing, Networking, Storage, and Analysis (November 2006)
10. Kogge, P.M., et al.: Exascale computing study: Technology challenges in achieving exascale systems. Tech. rep., University of Notre Dame CSE Department Technical Report, TR-2008-13 (September 2008)
11. Lin, P.T., Shadid, J.N., Sala, M., Tuminaro, R.S., Hennigan, G.L., Hoekstra, R.J.: Performance of a parallel algebraic multilevel preconditioner for stabilized finite element semiconductor device modeling. Journal of Computational Physics 228, 6250–6267 (2009)
12. Luszczek, P., Dongarra, J., Koester, D., Rabenseifner, R., Lucas, B., Kepner, J., Mccalpin, J., Bailey, D., Takahashi, D.: Introduction to the HPC Challenge (HPCC) benchmark suite. Technical Report (March 2005), http://icl.cs.utk.edu/projectsfiles/hpcc/pubs/hpcc-challenge-benchmark05.pdf
13. Rodrigues, A.: Programming Future Architectures: Dusty Decks, Memory Walls, and the Speed of Light, ch. 3, pp. 56–81. University of Notre Dame (2006)
14. Weaver, R., Gittings, M.: Massively parallel simulations with DOE's ASCI super-computers: An overview of the Los Alamos Crestone project. In: Adaptive Mesh Refinement - Theory and Applications. Lecture Notes in Computational Science and Engineering, vol. 41, pp. 29–56. Springer, Heidelberg (2005)

Impact of Kernel-Assisted MPI Communication over Scientific Applications: CPMD and FFTW

Teng Ma, Aurelien Bouteiller, George Bosilca, and Jack J. Dongarra

Innovative Computing Laboratory,
EECS, University of Tennessee
{tma,bouteill,bosilca,dongarra}@eecs.utk.edu

Abstract. Collective communication is one of the most powerful message passing concepts, enabling parallel applications to express complex communication patterns while allowing the underlying MPI to provide efficient implementations to minimize the cost of the data movements. However, with the increase in the heterogeneity inside the nodes, more specifically the memory hierarchies, harnessing the maximum compute capabilities becomes increasingly difficult. This paper investigates the impact of kernel-assisted MPI communication, over two scientific applications: 1) Car-Parrinello molecular dynamics(CPMD), a chemical molecular dynamics application, and 2) FFTW, a Discrete Fourier Transform (DFT). By focusing on the usage of Message Passing Interface (MPI), we found the communication characteristics and patterns of each application. Our experiments indicate that the quality of the collective communication implementation on a specific machine plays a critical role on the overall application performance.

1 Introduction

Enhanced by multi-core and many-core nodes, clusters of workstations are widely used for scientific computing, where MPI has been the de facto programming paradigm for scientific computing parallel applications. To fully exploit the potential of multi-core nodes, domain users can adopt different programming models, prominently a pure MPI approach, but also hybrid programming (e.g. MPI+OpenMP or MPI+multithreading), virtual shared memory systems (e.g. OpenMP only), HPF (high performance FORTRAN) and etc. Compared with other approaches, the pure MPI approach has the benefit of the portability, allowing application developers to implement the code once, and then run it everywhere. However, the overhead of MPI intra-node communications from excessive memory copy is a major concern. In the pure MPI approach, each MPI process is bound to a core for performance reasons. The most common approach for delivering messages between MPI processes, running on shared memory multi-core nodes, has been to establish a shared memory between the two processes. The sender copies a message into the shared memory zone and the receiver copies it out to the target buffer, resulting in a double copy for each point-to-point communication. This copy-in/copy-out approach wastes, not only CPU cycles,

Y. Cotronis et al. (Eds.): EuroMPI 2011, LNCS 6960, pp. 247–254, 2011.

but also memory bandwidth, especially for large messages and collective communication. With more cores and deeper memory hierarchies, it becomes more difficult, for a shared memory approach, to deliver optimal performance in a generic way.

Kernel-assisted memory copy can alleviate this issue by using system calls to offload the copy to the kernel. Because the kernel has a physical view of the memory space for both processes (the source and the destination), it can perform memory copies directly from the source memory to destination memory without an intermediate buffer. KNEM is a Linux kernel module that enables high-performance, inter-process, single-copy memory copies. It offers support for asynchronous and vector data transfers. MPI communities have realized the importance of integrating kernel assisted memory copy to MPI intra-node communications. Open MPI, since version 1.5, includes KNEM support in its shared memory point-to-point communications component. MPICH2, since version 1.1.1, uses KNEM in the DMA LMT to improve large message performance within a single node. The work in [1,2] has shown that KNEM-enabled MPI communication significantly improves the performance of some micro- and macro-benchmarks. However, the performance of real scientific applications using KNEM-enabled MPI communication has yet to be asserted. This paper focuses on the impact of KNEM-enabled MPI communication on scientific applications.

We selected two applications: CPMD [3] and FFTW [4]. These applications come from different scientific areas: from molecular dynamics to signal processing. They are widely known and used in the engineering and scientific computing communities. The CPMD code is a parallel plane wave/pseudo-potential implementation of density functional theory, particularly designed for ab-initio molecular dynamics [3]. FFTW, "Fastest Fourier Transform in the West", is one of the most popular libraries for computing discrete Fourier transforms (DFTs), developed by Matteo Frigo and Steven G. Johnson [4]. Both applications are developed around a core, involving both point-to-point and collective communications, and can be considered as lightly communication-intensive applications. Using a lightweight MPI profiling software (mpiP), we investigate the communication overhead distribution in each application, including the percentage of MPI runtime in the application runtime, the percentage of each MPI call runtime in the whole communication time, and the message size distribution.

The remainder of this paper is organized as follows: Section 2 introduces the related work about the shared memory and the kernel assisted approach. Section 3 depicts the two parallel applications used in this paper, followed by Section 4 where the experimental results of CPMD and FFTW are presented.

2   Related Work

With the increasing complexity of the node architecture, shared memory performance remains a critical corner-stone in MPI application performance. As such, significant efforts have been deployed to improve the MPI intra-node communication performance. Darius Buntinas *et al.* proposed single-copy communication to speed up large message point-to-point communication for MPICH2 (based on vmsplice and KNEM [1]). The KNEM assisted approach outperforms the standard transfer method in the MPICH2 implementation when no cache is shared between the processing cores, or when very large messages are being transferred. Even simply using KNEM assisted point-to-point communication underneath collective communication achieved a significant improvement [1,2]. Within Open MPI, a similar approach was implemented, with further emphasis on auto-tuning and performance portability [5]. KNEM based memory operation (with features such as persistent memory registration and copy direction control) has been leveraged from within the collective algorithm itself, allowing for most copies to happen in parallel in 'rooted' communications (e.g. one-to-all or all-to-one). Furthermore, the hardware features and the collective communication topology are mapped in order to minimize the volume of data transiting between distant memory hierarchies. Overall these improvements demonstrated substantial speedups of the communication operations on multicore systems [6]. However, the evaluation of the benefits of these approaches has been mostly centered around synthetic benchmarks.

From another perspective, several works have focused on determining the properties of parallel applications in the context of hierarchical systems [7,8]. Recent works have investigated the benefits of hybrid programming in the context of multicore nodes [9]. However, the conclusions of these studies are challenged by the performance now permitted by kernel assisted copies within MPI. Our present work focuses specifically on investigating the behavior of prominent application, taking into account this novelty.

3   Applications

Car-Parrinello Molecular Dynamics(CPMD) is a plane wave/pseudo-potential implementation of density functional theory, particularly designed for ab-initio molecular dynamic [3]. CPMD simulations use the most fundamental approaches to model condensed phases. Dynamic equations of motion are solved for the ions with the inter-ionic forces computed from the valence electron density, which is solved for at each time step using density functional theory. In the case of methane, a CPMD simulation consists of one C and four H ions and eight valence electrons per molecules. The ground state electron density is computed at each time step. And polarization and other short range forces are also taken into account in the CPMD [3]. CPMD provides several standard simulations, such as C-120, Si-64, water-32, etc. We selected the methan-fd-nosymm test in our experiments, that uses the finite-difference (FD) method, based on a discretization of the differential operator [10], without molecular symmetry.

FFTW, "Fastest Fourier Transform in the West", is one of the most popular libraries to compute discrete Fourier transforms (DFTs). FFTW can handle inputs with one or more dimensions, arbitrary size, and both real and complex data [4]. FFTW also features a MPI-based distributed implementation. To compute the FFT of a multi-dimensional array, each processor first transforms all the dimensions of the data that are completely local to it (rows). Then, the processors perform a transpose of the data in order to get the remaining dimension local to a processor (columns). This dimension is then Fourier transformed, and then the dataset is transposed, back to its original order.

4 Experiments

4.1 Experimental Conditions

Our experimental platform (named IG) is a 48 core AMD NUMA machine with 128GB of memory. The system is composed of 8 sockets with a six-core 2.8 GHz AMD Opteron 8439 SE, 5 MB L3 caches and 16 GB memory per NUMA node. The sockets are further divided as two sets of 4 sockets on two separate boards connected by a low performance interlink.

The Linux Red Hat 4.1.2 (2.6.35.7 kernel) operating system is used on the machine, with the KNEM (version 0.9.5) kernel memory copy module. The MPI implementation is Open MPI (trunk r24549), with mpiP (version 3.2.1) [11] to profile and record MPI usage. Inside Open MPI, two different setups are compared: the SM setup uses the tuned collective module [12] and the SM point-to-point Byte Transfer Layer (BTL); the KNEM setup uses the KNEM collective module and the SM/KNEM BTL [6] and is hence benefiting from kernel assisted memory copies for messages larger than 4KB. Because KNEM collective module implemented a subset of MPI collective operations: Broadcast, Gather(v), Scatter(v), Allgather(v), and Alltoall(v), operations not implemented in KNEM collective module such as Reduce, Allreduce, and etc. will use Open MPI's Basic collective module. The mapping between physical cores and MPI processes is identical for both setups, regardless of the underlying communication components.

The CPMD software (version 3.13.2) [3] is configured as 'LINUXMPI' with the BLAS/LAPACK libraries (LAPACK 3.3.0). We selected one simulation from CPMD's vibrational analysis tests: methan-fd-nosymm.inp. FFTW-3.2.alpha [4] is configured with MPI support. Our input for the FFT mpi-bench is [1500×1500 20 20] with the verification(-y), which stands for a 1500×1500 complex DFT, a 20 complex DFT, and a 20 complex DFT.

4.2 CPMD

Table 1 compares the execution time breakdown, into compute time and communication time, of the CPMD application on a large multicore node, between the two communication modes (KNEM and SM, differing in their use of kernel assisted memory copies). As expected, the computation execution time remains generally constant when changing the communication mode (201s versus 194s). The major performance difference between the two setups lies in the communication overhead (MPI time), which occupies 26.9% of the overall application runtime for the KNEM-enabled mode, while it rises to 54.1% when using the regular SM communication mode. The CPMD application makes extensive use of all-to-all collective communications, which enjoy a threefold speedup when using the KNEM-enabled approach, translating into a $1.5\times$ application speedup in the methan-fd-nosymm test case.

Table 1. Total application time and MPI time for CPMD's methan-fd-nosymm test between Open MPI's KNEM mode and SM mode, with 48 MPI processes on IG's 48 cores

	Total Application time(sec)	MPI time(sec)
KNEM-enabled	276	74.4
SM-enabled	423	229

Table 2. Sum of all processes' execution time for the 5 most used MPI functions in CPMD's methan-fd-nosymm using shared memory and KNEM (48 processes on IG's 48 cores)

KNEM-enabled			SM-enabled		
Call	Time(millisec)	MPI%	Call	Time(millisec)	MPI%
Alltoall	2.88e+06	90.32	Alltoall	1.02e+07	97.71
Bcast	1.15e+05	3.59	Bcast	1.08e+05	1.03
Allreduce	1.12e+05	3.52	Allreduce	1.05e+05	1.00
Barrier	7.01e+04	2.20	Allreduce	8.6e+03	0.08
Allreduce	7.49e+03	0.23	Recv	6.17e+03	0.06

Table 2 presents the accumulated time, over all processes, spent in the five most time consuming MPI functions. In both cases, using KNEM or SM, the AlltoAll operation takes more than 90% of the MPI execution time. However, compared with SM-based communication, the KNEM version reduces the Alltoall cost from 10,000s to about 2,900s. Based on the statistics gathered using mpiP, the average message size for each AlltoAll operation is 24KBytes, a size in the range where KNEM is beneficial to collective operation performance (bigger than 4KB). Here, Allreduce in the KNEM-enabled setup is worse than in the SM-enabled setup(1.12e5 vs 1.05e5), because allreduce in the KNEM setup actually triggers Open MPI's Basic collective module, which is a simple and basic implementation of collective operations without any optimization. It's not a surprise for a Tuned collective operation to outperform a Basic collective operation.

Figure 1 shows a strong scaling performance of CPMD's methan-fd-nosymm tests, using the KNEM-enabled and SM communication modes. In this experiment, the process i is bound to core i, for all modes. The KNEM-enabled MPI communication outperforms the SM MPI communications, regardless of processes in use and the number of NUMA nodes. The CPMD application benefits from a better scalability, when increasing the number of cores, with the KNEM-enabled MPI operations. The SM communications do not permit the application to scale to more than 24 processes (the limit where all processors are on the same system board, in this machine), because the SM communications are oblivious to the underlying hardware topology. On the other hand, the KNEM-enabled operations enable the application to benefit from the expected scalability for the CPMD application, even though the NUMA topology is extremely challenging on this platform.

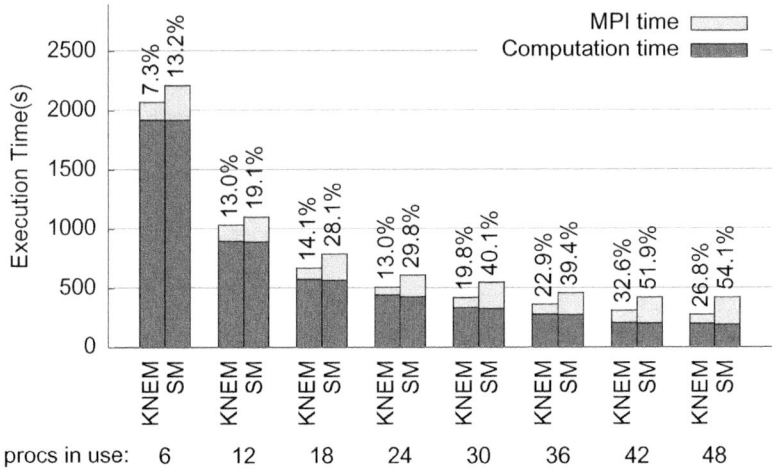

Fig. 1. Strong Scaling for CPMD's methan-fd-nosymm test over KNEM and shared memory. Processes are bound to IG's cores in a compact way (rank i is bound to core i).

4.3   FFTW

The next application is the Discrete Fourier Transform library, from FFTW. A detailed view of the contribution to the cumulative communication time of the 5 most used MPI functions of this application is presented in the table 3. One can notice that the most time consuming MPI call (the broadcast communication) enjoys nearly an order of magnitude improvement when using the KNEM collective component (from 6070s down to 959s). Even point-to-point communications see their performance improve, but less significantly. As an example, the sendrecv based on KNEM is 1.37× faster than the one based on shared memory. Similarly with the previous application, the average message size (see table 4) is larger than the message size where KNEM enabled communications become beneficial. For the broadcast collective, the average size is 14MB, and the average size for the sendrecv point-to-point communication is 16KB. The execution time of KNEM Scatterv is a little more than Tuned Scatterv here, because messages in

Table 3. Cumulative time of the five most used MPI calls in a 48 processes FFTW (1500x1500, 20, 20, with verification), running with KNEM or SM components, on the 48 cores of IG

KNEM-enabled Mode			SM-enabled Mode		
Call	Time (millisec)	MPI%	Call	Time (millisec)	MPI%
Bcast	9.59e+05	61.98	Bcast	6.07e+06	88.9
Sendrecv	3.52e+05	22.78	Sendrecv	4.82e+05	7.07
Scatterv	1.65e+05	10.69	Gatherv	1.39e+05	2.03
Gatherv	5.86e+04	3.79	Scatterv	1.33e+05	1.95
Comm_dup	4.33e+03	0.28	Comm_dup	1.81e+03	0.03

Table 4. Aggregate Sent Message Size for each MPI calls in FFTW. 48 processes on IG, one rank per core.

Call	Count	Total (bytes)	Avrg (bytes)	Sent%
Bcast	19488	2.8e+11	1.44e+07	94.20
Sendrecv	710888	1.14e+10	1.61e+04	3.84
Gatherv	19488	5.83e+09	2.99e+05	1.96
Bcast	288	1.47e+05	512	0.00
Bcast	288	1.47e+05	512	0.00

the FFTW's scatterv operation is smaller than KBytes and the overhead of trapping into kernel and distributing cookies offset the benefits of KNEM kernel copy.

Finally, the table 5 presents the total execution time, and the MPI contribution to that total, for both KNEM-enabled and SM communication modes. Thanks to the benefits on the broadcast operations, the KNEM-enabled communications induce a dramatic threefold reduction of the time spent communicating, from 143s to 38s. As FFTW is a communication intensive application, the decrease of the communication contribution to the execution time translates into a major improvement of the overall execution time (doubled performance). The last row of Table 5 indicates the performance of the OpenMP version of this same application. Although this version cannot run on distributed memory clusters, it is indicative of the performance attained by a tailored approach on this shared memory architecture. The introduction of KNEM-enabled communications have greatly reduced the efficiency gap between the OpenMP approach and the MPI approach on shared memory machines. However, the OpenMP code is still twice as fast; but it lacks the capability to span over multicore cluster.

Table 5. Total application time for the FFTW's application when using OpenMP(48 threads) or pure MPI over different communications: the KNEM-enabled or the shared memory-enabled communication with 48 processes (rank i bound to core i).

	Total Application time(sec)	MPI time(sec)
KNEM-enabled mode	107	38.1
SM-enabled mode	216	143
OpenMP mode	49.5	N/A

5   Conclusion

A lot of MPI users spend significant time porting their pure MPI applications to a hybrid model (usually MPI+OpenMP) to exploit the full potential of multi-core architectures. Inefficient shared memory-based communications are an important factor forcing domain users to look for alternative solutions inside the nodes. From the experiments presented in the previous Section, classical shared memory communications have certain difficulties to provide good point-to-point and collective performance, when the number of cores and consequently the complexity of the memory hierarchy increase. However, kernel-assisted single-copy approaches have the potential to alleviate this issue by offloading the memory copies into the kernel, reducing the number as well as their impact on the memory bus by a factor of two. Experiments show the KNEM-enabled MPI communication can increase application performance, and expose a better scalability than the classical shared memory approach when integrating more resources inside computing nodes. With more cores, increased core heterogeneity, and deeper memory hierarchies, kernel assisted MPI communication provides dependable performance, and offers a better alternative to hybrid approaches, while still retaining the simplicity of a single programming model.

References

1. Buntinas, D., Goglin, B., Goodell, D., Mercier, G., Moreaud, S.: Cache-Efficient, Intranode Large-Message MPI Communication with MPICH2-Nemesis. In: Proceedings of the 38th International Conference on Parallel Processing (ICPP-2009), pp. 462–469. IEEE Computer Society Press, Vienna (2009)
2. Moreaud, S., Goglin, B., Goodell, D., Namyst, R.: Optimizing MPI Communication within large Multicore nodes with Kernel assistance. In: CAC 2010: The 10th Workshop on Communication Architecture for Clusters, Held in Conjunction with IPDPS 2010. IEEE Computer Society Press, Atlanta (2010)
3. Hutter, J., Iannuzzi, M.: CPMD: parrinello Molecular Dynamics, http://www.cpmd.org/
4. Frigo, M., Johnson, S.: The Design and Implementation of FFTW3. Proceedings of the IEEE 93(2), 216–231 (2005)
5. Ma, T., Bosilca, G., Bouteiller, A., Dongarra, J.J.: Locality and topology aware intra-node communication among multicore CPUs. In: Keller, R., Gabriel, E., Resch, M., Dongarra, J. (eds.) EuroMPI 2010. LNCS, vol. 6305, pp. 265–274. Springer, Heidelberg (2010)
6. Ma, T., Bosilca, G., Bouteiller, A., Goglin, B., Squyres, J., Dongarra, J.: Kernel Assisted Collective Intra-node Communication Among Multicore and Manycore CPUs. Research report (2010)
7. Vetter, J.S., Mueller, F.: Communication characteristics of large-scale scientific applications for contemporary cluster architectures. J. Parallel Distrib. Comput. 63, 853–865 (2003)
8. Plaat, A., Bal, H.E., Hofman, R.F.H., Kielmann, T.: Sensitivity of parallel applications to large differences in bandwidth and latency in two-layer interconnects. Future Generation Computer Systems 17(6), 769–782 (2001)
9. Rabenseifner, R., Hager, G., Jost, G.: Hybrid MPI/OpenMP parallel programming on clusters of multi-core SMP nodes. In: Proceedings of the 2009 17th Euromicro International Conference on Parallel, Distributed and Network-based Processing, pp. 427–436. IEEE Computer Society Press, Washington, DC, USA (2009)
10. scholarpedia.org: Finite difference method, http://www.scholarpedia.org/article/Finite_difference_method)
11. Vetter, J.S.: mpiP: Lightweight, Scalable MPI Profiling, http://mpip.sourceforge.net/
12. Fagg, G.E., Bosilca, G., Pješivac-Grbović, J., Angskun, T., Dongarra, J.: Tuned: A flexible high performance collective communication component developed for Open MPI. In: Proccedings of DAPSYS 2006, pp. 65–72. Springer, Innsbruck (2006)

A Log-Scaling Fault Tolerant Agreement Algorithm for a Fault Tolerant MPI

Joshua Hursey, Thomas Naughton,
Geoffroy Vallee, and Richard L. Graham

Oak Ridge National Laboratory, Oak Ridge, TN USA 37831
{hurseyjj,naughtont,valleegr,rlgraham}@ornl.gov

Abstract. The lack of fault tolerance is becoming a limiting factor for application scalability in HPC systems. The MPI does not provide standardized fault tolerance interfaces and semantics. The MPI Forum's Fault Tolerance Working Group is proposing a collective fault tolerant agreement algorithm for the next MPI standard. Such algorithms play a central role in many fault tolerant applications. This paper combines a log-scaling two-phase commit agreement algorithm with a reduction operation to provide the necessary functionality for the new collective without any additional messages. Error handling mechanisms are described that preserve the fault tolerance properties while maintaining overall scalability.

Keywords: MPI, Fault Tolerance, Agreement Protocol, Run-through Stabilization, Algorithm Based Fault Tolerance.

1 Introduction

The lack of fault tolerance will soon become a limiting factor for application scalability in High Performance Computing (HPC) systems, in particular exascale systems. It is projected that the mean time to failure (MTTF), a measure of system reliability, will drop from days to hours or minutes in such HPC systems [2]. This indicates that process failure will be a normal event that the application must be prepared to handle to fully utilize next generation HPC systems. As a result, applications are looking to augment (or replace) their existing checkpoint/restart fault tolerance techniques with Algorithm Based Fault Tolerance (ABFT) techniques to improve the efficiency of application recovery.

Unfortunately, application developers are hindered by the lack of any, let alone scalable, resilience models necessary for ABFT in fundamental support libraries like the Message Passing Interface (MPI) [13]. The current MPI standard does not provide standard semantics in the presence of process failure except in the default, abort case (i.e., MPI_ERRORS_ARE_FATAL). Such semantics are left to be optionally defined by individual implementations.

The MPI Forum created the Fault Tolerance Working Group (FTWG) in response to the growing need for portable, scalable fault tolerant semantics and interfaces in the MPI standard. Fault tolerant agreement algorithms serve as a

Y. Cotronis et al. (Eds.): EuroMPI 2011, LNCS 6960, pp. 255–263, 2011.

fundamental building block for most fault tolerant applications and libraries [1]. These algorithms provide uniform agreement of a value (or set of values) even in the presence of process failure during the execution of the algorithm. The FTWG's run-through stabilization (RTS) proposal provides an interface to such an algorithm in the MPI_Comm_validate_all collective operation over communicators. A similar collective interface is also available for windows and file handles.

The MPI_Comm_validate_all collective operation must be able to be implemented in a scalable manner if it is to be relied upon in highly scalable, fault tolerant HPC applications. Most fault tolerant agreement algorithms struggle to scale well to large numbers of processes, while others propose overly complex algorithms that are difficult to implement in practice. These algorithms focus on the agreement of a single state (namely COMMIT or ABORT) after the execution of the transaction body. The MPI_Comm_validate_all collective operation must agree upon a set of failed processes constructed by the group. Using the existing agreement protocols would require a separate fault-aware reduction operation followed by a separate agreement protocol.

The presented algorithms combine the reduction operation with a two-phase agreement operation to construct the list of known failures during the *voting* phase and uniformly agree upon a single list during the *commit* phase. This paper describes algorithmic adjustments made to the *two-phase commit* agreement algorithm, for both linear-scaling and log-scaling variations, to provide this functionality. The log-scaling algorithm variation sustains a point-to-point message complexity of $O(2log(n))$. This paper describes the error handling mechanisms that preserve the fault tolerance guarantee of uniform agreement even in the presence of process failure, in addition to an optimization to the termination protocol.

2   Related Work

Applications are experimenting with the integration of fault tolerance techniques into their code to improve the efficiency of application recovery. ABFT techniques require specialized algorithms that are able to adapt to and recover from process loss [10]. ABFT techniques typically rely upon data encoding, algorithm redesign, and diskless checkpointing in addition to a fault tolerant message passing environment (e.g., MPI). Related to ABFT are natural fault tolerance techniques [5]. Natural fault tolerance techniques focus on algorithms that can withstand the loss of a process and still get an approximately correct answer, usually without the use of data encoding or checkpointing.

The FTWG's RTS proposal defines semantics and interfaces for the handling of *fail-stop* process failure [7]. *Fail-stop* (a.k.a. *crash fault*) process failures are failures in which a process is permanently stopped often due to a component crash event in the system [1]. For the detection of such failures, the proposal provides the application with a *perfect* failure detector. A *perfect* failure detector is both *strongly accurate* and *strongly complete* [4]. *Strong accuracy* means that no process is reported as failed before it actually fails. *Strong completeness* means that eventually every failed process will be known by all other processes.

Fault tolerant agreement algorithms play a central role in many fault tolerant applications, libraries, and distributed transaction processing services for database systems [1]. These collective algorithms provide uniform agreement of a state even in the presence of process failure during the execution of the algorithm. There are three often cited fault tolerant agreement algorithms: *two-phase commit*, *three-phase commit*, and *Paxos*. Agreement algorithms can be either *blocking* or *non-blocking*. A *blocking* algorithm may block, in some failure scenarios, in an undecided state until a peer process is restored and makes a decision from a write-ahead log file. A *non-blocking* algorithm does not require the restart of failed processes for the collective group to decide.

The *two-phase commit* algorithm is a *blocking* algorithm built from two linear reliable broadcast operations and one linear reliable gather operation [9]. An optional, termination detection algorithm can be used to reduce the opportunity for blocking when the coordinator fails. The linear nature of the reliable broadcast and gather operations allow for relative simplicity in the handling of process failures, but at the cost of poor scalability to large numbers of processes.

Multi-level, tree structured two-phase agreement algorithms have been explored in, and proven correct for transaction processing systems [14,15]. The algorithm presented in this paper combines the multi-level communication topology with the construction of the global list of failed processes. Combining operations reduces the message complexity required to do both consecutively.

The *three-phase commit* algorithm extends the two-phase commit algorithm by adding another round of messages to eliminate the need for blocking [17]. Since this algorithm adds another round of operations (one additional broadcast and gather) it further adds to the message complexity.

The *Paxos* algorithm is a non-blocking algorithm that uses *replicas* instead of a single coordinator to reach agreement [12]. This algorithm scales as well as the two-phase commit algorithm while still being a non-blocking algorithm. However, this algorithm has proven challenging to implement correctly in practice [3].

The FT-MPI project provided a first attempt at extending the semantics and interface of the MPI-1 standard to support ABFT [6]. FT-MPI extended the MPI communicator states and modified the MPI communicator construction functions. The FT-MPI project provided inspiration for the FTWG's RTS and Process Recovery proposals. The RTS proposal provides semantics similar to FT-MPI's blank communicator mode, where failed processes are replaced by MPI_PROC_NULL. Both projects have complementary semantics regarding point-to-point and collective operations. The main difference between these projects is in the handling of communicator and group objects. Upon process failure, FT-MPI destroys all MPI objects with non-local information (e.g., communicators and groups), except MPI_COMM_WORLD, requiring the application to manually recreate these objects after every failure. In contrast, the RTS proposal preserves all communicators and groups. Instead of providing a fault tolerant agreement protocol to the application (i.e., MPI_Comm_validate_all), the FT-MPI project provides it as a transparent component of the runtime environment which is used to determine the group membership for MPI_COMM_WORLD after each process failure. As such, FT-MPI requires that every process failure be recognized globally by all alive processes. In the RTS proposal, process failures can be recognized locally, and on a per-communicator basis. These differences allow the RTS proposal to more flexibly support libraries, and, by allowing for localized failure recognition, open the door to more scalable fault tolerance solutions.

3   Two-Phase Commit Algorithm

This section briefly describes the structure and fault management properties of the linear scaling two-phase commit algorithm [9]. The two-phase commit algorithm is discussed in terms of how it was implemented in Open MPI to support the MPI_Comm_validate_all collective operation.

The two-phase commit algorithm relies upon linear-scaling reliable broadcast and gather operations, in addition to an optional linear termination detection algorithm upon coordinator failure. The *coordinator* initiates the algorithm by broadcasting a *vote request* to all of the *participants*, skipping failed processes. Since the MPI_Comm_validate_all is a collective operation, all alive processes will eventually enter the operation, so the vote request round can be eliminated from the MPI implementation reducing the message complexity, called the *unsolicited vote* optimization [18]. The algorithm will decide either DECIDED or UNDECIDED (a.k.a, COMMIT, ABORT) along with a globally constructed list of process failures.

Upon entering the MPI_Comm_validate_all operation, the participants send their *vote* to the coordinator. In the Open MPI implementation, the vote is a bit-field of locally known failed processes at that rank. If the coordinator fails before a participant sends its vote, then the participant can safely decide UNDECIDED and exit the algorithm since the coordinator could not have made a decision without their contribution.

The coordinator gathers the contributions of each participant (skipping failed processes), and creates a *decision* – represented as list of globally known failed processes constructed from the local list at each process. The coordinator then broadcasts the decision to all alive participants.

If the coordinator fails after a participant sent its vote, but before the participant receives the decision message then the participant is in an *uncertain* state since it does not know if a decision was made before the coordinator failed. Without the optional termination detection algorithm, the uncertain participant would *block* and wait for the coordinator to be recovered. With the termination detection algorithm, the uncertain participant linearly asks all other participants if they have decided or not. If a peer participant has decided (either DECIDED or UNDECIDED), then the uncertain participant decides with them. Otherwise, if no other alive participant has decided, then this process blocks waiting for recovery.

The algorithm must keep at least two log entries in the volatile memory of each local process. A log entry contains the list being decided upon, the state of the agreement, and a monotonically increasing sequence number to distinguish rounds. One entry stores the last decision made by the group used to catchup a process in the termination protocol. The other entry maintains that state for the current round of the operation.

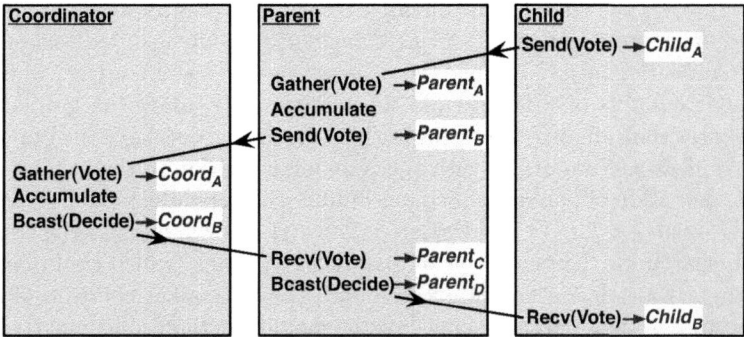

Fig. 1. Illustration of the three participants in the log-scaling two-phase commit algorithm. Error handling annotations highlighted in red italics (e.g., $\{Parent_C, Child_B\}$).

The RTS proposal does not provide the capability to restart processes, so if a process blocks it must call MPI_Abort. Progressing forward from an uncertain state may cause consistency issues with other communication contexts. In the Open MPI implementation, a runtime option is provided to allow the user to promote the uncertain state to an UNDECIDED decision, for the case where this semantic protection is not desired.

Ideally the MPI implementation would provide a non-blocking, fault tolerant agreement algorithm (e.g., three-phase commit), to avoid the uncertainty problem described above. This paper focuses on augmenting the two-phase algorithm for the sake of simplicity in explanation, and leaves the non-blocking algorithm extension to future work.

4   Log Scaling Two-Phase Commit Algorithm

The log-scaling two-phase commit algorithm uses a tree structure for both the broadcast and gather operations similar in communication structure to multi-level variations [14,15]. In practice, the *gather* is replaced by a *reduce*, but referred to in this section as a *gather* for consistency with Section 3.

The log-scaling two-phase commit algorithm differentiates between two types of participants: those that are not leaf elements, called *parent* participants; and those that are leaf elements, called *child* participants. The *coordinator* (at the root of the tree) is still ultimately responsible for making the final decision - construction of the final list of failed processes. The dependencies in the tree structure require additional error handling mechanisms to maintain the consistency of the algorithm and data in the presence of fail-stop process failures.

Figure 1 illustrates the three kinds of participants in the algorithm along with the basic communication pattern and error handling annotations (highlighted in red italics). This algorithm can be used with a tree of any depth, even though this figure shows only one level of parent participants.

In Figure 1, the primary error handlers of the coordinator and parents for the reliable gather operation are $\{Coord_A, Parent_A\}$ and for the broadcast operation are $\{Coord_B, Parent_D\}$. During the reliable gather and broadcast operations, parent participants of a failed child must recursively *adopt* the grandchildren. This ensures that all alive processes in the collective group have the opportunity to *vote* and *decide* uniformly with the remaining portion of the group.

Upon detection of parent participant failure, the dependent children are either in the $\{Parent_B, Child_A\}$ or $\{Parent_C, Child_B\}$ error handlers. The dependent children search for the nearest grandparent participant (which could be the coordinator). If an alive grandparent is found, the dependent child posts a query message to the new parent asking how it should continue participating in the algorithm. If the dependent child has not yet voted (in $\{Parent_B, Child_A\}$) then the new parent tells it to participate in the gather phase. If the child has voted (in $\{Parent_C, Child_B\}$), the new parent tells it to either participate in the gather phase (if the old parent failed before propagating the message up the tree), or the decision phase (if the old parent failed after propagating the message up the tree). If the new parent had successfully completed the collective operation, then it replies with the decision value and the child decides with the new parent. If the new parent also fails, the child queries the next alive parent.

If the coordinator fails, the termination detection algorithm is activated in each of the uncertain dependent parent participants of the coordinator, which may involve a parent at a lower level in the tree for cases where a first level parent failed. If a dependent parent participant has not voted (in $\{Parent_B, Child_A\}$) then it can decide UNDECIDED. If a dependent parent participant has voted (in $\{Parent_C, Child_B\}$) it becomes *uncertain* and enters the termination protocol. When a parent participant makes a decision, it must propagate that decision down the tree using the reliable broadcast operation.

As an optimization to the termination algorithm, instead of asking *all* participants in the collective group for a decision (as with the original two-phase algorithm), the uncertain participant only asks the parent participants dependent upon the coordinator. Notice that if the coordinator made a decision then the directly dependent parent participants of the coordinator would be the first to know, since they are in the broadcast group of the coordinator. Further, since children in the sub-tree below the parent only decide with their direct ancestors in the tree, those dependent children do not need to be queried since they cannot make a different decision than that of their parent.

The tree structure remains static between successful completions of the algorithm. Upon successful completion, the agreed upon list of failed processes is used to rebalance the tree structure. Rebalancing has been shown to improve the performance of collective operations after fail-stop process failure [11].

Even with the tree structured algorithm, there is still a possibility for blocking if the coordinator fails after the gather and before the broadcast operation. As mentioned in Section 3, in this one scenario the uncertain processes are aborted, by default, to protect the user.

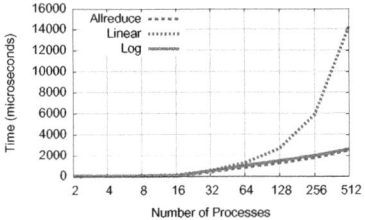
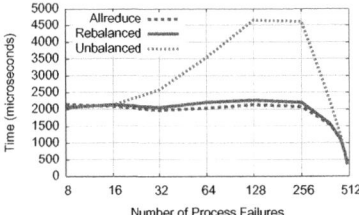

(a) Failure free scaling performance   (b) Performance of the log-scaling algorithm at 512 processes varying the number of failures

Fig. 2. Performance of fault tolerant log-scaling two-phase commit algorithm

5 Results

The following analysis used a prototype of the FTWG's RTS proposal based on the development trunk of the Open MPI implementation of the MPI standard [8]. Fault aware collectives are provided in the ftbasic component of the coll framework [11]. Three variations of the MPI_Comm_validate_all collective operation were implemented: a linear two-phase, a log-scaling two-phase, and a non-fault tolerant log-scaling allreduce. The allreduce algorithm is similar in form to the two-phase commit algorithm so it serves as an appropriate baseline for performance comparison. Though any tree topology can be used with the presented algorithm, a binomial tree is used in this implementation.

These tests used 32 nodes with each node containing four quad-core 2.0 GHz AMD Opteron processors. The Ethernet (tcp) and shared memory (sm) network drivers in Open MPI were used for these tests since they are the only fully supported interconnects provided by the prototype at this time. In the failure-full tests specific ranks are forcibly terminated before the performance testing by sending them the SIGKILL signal. This paper assumes fail-stop process failures, and once a process fails it is never restored. After a warmup phase, each data point is the average of 20 sets of an inner timing loop of 200 operations.

Figure 2(a) shows the failure-free performance of the three implementations. This figure illustrates the expected log-scaling performance of the algorithm presented in this paper. At 512 processes, the log-scaling algorithm shows a significant improvement over the linear-scaling algorithm, while staying within 3% of the baseline performance.

Figure 2(b) explores the performance impact on the log-scaling two-phase commit algorithm as the number of failures increases for a fixed sized job of 512 processes. The baseline performance in this figure is a failure-free run of the allreduce algorithm on a reduced sized communicator. As the number of failures increases the need to rebalance the validation tree upon successful agreement becomes readily apparent (2.6 times slower at 256 failures). The rebalanced performance is at worst 6% slower (at 64 failures) than the baseline allreduce algorithm. This difference indicates that there may be further room for improvement in the implementation.

6   Conclusion

Fault tolerant agreement algorithms play a foundational role in many fault tolerant applications, but existing algorithms often struggle with scalability to large numbers of processes. The MPI Forum's FTWG proposed, as part of the RTS proposal, a MPI_Comm_validate_all collective operation to encapsulate such algorithms. This operation provides uniform agreement of the set of known failed processes in the specified communicator (variations are also available for windows and file handles) even in the presence of process failures during the algorithm.

This paper explores an enhancement to the well established two-phase commit algorithm. By replacing the linear broadcast and gather operations with tree-based, log-scaling operations the point-to-point message complexity was reduced from $O(2N)$ to $O(2log(N))$. Further by combining the list construction operation with the two-phase commit protocol, no additional messages were added to fully support the MPI_Comm_validate_all collective operation. This paper describes the additional error handling mechanisms required to maintain the fault tolerance guarantees of uniform agreement even in the presence of failures. Using a prototype implementation in Open MPI, performance results showed scaling performance comparable to an MPI_Allreduce operation.

Since the two-phase commit algorithm is blocking in some situations, we are extending this design to non-blocking algorithms. Concurrently, we are exploring other methods for implementing this collective algorithm to further improve scalability and performance. Finally, we will consider dynamically constructed trees, instead of statically rebalanced trees, which may help manage the effects of process skew [16].

Acknowledgments. Research sponsored by the Mathematical, Information, and Computational Sciences Division, Office of Advanced Scientific Computing Research, U.S. Department of Energy, under Contract No. DE-AC05-00OR22725 with UT-Battelle, LLC.

References

1. Barborak, M., Dahbura, A., Malek, M.: The consensus problem in fault-tolerant computing. ACM Computing Surveys 25, 171–220 (1993)
2. Cappello, F., Geist, A., Gropp, B., Kale, L., Kramer, B., Snir, M.: Toward exascale resilience. International Journal of High Performance Computing Applications 23(4), 374–388 (2009)
3. Chandra, T.D., Griesemer, R., Redstone, J.: Paxos made live: An engineering perspective. In: Proceedings of the Twenty-sixth Annual ACM Symposium on Principles of Distributed Computing, PODC 2007, pp. 398–407. ACM, New York (2007)

4. Chandra, T.D., Toueg, S.: Unreliable failure detectors for reliable distributed systems. Journal of the ACM 43, 225–267 (1996)
5. Engelmann, C., Geist, G.A.: Super-scalable algorithms for computing on 100,000 processors. In: Sunderam, V.S., van Albada, G.D., Sloot, P.M.A., Dongarra, J. (eds.) ICCS 2005. LNCS, vol. 3514, pp. 313–321. Springer, Heidelberg (2005)
6. Fagg, G.E., Gabriel, E., Chen, Z., Angskun, T., Bosilca, G., Pjesivac-Grbovic, J., Dongarra, J.J.: Process fault-tolerance: Semantics, design and applications for high performance computing. International Journal for High Performance Applications and Supercomputing 19(4), 465–478 (2005)
7. Fault Tolerance Working Group: Run-though stabilization proposal, svn.mpi-forum.org/trac/mpi-forum-web/wiki/ft/run_through_stabilization
8. Gabriel, E., Fagg, G.E., Bosilca, G., Angskun, T., Dongarra, J., Squyres, J.M., Sahay, V., Kambadur, P., Barrett, B.W., Lumsdaine, A., Castain, R.H., Daniel, D.J., Graham, R.L., Woodall, T.S.: Open MPI: Goals, concept, and design of a next generation MPI implementation. In: Kranzlmüller, D., Kacsuk, P., Dongarra, J. (eds.) EuroPVM/MPI 2004. LNCS, vol. 3241, pp. 97–104. Springer, Heidelberg (2004)
9. Gray, J.: Notes on data base operating systems. In: Operating Systems, An Advanced Course, pp. 393–481. Springer, London (1978)
10. Huang, K.H., Abraham, J.A.: Algorithm-based fault tolerance for matrix operations. IEEE Transactions on Computers 33(6), 518–528 (1984)
11. Hursey, J., Graham, R.: Preserving collective performance across process failure for a fault tolerant MPI. In: 16th International Workshop on High-Level Parallel Programming Models and Supportive Environments (HIPS) Held in Conjunction with the 25th IEEE International Parallel and Distributed Processing Symposium (IPDPS), Anchorage, Alaska (May 2011)
12. Lamport, L.: The part-time parliament. ACM Transactions on Computer Systems (TOCS) 16, 133–169 (1998)
13. Message Passing Interface Forum: MPI: A Message Passing Interface. In: Proceedings of Supercomputing 1993, pp. 878–883. IEEE Computer Society Press, Los Alamitos (1993)
14. Mohan, C., Lindsay, B.: Efficient commit protocols for the tree of processes model of distributed transactions. ACM SIGOPS Operating Systems Review 19, 40–52 (1985)
15. Mohan, C., Lindsay, B., Obermarck, R.: Transaction management in the R* distributed database management system. ACM Transactions on Database Systems (TODS) 11, 378–396 (1986)
16. Raz, Y.: The dynamic two phase commitment (d2pc) protocol. In: Vardi, M.Y., Gottlob, G. (eds.) ICDT 1995. LNCS, vol. 893, pp. 162–176. Springer, Heidelberg (1995)
17. Skeen, D.: Nonblocking commit protocols. In: Proceedings of the 1981 ACM SIGMOD International Conference on Management of Data, SIGMOD, pp. 133–142. ACM, New York (1981)
18. Stonebraker, M.: Concurrency control and consistency of multiple copies of data in distributed Ingres. IEEE Transactions on Software Engineering SE 5(3), 188–194 (1979)

Fault Tolerance in an Industrial Seismic Processing Application for Multicore Clusters*

Alexandre Gonçalves[2], Matheus Bersot[1], André Bulcão[3], Cristina Boeres[1], Lúcia Drummond[1], and Vinod Rebello[1]

[1] Computer Science Department – Fluminense Federal University (UFF)
Niterói – RJ – Brazil
{mbersot,boeres,lucia,vinod}@ic.uff.br
[2] Federal Institute of Education, Science and Technology of Rio de Janeiro (IFRJ)
Rio de Janeiro – RJ – Brazil
alexandre.domingues@ifrj.edu.br
[3] Petrobras Research Center (CENPES)
Rio de Janeiro – RJ – Brazil
bulcao@petrobras.com.br

Abstract. Seismic processing applications are used to identify geological structures where reservoirs of oil and gas may be found. With oil companies seeking better precision over larger geographical regions, these applications require larger clusters to keep execution times reasonable. The combination of longer run times and clusters with greater numbers of components increases the probability of faults during the execution. To address this issue, this paper describes an application-level fault tolerance mechanism that considers node crashes and communication link failures. For this industrial application, experiments show that continued execution with the remaining resources is both feasible and efficient.

Keywords: fault tolerance, multicore clusters, RTM application.

1 Introduction

With a better cost-performance benefit than other parallel systems, computing clusters dominate the HPC market. But even in state-of-the-art environments, many applications require days or even months to run. Given that these systems are composed of thousands of components, any of which can fail, one of the key issues is the continued execution of applications in the face of hardware failures.

Applications designed for high-performance clusters generally use the MPI library. Although the MPI standard does not define provisions for fault tolerance, a number of academic implementations such as FT-MPI [6], MPI/FT [1] e MPICH-V [2] do offer this functionality. Nevertheless, it is hard to design applications that are portable across MPI implementations due to the need to conform to the specific requirements of each one.

In the case of many commercial MPI products, such as the Intel MPI library [8] that focuses on maximising performance, programmers must resort

---

* Supported by PRONEX E-26/110.552/2010, CNPq, FAPERJ, PETROBRAS.

Y. Cotronis et al. (Eds.): EuroMPI 2011, LNCS 6960, pp. 264–271, 2011.

to implementing application-level fault tolerance, where an application specific strategy can be adopted with the intent of minimising the adverse impact on fault-free performance. In this work, our target is an industrial application in seismic processing that determines the properties and position of various layers of subterranean rocks, with objective of identifying geological structures where reservoirs of oil and gas may potentially be found.

One of the standard data processing techniques used to create accurate geological images of subsurface rock strata is seismic surveying through geophysical depth migration. Shock waves generated at the surface travel at different speeds depending on rock types and patterns and thus are reflected at distinct angles and refracted differently as they meet new strata. Changes in the wave energy recorded at the surface by geophones are geometrically relocated in space to the location that the event occurred beneath the surface. The application tackled here employs the Reverse Time Migration (RTM) method [11] to solve the wave propagation equation, forward in time from the source and backward in time from the receiver. The RTM algorithm analyses a three-dimensional domain space that is partitioned amongst the available processors, with one process per core. Each iteration of the main loop calculates the energy at each point in the domain to the 10th order in space and 2nd order in time. Given that the solution for a point in 3D space depends not only on previous energy values at that point but also on the values of neighbouring points in the three dimensions, at the end of each loop, the energy values associated with data points in different processes are exchanged via messages.

This MPI application can require months of computational time even on a cluster with hundreds of processing cores and thus it would be prudent to take measures to protect it from failures. This paper describes the mechanisms for failure detection and recovery that have been aggregated to the RTM application and evaluates their impact on performance while using the Intel MPI library.

2 Fault Tolerance

MPI applications are not resilient to failures since the standard provides developers with only two options: either the execution is aborted or control is returned to the application without further assurance that new communications can occur. However, there are several approaches that address this issue and a few are summarised in Table 1, which highlights, whether faults are considered at the level of nodes or individual processes; the manner of fault detection; how the remaining processes are notified of faults; how the state is saved; and how failed processes are restored, for example, by creating new processes or using spare ones.

In production, the target application executes using Intel MPI and so was modified to incorporate the proposed application-level fault tolerant procedures. Monitor processes are distributed across servers to detect remote node or link failures, while application processes periodically save their state and are notified of faults by their local monitors through interrupts that initiate recovery procedures, as described in the rest of this section.

This work is based on a partially synchronous distributed system model [3] where an application is defined as a set of processes that communicate through messages. The target environment consists of a multicore computing cluster. The execution model considers a single application process, p_j, per core. We assume that multiple unrecoverable failures on nodes or communication links may occur, but only one fault at a time, and that the system is fully interconnected so that a link failure will not partition the network. After a failure, the application will use the remaining resources to continue its execution since maintaining spare resources idle would be inefficient in the case of long running applications.

Table 1. Fault Tolerance Services Available. C = checkpointing; ML = Message-logging; CC = Coordinated checkpointing; AC = Application-level checkpointing;

Implementation	Granularity	Detection	Fault Notification	Saving technique	Recovery
MPI-FT [10]	Node	Centralized	By message.	ML	New process
MPI/FT [1]	Process	Centralized	By message.	CC	New process
MPICH-V [2]	Process	Centralized	None	ML or CC	New process
FT-MPI [6]	Process	Not informed	By error code from MPI calls	AC	User defined
RADICMPI [5]	Node	Distributed	None	C + ML	New process
EasyGrid AMS [12]	Process	Distributed	None	AC + ML	New process

2.1   The Fault Monitoring Mechanism

In order to detect faults, application processes and a single monitor process m_i coexist in each node. The multicore node to which p_j and m_i are allocated to is denoted by $n(p_j)$ and $n(m_i)$, respectively. Each monitor m_i creates four threads: *Inspection()*, *Propagation()*, *Detection()* and *Heartbeat()* and coordinates the termination of the application. It manipulates the following lists that are shared among the threads: LS_i is a list of monitors that have supposedly failed; faulty application processes in LFP_i; and LFM_i with all monitor processes (and consequently, their nodes) that have failed during execution. Semaphores coordinate threads and enforce thread safe MPI behaviour.

Although perhaps particular to Intel MPI, the behaviour of blocking collective MPI operations had to be modified to allow the execution to continue with fewer processes, in case of a failure. Some MPI functions were re-implemented using the following techniques: the Dissemination algorithm [7] for MPI_Barrier; the Binomial Tree for MPI_Bcast and MPI_Reduce. The monitors assume the role of MPI_Finalize and detect the termination of the active MPI processes.

In asynchronous systems it is impossible to distinguish whether a processor is running very slowly or has stopped all together due to a crash failure [3]. There are different classes of failure detectors but none in the asynchronous model is implementable without making some synchrony assumptions. Based on the class $\diamond Q$, the threads in Algorithms 1 and 2 [9] were implemented to identify faulty nodes. The monitors are arranged so that $succ(m_i)$ and $pred(m_i)$ represent, respectively, the successor and predecessor of monitor m_i in the ring of active monitors.

Algorithm 1. Detection()

```
1  target_i ← succ(m_i);
2  LS_i ← ∅; ∀m_j : △_{i,j} ← timeout;
3  loop
4      wait(mutex_i);
5      send("Are you alive?", target_i);
6      received ← false;
7      signal(mutex_i);
8      sleep(△_{i,target_i});
9      wait(mutex_i);
10     if not received then
11         △_{i,target_i} ← △_{i,target_i} + 1;
12         LS_i ← LS_i ⋃{target_i};
13         target_i ← succ(target_i);
14     signal(mutex_i)
```

Algorithm 2. Heartbeat()

```
1  loop
2      recv(msg, m_j);
3      wait(mutex_i);
4      if msg = "Are you alive?" then
5          send("I am alive.", m_j);
6          if m_j ∈ LS_i then
7              LS_i ← LS_i − {m_j, ..., pred(target_i)};
8              target_i ← m_j; received ← true;
9      else if msg = "I am alive." then
10         if m_j = target_i then
11             received ← true;
12         else if m_j ∈ LS_i then
13             LS_i ← LS_i − {m_j, ..., pred(target_i)};
14             target_i ← m_j; received ← true;
15     signal(mutex_i);
```

Detection() in Algorithm 1 sends, after at least $\triangle_{i,target_i}$ time steps, a heartbeat request message to its neighbouring monitor, $target_i$, to discover if that node has failed. If after a further $\triangle_{i,target_i}$ time units, m_i has not received an answer from $target_i$ (via *Heartbeat()*), $target_i$ will be placed under suspicion of having failed and included in LS_i. Since a lack of response may be due to the round trip latency of the messages being longer than $\triangle_{i,target_i}$, the timeout is incremented to reduce the possibility of future incorrect presumptions.

Upon receiving a message *msg* from monitor m_j, *Heartbeat()* in Algorithm 2 either replies to the sender with a heartbeat or records the receipt of such a message. If the sender m_j is in LS_i, all the monitors between m_j and $pred(target_i)$ are removed, since m_j is in fact alive and the other monitors will be tested by m_j and/or its successors. This mechanism can therefore correct false positives made by unreliable failure detectors. On the other hand, if a monitor in this path has actually failed, this will eventually be detected again.

In Algorithm 3, *Inspection()*, at every *inspection_timeout* time units, m_i checks for a new failed monitor (that is, a node) and, if a failure occurred, updates LFM_i and broadcasts this information to synchronise and update the remaining monitors. If *Propagation()* (Algorithm 4) receives a new list of faulty monitors from another monitor, m_i updates its own LFM_i and LFP_i. An interrupt signal and LFP_i are sent to all application processes present on $n(m_i)$. In turn, an acknowledgement message, sent by each p_j on $n(m_i)$ should be received by m_i to confirm the receipt of the failure information.

Algorithm 3. Inspection()

---
1 **loop**
2      $sleep(inspection\_timeout)$; $wait(mutex2_i)$; $wait(mutex_i)$;
3   **if** $LS_i \neq \emptyset$ **then**
4   **if** $LFM_i = \emptyset$ **then**
5              $LFM\_aux_i \leftarrow LS_i$; $LFM_i \leftarrow LS_i$;
6   **else**
7              $LFM\_aux_i \leftarrow (LS_i - LFM_i)$; $LFM_i \leftarrow LFM_i \cup LFM\_aux_i$;
8      $signal(mutex_i)$;
9   **if** $LFM\_aux_i \neq \emptyset$ **then**
10         $\forall m_j$, **if** $m_j \notin LFM_i$ **then** $send(LFM_i, m_j)$;
11     $signal(mutex2_i)$;

Algorithm 4. Propagation()

---
1 **loop**
2      $recv(LFM_k, m_k)$; $wait(mutex2_i)$;
3      $LFM_i \leftarrow LFM_i \cup (LFM_k - LFM_i)$;
4      $\forall p_j$, **if** $n(p_j) = n(m_i)$ **then** $sendInterruptionSignal(p_j)$;
5      $LFP_i \leftarrow IdentifyFaultyApplicationProcesses(LFM_i)$;
6      $\forall p_j$, **if** $n(p_j) = n(m_i)$ **then** $send(LFP_i, p_j)$;
7      $\forall p_j$, **if** $n(p_j) = n(m_i)$ **then** $recv(\text{END\_PROPAGATION}, p_j)$;
8      $signal(mutex2_i)$;

2.2   Failure Recovery

In this work, crash faults are tackled with checkpoints and rollback recovery techniques within the algorithm. The recovery process consists of restoring the application to a consistent global state prior to the failure. A global checkpoint is a set of local checkpoints, one from each process of the application and it is consistent if, for the given p local checkpoints, there are no messages (or a causal chain of messages) sent by p_i after its local checkpoint that must be received by p_j before its corresponding checkpoint.

In RTM, application processes execute a loop where during each iteration, messages are exchanged with its neighbours (determined by the domain decomposition), before moving on to the next iteration. Thus, recording checkpoints at the end of each iteration guarantees that the local checkpoints form a consistent global state. The checkpoints contain information relating to 16 matrixes:

2 three dimensional matrixes representing wave fields and 14 bi-dimensional energy matrixes. To create a checkpoint, all of these matrixes are copied into a unique three dimensional matrix and then compressed using zlib [4]. Initial tests showed that application runtime increases in the order of two to three times when uncompressed matrix is used. In our experiments, each local checkpoint is saved in a different file in a common repository managed by NFS. Only the most recent checkpoint needs to be kept for each process. Since the remaining active processes divide the original domain space amongst themselves, each process must load the matrixes that correspond to the new size and position of the local domain space. Fault recovery requires the remaining application processes to read and uncompress the checkpoint files necessary to obtain the wave field and energy data of their newly assigned subdomain.

3 Experimental Results

The original application and the fault tolerance procedures were implemented in C++ and Fortran together with the Intel MPI library. The experiments were run on a 40 nodes cluster, each node containing two Intel Xeon E5430 2.66GHz Quad core processors with 12MB L2 cache each and 16GB RAM memory per node, running RHEL 5.3 and NFS, and interconnected by a Gigabit Ethernet network. Tests were run over a real problem instance, with each process initially receiving a $210 \times 210 \times 832$ wave field matrix.

Table 2. The overhead of the fault tolerance mechanisms with checkpoint intervals of 250 and 500 iterations (ET_{250} and ET_{500})

N $(n_x * n_y)$	Matrix Dimension	ET (s)	ET_{250} (s)	O_{250} (%)	ET_{500} (s)	O_{500} (%)
24 (8×3)	$1600 \times 600 \times 800$	3609.35	3648.37	1.08	3627.35	0.50
32 (8×4)	$1600 \times 800 \times 800$	3560.65	3627.33	1.87	3595.32	0.97
64 (8×8)	$1600 \times 1600 \times 800$	3812.89	3987.31	4.57	3915.06	2.68
128 (8×16)	$1600 \times 3200 \times 800$	3956.72	4165.27	5.27	4063.90	2.71
256 (8×32)	$1600 \times 6400 \times 800$	4074.69	4365.47	7.14	4221.26	3.60

Table 2 shows the number of processes N where n_x and n_y are the number of cores per node and of nodes, the total dimension of the problem and ET, the average of five executions of the original code without any fault tolerance mechanisms. The first set of tests analyse the monitoring and checkpointing overhead in a scenario without failures. The following parameters were used: *timeout* = 60 seconds; *inspection_timeout* = 180 seconds; and checkpoints at two different intervals, every 250 and 500 iterations, respectively. In all tests, RTM executed 3.077 iterations, resulting in 12 checkpoints for the 250 iteration interval and 6, for 500. As expected, the application's runtime and the overheads of the fault tolerance procedures grew with the number of processes and such overheads were smaller for a checkpoint interval of 500 than for 250 iterations. From these results, it is seen that the performance degradation caused by monitoring and checkpointing was not substantial, ranging from about 1% and 0.5 % with 24 processes up to 7.14% and 3.6% with 256 processes, for 250 and 500 iterations, respectively. Overheads increased only slightly when more than 64 processes were employed, indicating that the proposed approach may be scalable.

The next experiment considers the occurrence and recovery from a single fault at three different points during the execution of the RTM application: (i) at the beginning, immediately after the first checkpoint; (ii) in the middle of the execution, after the sixth and third checkpoint, for 250 and 500 iterations, respectively, and (iii) at the end after the last checkpoint. Let ET_m and ET_{m-1} be the fault free runtimes of the application on m and $m - 1$ nodes, respectively, and let n_t be the total number of checkpoints and n_c be the number of checkpoints prior to the single failure, respectively, taken by fault tolerant version (FT-RTM). Then, $MET = \frac{n_c}{n_t+1}ET_m + \frac{n_t+1-n_c}{n_t+1}ET_{m-1}$ is a lower bound for the minimum execution time of FT-RTM with a single failure. The results in Table 3 are again the average time, in seconds, of five executions, that showed a standard deviation of less than 1%. The results for 500 iterations presented similar behaviour but with smaller overheads and were omitted due to space limitations. In all cases, the overhead for a single failure was less than 18% and the detection overhead was about 3,8% of 1-failure runtime, thus showing that the proposed approach is feasible and more attractive than restarting the application.

Table 3. Average execution times (in seconds) with and without failure for scenarios (i), (ii) and (iii) with checkpoint intervals of 250 iterations and their overheads

	N	24	32	64	128	256
	MET	4935.68	4521.16	4245.92	4497.13	4254.16
Scenario (i)	1 failure	5390.90	4894.97	4639.15	4953.95	4993.33
	Overhead (%)	9.22	8.27	9.26	10.16	17.38
	MET	4383.04	4120.95	4065.49	4271.96	4179.38
Scenario (ii)	1 failure	4731.40	4441.49	4509.76	4743.41	4786.03
	Overhead (%)	7.95	7.78	10.93	11.04	14.52
	MET	3719.88	3640.69	3848.97	4001.76	4089.64
Scenario (iii)	1 failure	4001.30	3957.74	4324.22	4559.41	4773.69
	Overhead (%)	7.57	8.71	12.35	13.94	16.73

During the experimental evaluation, it was observed that when a failure occurred, TCP would continuously re-transmit an undelivered message, previously sent by an MPI process to one that has failed, up to a maximum number of tries, given by the variable *tcp_retries2*. After that, the error would propagate from TCP to MPI, causing the application to abort with no chance of recovery. In order to adopt the proposed fault tolerance procedures, it was necessary to change the value of *tcp_retries2*, the default being 15, to a larger value, to allow enough time to detect and recover from the fault.

4   Conclusion and Future Work

The fault tolerance techniques presented here can be employed in a variety of scientific areas that require domain decomposition simulation. Through an extensive experimental analysis, the proposed approach showed to be applicable in practical terms due to its small impact, since the overheads are reasonably low and the mechanisms appear to be scalable. However, further analysis will be carried out with an increasing number of nodes. In this situation, recording all checkpoints in a common repository may become a bottleneck particularly when the number of processes is high and/or checkpointing becomes more frequent. In order to address this problem, all processes of a node could send their compressed checkpoints to a local leader process that would assume the role of writing them to disk. A preliminary evaluation concluded that the scheme is advantageous for 128 and 256 processes, thus indicating that for large systems such a procedure might reduce the bottleneck of using a common fault free repository.

References

1. Batchu, R., Dandass, Y., Skjellum, A., Beddhu, M.: MPI/FT: a model-based approach to low-overhead fault tolerant message-passing middleware. Cluster Computing 7(4), 303–315 (2004)
2. Bouteiller, A., Herault, T., Krawezik, G., Lemarinier, P., Cappello, F.: MPICH-V project: A multiprotocol automatic fault-tolerant MPI. International Journal of High Performance Computing Applications 20(3), 319–333 (2006)
3. Chandra, T.D., Toueg, S.: Unreliable failure detectors for reliable distributed systems. J. ACM 43(2), 225–267 (1996)
4. Deutsch, P., et al.: Zlib compressed data format specification version 3.3 (1996)
5. Duarte, A., Rexachs, D.I., Luque, E.: An intelligent management of fault tolerance in cluster using RADICMPI. In: Mohr, B., Träff, J.L., Worringen, J., Dongarra, J. (eds.) PVM/MPI 2006. LNCS, vol. 4192, pp. 150–157. Springer, Heidelberg (2006)
6. Fagg, G.E., Dongarra, J.: FT-MPI: Fault Tolerant MPI, Supporting Dynamic Applications in a Dynamic World. In: Dongarra, J., Kacsuk, P., Podhorszki, N. (eds.) PVM/MPI 2000. LNCS, vol. 1908, pp. 346–353. Springer, Heidelberg (2000)
7. Hoefler, T., Mehlan, T., Mietke, F., Rehm, W.: A Survey of Barrier Algorithms for Coarse Grained Supercomputers. Chemnitzer Informatik Berichte 04(03) (2004)
8. Intel: Intel MPI Library Reference Manual (2011),
   http://software.intel.com/en-us/articles/intel-mpi-library-documentation
9. Larrea, M., Arévalo, S., Fernández, A.: Efficient algorithms to implement unreliable failure detectors in partially synchronous systems. Dist. Comp., 847–847 (1999)
10. Louca, S., Neophytou, N., Lachanas, A., Evripidou, P.: MPI-FT: Portable Fault Tolerance Scheme for MPI. Parallel Processing Letters 10(4), 371–382 (2000)
11. Ortigosa, F., Araya-Polo, M., Rubio, F., Hanzich, M., Cruz, R., Cela, J.: Evaluation of 3d RTM on HPC platforms. SEG Expanded Abstracts 27(1), 2879–2883 (2008)
12. da Silva, J.A., Rebello, V.E.F.: Low Cost Self-healing in MPI Applications. In: Cappello, F., Herault, T., Dongarra, J. (eds.) PVM/MPI 2007. LNCS, vol. 4757, pp. 144–152. Springer, Heidelberg (2007)

libhashckpt: Hash-Based Incremental Checkpointing Using GPU's

Kurt B. Ferreira[1,3], Rolf Riesen[2], Ron Brighwell[1],
Patrick Bridges[3], and Dorian Arnold[3]

[1] Scalable System Software
Sandia National Laboratories[*]
{kbferre,rbbrigh}@sandia.gov
[2] IBM Research, Ireland
rolf.riesen@ie.ibm.com
[3] Department of Computer Science
University of New Mexico
{kurt,bridges,darnold}@cs.unm.edu

Abstract. Concern is beginning to grow in the high-performance computing (HPC) community regarding the reliability guarantees of future large-scale systems. Disk-based coordinated checkpoint/restart has been the dominant fault tolerance mechanism in HPC systems for the last 30 years. Checkpoint performance is so fundamental to scalability that nearly all capability applications have custom checkpoint strategies to minimize state and reduce checkpoint time. One well-known optimization to traditional checkpoint/restart is incremental checkpointing, which has a number of known limitations. To address these limitations, we introduce libhashckpt; a hybrid incremental checkpointing solution that uses both page protection and hashing on GPUs to determine changes in application data with very low overhead. Using real capability workloads, we show the merit of this technique for a certain class of HPC applications.

1 Introduction

Disk-based coordinated checkpoint/restart has been the dominant fault tolerance mechanism in high performance computing (HPC) systems for at least the last 30 years. In current large distributed-memory HPC systems, this approach generally works as follows: periodically all nodes quiesce activity, write all application and system state to stable storage, and then continue with the computation. In the event of a failure, the stored checkpoints are read from stable storage to return the application to a known-good state.

---

[*] Sandia National Laboratories is a multi-program laboratory operated by Sandia Corporation, a wholly owned subsidiary of Lockheed Martin Corporation, for the U.S. Department of Energy's National Nuclear Security Administration under contract DE-AC04-94AL85000.

Y. Cotronis et al. (Eds.): EuroMPI 2011, LNCS 6960, pp. 272–281, 2011.

Checkpoint performance impacts scalability of large-scale applications to such a degree that many capability applications have their own custom *application-specific* checkpoint mechanism to minimize the saved checkpoint state and therefore the time to checkpoint (this time is also referred to checkpoint commit time). While this approach minimizes the application state that must be written to disk, it requires intimate knowledge of the application's computation and data structures, and is typically difficult to generalize to other applications.

One well-known and generalized optimization of traditional checkpoint/restart is *incremental checkpointing*. Incremental checkpointing [6,8,17] attempts to reduce the size of a checkpoint, and therefore the time to write a checkpoint, by saving only differences in state from the last checkpoint.

Current incremental methods have failed to achieve dramatic decreases in checkpoint size because of a reliance on page protection mechanisms to determine which address ranges have been written, or *dirtied*, during the checkpoint interval [8]. Relying solely on page-based mechanisms forces such an approach to work at a granularity of the operating systems page size. Even if only one byte in a page is written, the entire page is marked as dirty and must be saved. Furthermore, if identical values are written to a location, that page is still marked as dirty. These problems are also compounded by the increasing maximum page sizes of modern processors and the increased performance for HPC applications on these larger page sizes.

To address these limitations, we introduce libhashckpt: a hybrid incremental checkpointing approach that uses page protection mechanisms, a hashing mechanism, and MPI hooks to determine the locations within a page that have changed. To reduce the overhead of the hash calculation, libhashckpt also uses graphics processing units (GPU) to offload the hash calculation. Using real HPC workloads, we compare the performance of this technique against page protection-based incremental systems and highly optimized, application-specific checkpoint techniques. Our results show that our approach is able to dramatically reduce system checkpoint sizes compared to previous incremental checkpointing systems, in some cases approaching the checkpoint sizes of hand-tuned application-specific checkpointing systems.

2   Approach

2.1   Overview

The hash-based incremental checkpointing mechanism in libhashckpt works as follows. While the application is running, the library uses the page-protection mechanism to mark those virtual memory pages that have been written in the checkpoint interval as potentially dirty. To support MPI applications, the library also intercepts receive calls and marks message buffers as dirty, identifying them as candidates to be checked by the hashing mechanism. These message buffers require marking because changes in memory from user-level network hardware is not subject to the processor's page protection mechanisms.

When a checkpoint is requested, the library hashes all blocks corresponding to potentially dirty pages, comparing the key with previously stored values, if they exist. If no key exists, or if the key has changed, the block is marked to be included in the checkpoint and excluded otherwise. If the node contains a GPU, potentially dirty blocks are copied down to the GPU and the computed keys are copied up to host memory. Finally, once the hash calculation has completed, all blocks that have been marked as changed by the library are then saved to stable storage for later retrieval, if needed.

2.2   Library Implementation Details

libhashckpt is based on the libckpt library [17], now referred to as clubs [2]. Clubs is a transparent, user-level, checkpoint library for Unix based systems. It contains a number of optimizations including:

- Virtual memory page-protection based incremental checkpointing;
- Forked checkpointing; and,
- User-directed checkpointing which allows the user to include or exclude portions of the processes address space in the checkpoint.

We added the following functionality to this library. Firstly, we added a framework for calculating and storing hash keys of arbitrary block size. The block size can be adjusted to be larger or smaller than the native page size. We also modified the library to intercept MPI receive calls using the MPI profiling layer found in most modern MPI libraries. Finally, we added an engine for offloading this hash calculation to graphics processing units, if any are present.

2.3   Applications and Platform

To evaluate the merit of our hash-based checkpointing library, we present results from two key HPC applications; CTH [9] and LAMMPS [18,19]. These applications represent important HPC modeling and simulation workloads. They use different computational techniques, are frequently run at very large scale for weeks at a time, and are key simulation applications for the US Department of Energy. Also, each of these applications contain highly-optimized application-specific checkpoint mechanisms that will be used for comparison with the methods outlined in this paper.

These application tests were conducted on the Cray Red Storm system at Sandia National Laboratories. For these application runs, the hashing was performed by a spare on-node CPU core as Red Storm system does not contain GPUs. For the GPU results in this paper, we compare the performance of the Opteron processor on Red Storm [5] against that of a NVIDIA Tesla C1060 GPU.

3   Results

In this section, we outline the performance of libhashckpt. First, we examine the results of hashing versus page-based protection mechanisms for determining the percentage of application memory that has actually changed. Following this, we examine the performance of this library with the two aforementioned simulation workloads, comparing this hash-based approach with both standard page protection-based incremental checkpointing and each application's specific checkpoint mechanism. Finally, we examine the performance advantage of computing the MD5 [12] hash used by libhashckpt using a GPU versus a CPU and use a simple model to outline the viability of this method.

With this hash-based approach *aliasing* is a concern. Aliasing, also referred to as collisions, comes about when modifications to a block are just such that the key values are identical. The danger with aliasing is the library will not save modified application data, thereby corrupting the application in the event of a restart. Previous studies have shown the likelihood of aliasing to be higher in practice then expected theoretically for a number of hash functions. Specifically, with the hash signature functions CRC32 and XOR, the probability of collision has been shown to be too high to be considered safe [7]. Secure hash signatures like MD5 and SHA1, however, have been shown to behave in practice as expected theoretically, and therefore reliable enough to be used in a hash-based approach [13].

3.1   Hash-Based Dirty Data Detection

The key feature that libhashckpt hopes to exploit is finer-grained detection of dirtied blocks than is currently possible using mechanisms based solely on page protection mechanisms. To examine the overall potential of such a hash-based approach, we first used libhashckpt to examine what portion of an application's memory actually changed (using fine-grained hashing) versus the percentage that a pure page protection-based mechanism would indicate was changed.

Figure 1 shows the percentage of memory that our hash-based mechanism indicates actually changed at each 15 minute checkpoint interval versus the percentage that a page protection mechanism indicates may have changed. For each of these tests, we use a 512 byte block size on an operating system with 4KB pages. Therefore each machine page contains 8 hash blocks. In Figure 1(a), we see that, while nearly all the allocated memory is written in a checkpoint interval, a very small percentage of that memory actually changes. This small percentage of change is an artifact of the simulation problem. The application uses thresholding such that, in a small simulation-time interval, sections of the simulation do not change. In contrast, for LAMMPS in Figure 1(b), the amount of data changed is nearly identical to the data written. This is because the largest data structure in LAMMPS is the neighbor structure, which continuously changes as atoms move around.

These results demonstrate the potential accuracy advantage a hash-based incremental checkpointing approach can provide over a purely page protection-based mechanism. On the other hand, these results also show that the potential benefits are also highly application-dependent.

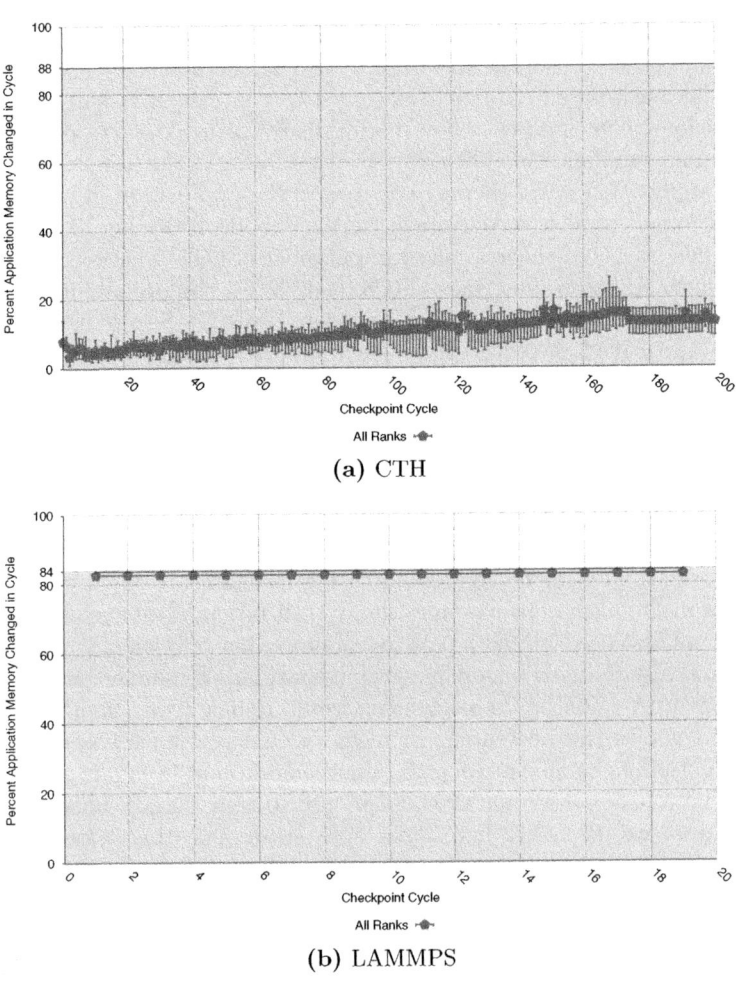

(a) CTH

(b) LAMMPS

Fig. 1. Average percent of allocated memory changed detected using a hash-based incremental checkpointing mechanism for the CTH and LAMMPS. The shaded region represents the average percent of memory written to using a page-protection based mechanisms. Errorbars are shown for CTH but omitted for LAMMPS as the per-process variation is ±0.5%.

3.2   Checkpoint File Size Comparison

Based on the results in the previous section, we then examined the resulting difference in checkpoint sizes between the two incremental checkpointing approaches (pure page protection vs. libhashckpt's hybrid page protection/hashing scheme). We also compared the size of these checkpoints with those generated by the application-specific mechanisms. These application specific methods are highly optimized, and, for the purpose of this work, we view these checkpoint sizes as a file size optimum.

Table 1. Per-process checkpoint size for CTH and LAMMPS. This table contains the size of the checkpoint using standard page protection-based system-level incremental checkpointing (VM CKPT), libhashckpt's hybrid approach, and an application-specific checkpointing approach (App CKPT). For the latter two columns the number in parenthesis is the percent reduction in size when compared to a system-based incremental checkpoint. The VM CKPT and Hash CKPT checkpoints contains data from both the application as well as other libraries linked with the application, for example MPI library data and its associated buffers.

Application	VM CKPT (MB)	Hash CKPT (MB)	App CKPT (MB)
CTH	513	35 (93%)	26 (95%)
LAMMPS	2735	2670 (2.3%)	608 (78%)

Table 1 shows a comparison in per-process checkpoint sizes for our two applications. We see that for CTH, libhashckpt's hash-based method dramatically reduces the size of system-based incremental checkpoints based solely on a page protection mechanism. Custom application-specific checkpointing mechanism does better still, but our hybrid scheme results in checkpoints that are only 35% larger than this highly-optimized approach. One reason our hash-based library is larger than the application-specific method has to do with the fact that the application checkpoint contains *only* application data, while the other methods shown save state from the application as well as the libraries linked with the application, most notably the MPI library and its associated data and buffers.

In contrast to CTH, the hash- and page-based schemes are nearly identical in size for LAMMPS, with application-specific checkpointing routines offering a 75% reduction in checkpoint sizes. This is because the application-specific checkpointing mechanism in LAMMPS can completely avoid writing neighbor structures to checkpoints because they can be reconstructed at application restart, while system-based methods do not have the application-specific knowledge needed to do this.

3.3   GPU Performance

Figure 2 compares GPU vs CPU performance of an MD5 calculation for varying block sizes. The GPU numbers presented in this plot represent the best measured for a block size varying the number of threads and the size of the overlap of the concurrent copy down to the card and computation. Also, these GPU numbers include the time to copy data down to the GPU as well as the time to copy computed keys to host memory. The CPU numbers use the Libgcrypt MD5 implementation. From this figure, we see that the GPU greatly outperforms the CPU implementation.

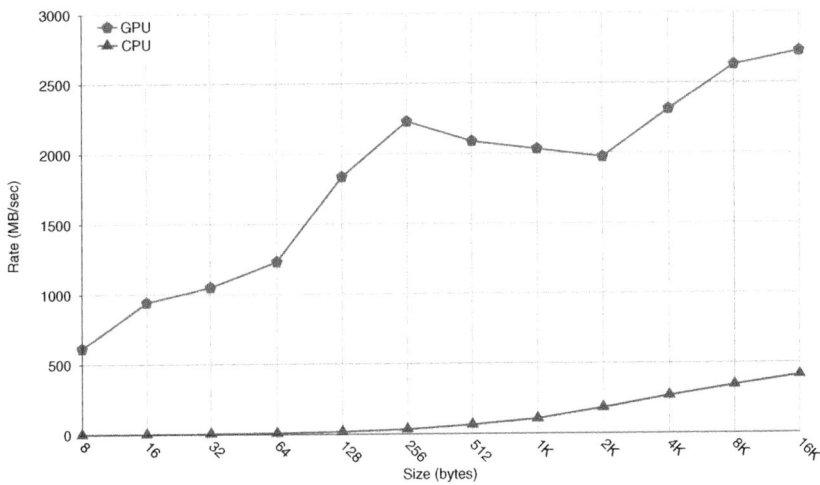

Fig. 2. A comparison of MD5 hashing rates for CPU and GPU. Note, the GPU rate includes both the copying of data to be checksummed down to the cards local memory as well as the copying of the computed keys from the card to host memory. The GPU data is the best recorded for a block size varying the number of threads and the amount of overlap in copy and computation. The CPU numbers are using the Libgcrypt [1] MD5 hashing algorithm.

In addition, with a per-process rate between 600 and 2600 MB/sec, the GPU-based data rates greatly exceed the per-process rate to stable storage for many large scale systems. In the next section we construct a simple model to further illustrate the viability of this approach.

3.4 Viability of Hash-Based Incremental Checkpointing

To evaluate the viability of this method we will compare the performance of this hash-based mechanism with that of a strictly page-based approach. This hash-based approach will outperform the page-based approach when the reduction in the checkpoint size for the hash method outweighs the cost of computing the hashes of the modified pages. More specifically, this approach is viable when:[1]

---

[1] Plank et al pose a similar concept [16].

$$\frac{|checkpoint|}{\beta_{hash}} + \frac{(1 - compression) \times |checkpoint|}{\beta_{ckpt}} < \frac{|checkpoint|}{\beta_{ckpt}} \qquad (1)$$

where $|checkpoint|$ is the size of page-based checkpoint, $compression$ is the percent reduction of hash-based approach in comparison to the page-based method, β_{hash} is the hashing rate, and β_{ckpt} is the rate of checkpoint commit. This equation can be reduced to:

$$\frac{\beta_{ckpt}}{\beta_{hash}} < compression \qquad (2)$$

Using the CTH data presented previously in this paper, $compression$ is 83% and the β_{hash} mean is around 2.0GB/s. Therefore, if a machine has a per-process checkpoint commit speed is less then 1.66GB/s then the hash-based approach will have a lower overhead than the strictly page-based approach. Even with many optimizations and high performance parallel file systems that stripe large writes simultaneously across many disks and file servers, it is difficult to achieve per-process disk commit bandwidth of this magnitude for many large scale systems. A per-process commit rate greater than this 1.66GB/sec value and the page based approach will have lower overheads. For LAMMPS, the compression is 2.4%, therefore the per-process checkpoint commit breakpoint speed is much lower at 48MB/sec; a value more easily reached by current parallel I/O systems.

4   Related Works

Checkpoint/restart is a well-known method for application fault-tolerance for large-scale distributed and parallel systems that has been studied extensively for over thirty years [8]. A number of optimizations has been suggested including; forked or copy-on-write checkpointing [10], checkpointing to remote nodes [20], communication-induced checkpointing [15], compiler-assisted checkpointing [4], incremental checkpointing [6,11], and probabilistic or hash-based checkpointing [14,3]. However, none of these methods have yet matched the performance of application-specific methods and are therefore not widely accepted by most capability workloads.

Most closely related to this work, Agarwal et al. [3] investigated the performance characteristics of a hash-based adaptive incremental checkpointing library. Similar to this work, the authors use an MD5 hash to determine the portions of an application address space that have changed in a checkpoint interval. In contrast to this work, we evaluate the merit of this hash-based technique on actual HPC capability workloads. In addition, we show how GPUs can be used to significantly reduce the overhead of the hash computations. This overhead is important as the computation overhead must be kept significantly lower than the rate to save to stable storage. Also, we compare the merit of this technique with an optimal application-specific checkpoint mechanism. Finally, our work varies from this previous work as we show that, while this technique may be appropriate for some applications, there are classes of HPC applications for which this method is clearly not appropriate.

5   Conclusions and Future Work

In this paper, we introduced libhashckpt, an incremental checkpointing library that uses hashing to save only the changed state of an application in a checkpoint interval. To significantly decrease the overhead of the hash calculation, libhashckpt can utilize GPUs. Using this library, we compare the checkpoint file sizes of this hash-based method with that of a standard page-protection mechanism and a highly optimized application-specific mechanism. Using real capability HPC workloads we show that, for a certain class of applications, this hash-based method can reduce the checkpoint file size to be around 15% of that of a page-based approach. In addition, this method can create checkpoint files which are only 35% larger than that of a manually-coded, application-specific method. Finally, we introduced a simple model to illustrate this proposed techniques viability for real-world HPC workloads.

There are several avenues of future work related to this research. First, we would like to analyze more applications in order to evaluate the merit of this technique to a broader set of large-scale applications. In addition, we would like to investigate other hash and checksum algorithms. For this study we used a cryptographically secure hash (MD5), but this algorithm may be overkill for determining block changes and other collision resistant, yet less computationally intense, hash signatures may have lower overheads. Lastly, we need to compare this method with other checkpoint optimization techniques, such as compiler-assisted incremental checkpoint methods.

References

1. libgcrypt web page (July 2010), http://directory.fsf.org/project/libgcrypt/
2. Libckpt web page (2011),
   http://web.eecs.utk.edu/~plank/plank/www/libckpt.html
3. Agarwal, S., Garg, R., Gupta, M.S., Moreira, J.E.: Adaptive incremental checkpointing for massively parallel systems. In: Proceedings of the 2004 International Conference on Supercomputing, St. Malo, France (2004)
4. Bronevetsky, G., Marques, D., Pingali, K., McKee, S.A., Rugina, R.: Compiler-enhanced incremental checkpointing for openmp applications. In: IPDPS, pp. 1–12. IEEE, Los Alamitos (2009)
5. Camp, W.J., Tomkins, J.L.: Thor's hammer: The first version of the Red Storm MPP architecture. In: Proceedings of the SC 2002 Conference on High Performance Networking and Computing, Baltimore, MD (November 2002)
6. Chen, Y., Plank, J.S., Li, K.: CLIP: a checkpointing tool for message-passing parallel programs. In: Proceedings of the 1997 ACM/IEEE conference on Supercomputing (CDROM), pp. 1–11. ACM, New York (1997),
   http://doi.acm.org/10.1145/509593.509626

7. Elnozahy, E.N.: How safe is probabilistic checkpointing? In: Proceedings of the The Twenty-Eighth Annual International Symposium on Fault-Tolerant Computing, FTCS 1998, pp. 358–363. IEEE Computer Society, Washington, DC (1998), http://portal.acm.org/citation.cfm?id=795671.796882

8. Elnozahy, E.N., Alvisi, L., Wang, Y.M., Johnson, D.B.: A survey of rollback-recovery protocols in message-passing systems. ACM Computing Surveys 34(3), 375–408 (2002)

9. Hertel Jr., E.S., Bell, R., Elrick, M., Farnsworth, A., Kerley, G., McGlaun, J., Petney, S., Silling, S., Taylor, P., Yarrington, L.: CTH: A Software Family for Multi-Dimensional Shock Physics Analysis. In: Proceedings of the 19th International Symposium on Shock Waves, Held at Marseille, France, pp. 377–382 (July 1993)

10. Feldman, S.I., Brown, C.B.: Igor: a system for program debugging via reversible execution. In: Proceedings of the 1988 ACM SIGPLAN and SIGOPS Workshop on Parallel and Distributed Debugging, PADD 1988, pp. 112–123. ACM, New York (1988), http://doi.acm.org/10.1145/68210.69226

11. Gioiosa, R., Sancho, J.C., Jiang, S., Petrini, F.: Transparent, incremental checkpointing at kernel level: a foundation for fault tolerance for parallel computers. In: Proceedings of the 2005 ACM/IEEE Conference on High-Performance Computing and Networking, Seattle, WA, USA (2005)

12. Menezes, A.J., Vanstone, S.A., Oorschot, P.C.V.: Handbook of Applied Cryptography, 1st edn. CRC Press, Inc., Boca Raton (1996)

13. Chang Nam, H., Kim, J., Hong, S.J., Lee, S.: A secure checkpointing system. In: Proceedings of Pacific Rim International Symposium on Dependable Computing, pp. 49–56 (2001)

14. Nam, H.C., Kim, J., Hong, S., Lee, S.: Probabilistic checkpointing. In: Twenty-Seventh Annual International Symposium on Fault-Tolerant Computing, FTCS-27, June 1997, pp. 48–57 (1997)

15. Netzer, R.H.B., Xu, J.: Necessary and sufficient conditions for consistent global snapshots. IEEE Trans. Parallel Distrib. Syst. 6, 165–169 (1995), http://dx.doi.org/10.1109/71.342127

16. Plank, J.S., Li, K.: ickp: A consistent checkpointer for multicomputers. Parallel & Distributed Technology: Systems & Applications 2(2), 62–67 (1994)

17. Plank, J.S., Beck, M., Kingsley, G., Li, K.: Libckpt: transparent checkpointing under unix. In: Proceedings of the USENIX 1995 Technical Conference Proceedings, TCON 1995, pp. 18–18. USENIX Association, Berkeley (1995), http://portal.acm.org/citation.cfm?id=1267411.1267429

18. Plimpton, S.J.: Fast parallel algorithms for short-range molecular dynamics. Journal Computation Physics 117, 1–19 (1995)

19. Sandia National Laboratory: LAMMPS molecular dynamics simulator April 10 (2010), http://lammps.sandia.gov

20. Zandy, V.C., Miller, B.P., Livny, M.: Process hijacking. In: Proceedings of the 8th IEEE International Symposium on High Performance Distributed Computing, HPDC 1999, p. 32. IEEE Computer Society, Washington, DC (1999), http://portal.acm.org/citation.cfm?id=822084.823234

Noncollective Communicator Creation in MPI*

James Dinan[1], Sriram Krishnamoorthy[2], Pavan Balaji[1], Jeff R. Hammond[1],
Manojkumar Krishnan[2], Vinod Tipparaju[3], and Abhinav Vishnu[2]

[1] Argonne National Laboratory, Argonne, Illinois
{dinan,balaji}@mcs.anl.gov, jhammond@alcf.anl.gov
[2] Pacific Northwest National Laboratory, Richland, Washington
{sriram,manoj,abhinav.vishnu}@pnl.gov
[3] Oak Ridge National Laboratory, Oak Ridge, Tennessee
tipparajuv@ornl.gov

Abstract. MPI communicators abstract communication operations across application modules, facilitating seamless composition of different libraries. In addition, communicators provide the ability to form groups of processes and establish multiple levels of parallelism. Traditionally, communicators have been collectively created in the context of the parent communicator. The recent thrust toward systems at petascale and beyond has brought forth new application use cases, including fault tolerance and load balancing, that highlight the ability to construct an MPI communicator in the context of its new process group as a key capability. However, it has long been believed that MPI is not capable of allowing the user to form a new communicator in this way. We present a new algorithm that allows the user to create such flexible process groups using only the functionality given in the current MPI standard. We explore performance implications of this technique and demonstrate its utility for load balancing in the context of a Markov chain Monte Carlo computation. In comparison with a traditional collective approach, noncollective communicator creation enables a 30% improvement in execution time through asynchronous load balancing.

1 Introduction

MPI communicators [6] provide communication contexts that differentiate both point-to-point and collective operations. This functionality enables the programmer to isolate communication between application modules by effectively sandboxing communication in different communicators. This has enabled the development of large applications composed of independently developed modules and libraries. In addition to this primary function, communicators also provide the ability to form groups of MPI processes and perform communication, especially collective communication, within these groups. Such process groups enable the programmer to express multiple levels of parallelism within MPI applications, a capability that has been shown to be increasingly important as computing system size increases.

* This work was supported through a resource grant from the Argonne Leadership Computing Facility (ALCF) and by the U.S. Department of Energy under contracts DE-AC02-06CH11357, DE-AC05-00OR22725, and DE-ACO6-76RL01830.

Y. Cotronis et al. (Eds.): EuroMPI 2011, LNCS 6960, pp. 282–291, 2011.

At the MPI implementation level, the key ingredient in a communicator is a context id. All processes participating in a communication operation identify the communicator using its context id, often an integer. The context id essentially serves as another tag, in addition to any user-provided communication tag, in matching communication operations. As such, consensus on the context id is required in order to correctly match communication operations.

MPI supports collective creation of communicators, where all processes in the parent communicator participate in the creation of the child communicator. However, the recent push towards petascale and beyond has brought forth new application architecture idioms and programming model use cases that highlight the need for noncollective creation of communicators. For example, in applications where a small subset of processes dynamically cooperate to make progress on a work component, this subset of processes might want to create a communicator without synchronizing with the remaining processes in the system. Similarly, when a process fails, recreating the communicator should be possible without involving the failed process. However, current MPI communicator creation operations such as MPI_Comm_dup, MPI_Comm_split, and MPI_Comm_create do not allow for such flexibility.

This collective mode of creation is so widely taught and practiced that noncollective creation of communicators was considered impossible within the MPI standard. In this paper, we present a new communicator creation algorithm that constructs a communicator collectively only on the group of processes that will be members in the new communicator. This algorithm is portable and uses only functionality provided by the current MPI standard. In short, our algorithm works around the MPI API's limitation by hierarchically constructing and merging intercommunicators into intracommunicators.

We present key use cases from a variety of domains that motivate the need for communicator creation that is not collective on a parent communicator. In addition, we evaluate the overhead of this implementation as compared with the traditional collective creation directly supported in the MPI API. We evaluate the benefits of this approach to asynchronous dynamic load balancing through a Markov chain Monte Carlo benchmark kernel. Compared with a traditional collective approach to load balancing, noncollective communicator formation enables a 30% improvement in execution time.

This paper is organized as follows. In Section 2 we present the current state of MPI communicators and motivate the need for noncollective communication creation. In Section 3 we present our noncollective communicator creation algorithm. In Section 4 we present an empirical evaluation of the overhead and performance impact of noncollective communicator creation. Section 5 contains a discussion of how this functionality can be incorporated into the MPI standard to improve performance. We summarize our conclusions in Section 6.

2   Need for Noncollective Communicator Creation

The processes cooperating in a subcomputation of a program are said to form a process group. In MPI, such groups can be conveniently specified using MPI_Group objects. These objects, created using local operations, specify the participation and ordering of processes in a group. While MPI groups allow querying for membership, they are not sufficient for communication operations. Such operations require the creation of an MPI communicator, which backs the group information with one or more context ids.

The widely used interfaces for MPI communicator creation are MPI_Comm_create, MPI_Comm_dup, and MPI_Comm_split. MPI_Comm_dup and MPI_Comm_split result in valid communicator handles on all processes in the parent group and hence are naturally collective on all member processes in the parent communicator. MPI_Comm_create, on the other hand, takes an MPI_Group object and creates a communicator on the subset of processes specified by the group. While the outcome is useful only for the processes participating in the subcommunicator, it is specified to be collective on the parent communicator. This has resulted in the common belief that MPI communicator creation requires full cooperation of all processes in the parent communicator. In the remainder of this section, we present several case studies where a communicator creation operation that is not collective over the parent communicator is required to enable a certain capability (e.g., collective communication after one or more process failures) or is helpful to improve performance.

2.1 Fault Tolerance

Several solutions have been proposed to provide fault tolerance for MPI programs. All approaches must address the reconstruction of a communicator that can be used for continued program execution. Proposed approaches include the use of explicit intercommunicators [4] and an MPI extension to introduce dynamic communicators that support grow and shrink operations [3]. The MPI standard leaves the behavior of an MPI implementation following process or network failures undefined, and several implementations allow for specific communication operations to proceed in such cases. For example, if a process has failed, point-to-point communication between remaining processes is not affected; all communication with a failed process would return an error.

Supporting collective operations after a failure has occurred is more challenging, as all communicators that contain a failed process can no longer be used. A collective operation on such a communicator can return an error. Furthermore, since all operations to create new communicators are collective, the application cannot create a new communicator that excludes the failed process, thus making collective operations unusable after a process failure has occurred.

With the algorithm we present in this work, a new communicator can be rebuilt by the application after a failure without introducing the complexity associated with intercommunicators or an extension to the MPI standard. Our approach relies on the observation that MPI_COMM_SELF is well defined on all live processes, irrespective of the state of any other communicators.

2.2 Global Arrays

Global Arrays [9] is a global address space programming model that provides a global view of multidimensional, shared arrays distributed across the memory of multiple processes. Much of GA's functionality is implemented on top of the remote memory operations provided by the Aggregate Remote Memory Copy Interface (ARMCI) [7]. Global Arrays and ARMCI were designed to be fully interoperable with MPI and employ MPI for process management, message passing, and collective operations.

Support for process groups in GA was initially built using MPI communicators. Subsequent application use cases motivated GA to support process groups that are collectively constructed only on the processes that are members of the new group. The implementation of these alternative process groups was not backed by an MPI communicator. The lack of an MPI communicator for each process group necessitated alternative pathways for functionality in the implementation that did not rely on communicators, primarily in supporting two-sided and collective communication. This design was based on the widely held assumption that MPI cannot support the needed mode of communicator creation. While efficient and practical, this broke the interoperability between ARMCI and MPI. GA has henceforth supported both functionalities, letting the user trade MPI interoperability for increased flexibility. The work presented in this paper resolves this dichotomy.

2.3   Dynamic Load Balancing and Multilevel Parallelism

Several applications have stressed the need for flexible management of process groups. Flexible process groups have been used in mixed quantum-mechanical and molecular mechanical calculations (QM/MM) [5] that couple classical force calculations for long-range interactions with short-range quantum mechanical corrections. The work per task performing a quantum mechanical calculation can vary widely and can only be approximately estimated a priori, making static load balancing difficult. One approach [8] employed a dynamic load balancing scheme in which the each QM task specified the number of processes that form a group to execute that task. Idle processes are identified and batched into a group to execute the next available task. This approach required idle processes to form a group while other processes are actively executing other tasks.

Dynamical nucleation theory Monte Carlo (DNTMC) [10,11] simulations are used for determining molecular nucleation rate constants and chemical properties. One of the main components of these algorithms involves many parallel Markov chain walkers to accelerate the exploration of the potential energy surface of interest. The walkers, each of which is executed in parallel on a subgroup, are all periodically synchronized to collect statistics and restart information, determine convergence, and steer for the simulation. One of the major concerns of this model was the load imbalance that can occur between the individual Markov chains. The reason behind this imbalance is the variable time for individual energy evaluations, which depends on the overall molecular cluster configuration and method being used for the evaluation. An alternative method currently under development allows a group that has completed its assigned work to help another group. The two groups merge to form a larger group and accelerate the lagging Markov chain calculation. This approach requires localized creation of groups with participation from only processes contributing to the particular work of interest.

Nonequilibrium umbrella sampling (NEUS) [2] is a technique for obtaining transition rates for rare events. Its computational profile is similar to DNTMC, although load imbalance can emerge from many different sources, as the walkers evaluate multistep dynamic trajectories rather than an energy evaluation. Because of the scalability of the underlying molecular dynamics simulations and the possibility of large variation in the execution time of each trajectory (the termination criteria depend greatly on the physics), NEUS can and should dynamically adjust the number of nodes assigned to each task.

3   Noncollective Formation of MPI Communicators

As discussed in Section 2, the routines provided by MPI for communicator creation (e.g., MPI_Comm_create) are collective over an existing parent communicator. In this section, we define a new group-collective communicator creation model where communicator creation is collective over only the processes that will be members in the resulting communicator. In addition, this algorithm does not require a parent communicator that is valid for collective communication. This is useful when a parent communicator (e.g., MPI_COMM_WORLD) has become invalid for collective communication because of a failure, when all processes in the parent communicator cannot be recruited to participate in communicator creation, and for performance when the output communicator is much smaller than the parent communicator.

The group-collective communicator creation algorithm is given in Algorithm 1. This algorithm accepts as input the MPI group corresponding to the new communicator, an existing communicator that contains all ranks in $group$, and a tag that can be safely used by this operation for communication on $comm$. The algorithm is collective only on processes that are members of $group$, and $group$ must be identical on all ranks. If desired, a check for $grp_rank = $ MPI_UNDEFINED can be used to filter out callers that are not in $group$, returning MPI_COMM_NULL on these processes. As output, a new communicator is produced where the ranks are ordered according to $group$'s ordering. The algorithm performs $\log |group|$ intercommunicator creation and merge steps to form the final intracommunicator.

The first step in this algorithm is to translate $group$'s ranks, $\{0..|group| - 1\}$, to the corresponding ranks in $comm$. In most MPI implementations, this step requires $O(|group| \cdot |comm|)$ steps except when translating to MPI_COMM_WORLD, whose translation table is cached, yielding a complexity of $O(|group|)$.

The output communicator, $comm'$, is initially assigned MPI_COMM_SELF. This communicator is then recursively merged between pairs of adjacent groups until a single communicator remains. If the current group identity is even, the group attempts to create an intercommunicator with the group to its right. This operation requires a tag that MPI can use internally to create the intercommunicator. The tag argument to the group-collective communicator creation algorithm is particularly important when multiple threads invoke this routine concurrently; the user must supply tags such that each operation can be uniquely identified. If no right neighbor group exists (i.e., $size$ is not a power of two), the group skips this round and will participate as a right neighbor in a future round. If an intercommunicator is created, it is then merged into an intracommunicator and stored in $comm'$. A high/low argument to MPI_Intercomm_merge is used to ensure that the rank ordering given in $pids$ is preserved.

Algorithm 1. Group-collective communicator creation algorithm.

INPUT: $group, comm, tag$
OUTPUT: $comm'$
REQUIRE: $group$ is ordered by desired rank in $comm'$ and is identical on all callers
LET: $grp\_pids[0..|group| - 1] = \mathbb{N}$ and $pids[\,]$ be arrays of length $|group|$

MPI_Comm_rank($comm$, &$rank$)
MPI_Group_rank($group$, &grp_rank), MPI_Group_size($group$, &grp_size)
MPI_Comm_dup(MPI_COMM_SELF, &$comm'$)

MPI_Comm_group($comm$, &$parent\_grp$)
MPI_Group_translate_ranks($group$, grp_size, grp_pids, $parent\_grp$, $pids$)
MPI_Group_free(&$parent\_grp$)

for ($merge\_sz \leftarrow 1$; $merge\_sz < grp\_size$; $merge\_sz \leftarrow merge\_sz \cdot 2$) **do**
  $gid \leftarrow grp\_rank/merge\_sz$, $comm\_old \leftarrow comm'$
  if gid mod $2 = 0$ **then**
    if $((gid + 1) \cdot merge\_sz < grp\_size$ **then**
      MPI_Intercomm_create($comm'$, 0, $comm$, $pids[(gid + 1) \cdot merge\_sz]$, tag, &ic)
      MPI_Intercomm_merge(ic, 0 /* LOW */, &$comm'$)
    end if
  else
    MPI_Intercomm_create($comm'$, 0, $comm$, $pids[(gid - 1) \cdot merge\_sz]$, tag, &ic)
    MPI_Intercomm_merge(ic, 1 /* HIGH */, &$comm'$)
  end if
  if $comm' \neq comm\_old$ **then**
    MPI_Comm_free(&ic)
    MPI_Comm_free(&$comm\_old$)
  end if
end for

4   Experimental Evaluation

We have evaluated the cost of our group-collective communicator creation method relative to the cost of the parent-collective MPI_Comm_create routine. In addition, we
present a Markov chain Monte Carlo benchmark kernel to explore the performance implications of group-collective communicator creation to load balancing. Experiments
were conducted on a Blue Gene/P system using IBM MPI, which is a derivative of
MPICH2. A node in this system contains a 4-core 850 MHz PowerPC 450 processor
with 2 GB of memory. Racks consist of 1024 nodes and the total number of racks is 40,
yielding 163,840 total processing cores. Because of a bug in the MPI implementation's
intercommunicator creation routine, we have been forced to limit our experimentation
to two racks, or 8,192 cores.

4.1   Group Creation Cost

In Figure 1 we present the costs of group- and parent-collective communicator creation over a range of output group sizes. All experiments in this figure were run on

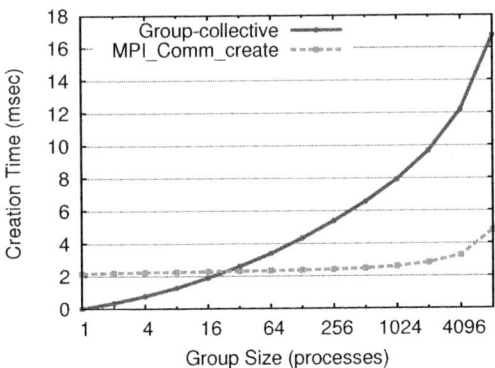

Fig. 1. Communicator creation cost for group-collective versus MPI_Comm_create

8,192 cores. In the case of MPI_Comm_create, collective communication was performed across all ranks in the parent group (MPI_COMM_WORLD for this experiment) regardless of the output group size. This explains the flat cost of MPI_Comm_create relative to the output group size.

In comparison, the group-collective communicator creation must perform $\log |group|$ collective communication steps; the size of the groups involved in this collective communication increases exponentially at each step because of the recursive merging nature of the algorithm. For small groups, we see that this approach is significantly faster than MPI_Comm_create. The cost increases well beyond the cost of MPI_Comm_create; however, as we demonstrate in the next section, this cost can be amortized by potential benefits to the application.

4.2   MCMC Load-Balancing Example

Markov chain Monte Carlo (MCMC) simulations are typically composed of walkers that explore a state space with sequential state transitions. The Monte Carlo transition from one state to the next is tested to determine whether the state is valid; if it is not, it is rejected, and another transition attempt is made. In addition, the amount of computation involved in calculating acceptance can vary across states with respect to the input data. Because of these factors, load balancing MCMC applications is extremely challenging. Often, the work performed by a walker can be parallelized and executed on a group of processes. In our current work with the DNTMC application [11], we have developed a load-balancing solution that reassigns idle processes to active walker groups in order to accelerate that walker.

For this work, we have developed a benchmark kernel that is representative of such MCMC simulations. This benchmark creates a set of initial walker groups of size G and assigns each group a workload. The workload is composed of S work items, corresponding to S state transitions in the Markov chain; processing of each item requires $T/group\_size$ milliseconds; for simplicity, all state transitions are accepted. When a group finishes processing its S work items, it merges with the group to its right. Likewise, groups must periodically check for incoming merge requests; when one arrives,

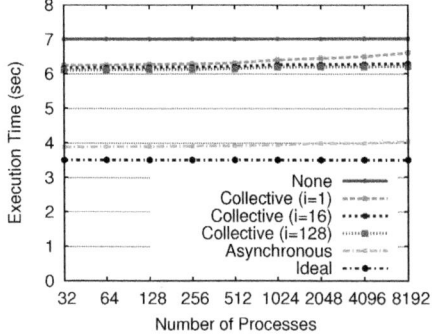

Ld. Bal.	i	Avg.	St. Dev.	Min	Max
None		0.00	0.00	0	0
Async.		14.38	3.54	5	26
Collect.	1	5.38	2.12	2	8
Collect.	2	5.38	2.12	2	8
Collect.	4	5.38	2.12	2	8
Collect.	8	5.38	2.12	2	8
Collect.	16	5.38	2.12	2	8
Collect.	32	5.38	2.12	2	8
Collect.	64	3.75	1.20	2	5
Collect.	128	2.62	0.48	2	3

Fig. 2. Markov chain Monte Carlo benchmark weak scaling up to 8192 cores with none, asynchronous, and collective load balancing

Table 1. Average number of regrouping operations performed per process for the experiment in Figure 2 on 8,192 cores

the old group is freed, and a new group is created. We have implemented this algorithm using both group- and parent-collective communicator creation. In the group-collective case, point-to-point merge requests are sent and result in a merge operation that involves only the merging processes. In the parent-collective case, all processes must perform periodic collective exchange of load information followed by regrouping. This collective load balancing is performed every i work units.

In Figure 2 we present data for a weak scaling experiment with the MCMC benchmark kernel. In this experiment G was four processes, T was 100 ms, and S was $10 \cdot R$ mod 32, where R is the group leader's rank. This resulted in a cyclic work distribution of $0, 40, 80, 120, 160, 200, 240, 280, 0, \ldots$. In the baseline case, regrouping is disabled, and the execution time is bounded by the time required to process the longest Markov chain: $S \cdot T/G$ or $280 \cdot 100ms/4 = 7sec$. The ideal execution time is also shown; this is the calculated execution time with perfect load balancing. Because we have chosen a cyclic, triangular workload, the ideal time is half of the baseline execution time.

Collective load balancing with load balancing intervals of $i = 1, 16, 128$ steps are shown and result in a roughly 15% improvement in execution time compared with no load balancing. Asynchronous group-collective load balancing yields over a 40% improvement in execution time compared with the baseline and over a 30% improvement compared with collective load balancing. The gap between ideal and asynchronous load balancing is due to the interval at which load balancing is performed. Polling for load balancing requests is performed once after each step in the Markov chain. The time between polling operations is the step execution time, $T/group\_size$. For the cyclic work distribution with period $P = 8$, this results in an overhead of up to $(P - 1) \cdot T/group\_size$ for each group.

Table 1 shows the number of regroupings that occurred for each load-balancing configuration on 8,192 cores. We can see from this data that the collective scheme results in a regular load-balancing pattern. In contrast, the asynchronous scheme takes advantage of more fine-grained load-balancing opportunities, leading to a significantly higher average number of regroupings over all processes.

5   Discussion

Intercommunication creation and merge steps perform an all-reduce operation which requires $O(\log p)$ communication steps. In the group-collective communicator creation algorithm, intercommunicator creation and merge steps are repeated $\log p$ times, yielding a time complexity of $O(\log^2 p)$. In comparison, the standard MPI communicator creation routine performs a single all-reduce step and has time complexity $O(\log p)$. The additional $\log p$ cost associated with group-collective communicator creation can be eliminated by extending MPI to provide a direct method for group-collective communicator formation.

5.1   Group-Collective Communicator Creation

The simplest method by which MPI can provide more efficient support for group-collective communicator creation is to include a group-collective communicator creation routine in the MPI standard. This would allow MPI implementors to provide a direct method for backing the provided group with a context ID, for example via a point-to-point all-reduce. Such a routine would take the form:

```
int MPIX_Group_comm_create(MPI_Comm in, MPI_Group grp, int tag, MPI_Comm *out)
```

In this routine, the input intracommunicator and tag are used to create the output intracommunicator. A communicator and tag are necessary to provide MPI with a safe conduit for noncollective communication; this is similar to the mechanism used by MPI's intercommunicator creation routines. The tag plays an important role in ensuring safety of this routine in the presence of threads. Creation of the new communicator is collective over members of the input group, and the input group must be a subset of the input communicator's group. We have included an implementation of this routine using the portable algorithm presented in this paper as an extension in version 1.4 of the MPICH2[1] MPI distribution. We are working toward an integrated implementation that uses MPICH2's internal API to eliminate the overheads identified in this algorithm.

5.2   Generalized Multicommunicators

An alternative to group-collective communicator creation would be to accomplish communicator creation with a single multicommunicator creation and merging step, eliminating a factor of $\log p$ from the creation cost. We present the concept of a multicommunicator as generalization of the current MPI communicator. In the current standard, an MPI intracommunicator is defined to contain a single MPI group. An intercommunicator is defined to contain two nonoverlapping MPI groups. A multicommunicator would be capable of containing an arbitrary number of nonoverlapping groups.

Multiple groups within a single communicator present a significant programmability challenge and significant difficulty in mapping multicommunicators to existing MPI routines. For the purpose of incorporating these generalized communicators with existing MPI functionality, the multicommunicator can be flattened into an intercommunicator. This flattening would merge all nonlocal groups into a single remote group and produce a new intercommunicator. Thus, group-collective communicator formation could be achieved in three steps: multicommunicator creation, flattening into an intercommunicator, and merging of the intercommunicator into an intracommunicator.

6 Conclusion

We have presented an algorithm for MPI communicator creation that is collective over the output group and utilizes only functionality in the current MPI standard. This type of group-collective communicator creation is a key capability for fault tolerance, multi-level parallelism, and load balancing. We have measured the overhead of our technique and demonstrated its effectiveness on a Markov chain Monte Carlo benchmark kernel. Compared with a traditional collective approach, group-collective communicator creation yields a 30% improvement in execution time to the MCMC benchmark through improved load balance.

References

1. MPICH2 Project Website (June 2011),
   http://www.mcs.anl.gov/research/projects/mpich2/
2. Dickson, A., Maienschein-Cline, M., Tovo-Dwyer, A., Hammond, J.R., Dinner, A.R.: Flow-dependent unfolding and refolding of an RNA by nonequilibrium umbrella sampling. ArXiv e-prints (1104.5180), cond–mat.stat–mech (April 2011)
3. Graham, R.L., Keller, R.: Dynamic communicators in MPI. In: Ropo, M., Westerholm, J., Dongarra, J. (eds.) PVM/MPI. LNCS, vol. 5759, pp. 116–123. Springer, Heidelberg (2009)
4. Gropp, W.D., Lusk, E.: Fault tolerance in MPI programs. International Journal of High Performance Computer Applications 18(3), 363–372 (2004)
5. Kamiya, M., Hirata, S., Valiev, M.: Fast electron correlation methods for molecular clusters without basis set superposition errors. The Journal of Chemical Physics 128(7), 74103 (2008)
6. MPI Forum: MPI: A Message-Passing Interface Standard. Version 2.2 (September 4, 2009)
7. Nieplocha, J., Carpenter, B.: ARMCI: A portable remote memory copy library for distributed array libraries and compiler run-time systems. In: Rolim, J.D.P. (ed.) IPPS-WS 1999 and SPDP-WS 1999. LNCS, vol. 1586, pp. 533–546. Springer, Heidelberg (1999), doi:10.1007/BFb0097937
8. Nieplocha, J., Krishamoorthy, S., Valiev, M., Krishnan, M., Palmer, B., Sadayappan, P.: Integrated data and task management for scientific applications. In: Bubak, M., van Albada, G.D., Dongarra, J., Sloot, P.M.A. (eds.) ICCS 2008, Part I. LNCS, vol. 5101, pp. 20–31. Springer, Heidelberg (2008)
9. Nieplocha, J., Palmer, B., Tipparaju, V., Krishnan, M., Trease, H., Aprà, E.: Advances, applications and performance of the global arrays shared memory programming toolkit. Int. J. High Perform. Comput. Appl. 20(2), 203–231 (2006)
10. Schenter, G.K., Kathmann, S.M., Garrett, B.C.: Dynamical nucleation theory: A new molecular approach to vapor-liquid nucleation. Physical Review Letters 82(17), 3484 (1999)
11. Windus, T.L., Kathmann, S.M., Crosby, L.D.: High performance computations using dynamical nucleation theory. Journal of Physics: Conference Series 125(1), 12017 (2008)

Evaluation of Interpreted Languages with Open MPI

Matti Bickel, Adrian Knoth, and Mladen Berekovic

Institute of Computer Science
Friedrich-Schiller-University Jena, Germany
{Matti.Bickel,Adrian.Knoth,Mladen.Berekovic}@uni-jena.de

Abstract. High performance computing (HPC) seems to be one of the last monopolies of low-level languages like C and FORTRAN. The de-facto standard for HPC, the Message Passing Interface (MPI), defines APIs for C, FORTRAN and C++ only. This paper evaluates current alternatives among interpreted languages, specifically Python and C#. MPI library wrappers for both languages are examined and their performance is compared to native (C) Open MPI using two benchmarks. Both languages compare favorably in code and performance effectiveness.

1 Introduction

Current parallel computing frameworks such as the Message Passing Interface (MPI) or OpenMP only offer language bindings for C, C++ and FORTRAN. To use object oriented principles in today's high-performance world, one is left with C++ as the only choice, which is non-satisfactory wrt modern features and consequently productivity gain present in languages like Python or C#.

In addition, unmanaged languages like C or C++ do not cater to "rapid prototyping", the ability to quickly implement and test an algorithm, enabling programmers to explore possible solutions and to come up with a sound solution faster than using a language like C with its known deficiencies, e.g., manual resource management and cumbersome string handling. After an initial solution is found, the program can be refined and adapted to changing needs. If performance is of concern, the program or parts of it can be ported to lower languages like C.

The benefits of interpreted languages have led to various integration attempts with the de-facto standard for intra-cluster communication, the Message Passing Interface (MPI), targeting languages like Ruby[1], Python[2], or Perl[3].

However, some of these projects have not been touched in years, lack support for the MPI-2.1[10] standard or are hardly documented.

Section 2 summarizes the current state, maturity and recent activity of existing MPI language bindings. Despite the advantages of interpreted languages,

---

[1] http://www.mcs.anl.gov/research/projects/mpi/mpi_ruby/

[2] http://www.boost.org/doc/libs/1_39_0/doc/html/mpi/python.html

[3] http://search.cpan.org/~josh/Parallel-MPI-0.03/MPI.pm

Y. Cotronis et al. (Eds.): EuroMPI 2011, LNCS 6960, pp. 292–301, 2011.

their nature suggests a performance penalty. While others have already provided an estimation for this penalty from synthetic benchmarks (Section 3), we complement their findings with an evaluation of a real-world application (Section 4).

```
# from http://tat.wright.name/game-of-life/
def iterate():
    global board  # need to reassign to board
    # find number of neighbours each square has
    neighbour_count = numpy.zeros_like(board)
    neighbour_count[1:, 1:] += board[:-1, :-1]
    neighbour_count[1:, :-1] += board[:-1, 1:]
    neighbour_count[:-1, 1:] += board[1:, :-1]
    neighbour_count[:-1, :-1] += board[1:, 1:]
    neighbour_count[:-1, :] += board[1:, :]
    neighbour_count[1:, :] += board[:-1, :]
    neighbour_count[:, :-1] += board[:, 1:]
    neighbour_count[:, 1:] += board[:, :-1]

    # a live cell is killed if it has fewer than 2
    # or more than 3 neighbours.
    part1 = ((board == 1) & (neighbour_count < 4) & (neighbour_count > 1))

    # a new cell forms if a square has exactly three members
    part2 = ((board == 0) & (neighbour_count == 3))

    # convert to integer from boolean
    board = numpy.cast[numpy.int8](part1 | part2)
    return board
```

Fig. 1. Python code to implement the update function for Conway's *Game of Life*. Array slicing is used to calculate the neighbourhood of all cells at once without utilizing explicit (for-)loops.

2 MPI Language Bindings

The Message Passing Interface is a *specification* produced by the MPI Forum and is under continuous development. In September 2009, version 2.2 of the standard was released to the public. The MPI standard is extensively covered in the literature, for example in [11], [5] and [7] and explained function by function in [9] for MPI-1 and [4] for MPI-2, respectively.

There are two major free implementations available today: Open MPI and MPICH. All of the findings in this report are based on Open MPI but are applicable to MPICH and any other conforming MPI implementation as well.

Research on MPI continues in many areas, including interoperability with OpenMP, performance optimizations at various message sizes and with various physical node layouts. With regard to language bindings, the MPI forum moved to deprecate[4] the C++ API specification from the current MPI-2.2 standard and might remove it completely in MPI-3.0. The forum argued that "better C++ MPI bindings have emerged elsewhere (e.g., Boost.MPI), which explicitly accepts language binding work outside of the standard. This might indicate little or no desire to work on additional and maybe more complex bindings in the future.

Nevertheless, several MPI bindings for interpreted languages exist. Examples include mpi-ruby [6], initially developed by Emil Ong, mpi.net[5] or Parallel::MPI [12] for Perl. However, most of these bindings are orphaned. They have not been in (public) active development for years, miss documentation or an active community. The mpi.net project's last release was 1.0.0 in 2008, and patches sent in February 2009 to the mailing list to fix its Linux support with Open MPI have not been merged into a new release. The Perl script used by this project to build a glue layer between Open MPI and C# appears to be failing with recent Open MPI releases (version 1.3 and newer). The Parallel::MPI project last released in 2002 and no updates have been made to the CPAN site ever since.

A notable exception to this list of potentially discontinued projects is mpi4py project[6], which released a new version (1.2.2) in September 2010 (making it the only studied project with code changes in 2009 or later) and continues to receive updates. It also features an active Usenet group and mailing list.

The package itself is written in a C library wrapping language, Cython, a dialect of Python. It is a sub-project of Scientific Python (SciPy), which "is open-source software for mathematics, science, and engineering" (from the SciPy site). Thus, it's a sister project to NumPy, a package for fast multi-dimensional array operations, including Fourier transforms, linear algebra and random number generation. In fact, mpi4py integrates seamlessly with NumPy, using its array class with minimal overhead compared to C (see Section 4).

C# was created by Microsoft Corporation in 2000 and named ISO/EIC Standard 23270 in 2003. In 2006, the language's standards body Ecma has released version 3.0 of the language specificationDue to its support by Microsoft, Hewlett-Packard, Intel and Novell, C# has become a widely used programming language, currently ranking 4th in the TIOBE index as of May 2011.

Mpi.net was written by the authors of Boost.MPI and originally targeted the Microsoft MPI implementation. In its current version, it uses a Perl script to build a wrapper around the most common MPI implementations. Most of the work went into object serialization and providing a C#-like interface. So instead of explicitly calling MPI_Init() in the program, the MPI part may be wrapped in a using directive, automatically ensuring proper setup and deconstruction of the MPI environment. Thus, MPI programs might start like this:

```
using (new MPI.Environment(ref args)) {
    Intracommunicator c = Communicator.world;
    int worldSize = c.Size;
    int myRank = c.Rank;
```

---

[4] https://svn.mpi-forum.org/trac/mpi-forum-web/ticket/150
[5] http://www.osl.iu.edu/research/mpi.net
[6] http://mpi4py.scipy.org

```
ReceiveRequest r =
        c.ImmediateReceive<Obj>(1, 0);
    r.Wait();
}
```

3   Related Work

After looking at the state of various MPI library wrappers for interpreted languages in Section 2, mpi4py and mpi.net are studied in more detail. The availability of documentation and recent scientific work on them was indeed a key component in the decision to choose these specific two libraries for comparison. Hence, this section will point out issues raised in [1] and [3], as well as compare their performance testing results to those presented in this work in Section 4.

One issue highlighted by Gregor and Lumsdaine [3] is serialized data (i.e. class objects). They state:

> MPI.NETs protocol for transmitting messages of arbitrary length via the native MPI interface introduces significant overhead to the already expensive point-to-point operations for serialized data, due to the higher message volume and the use of synchronous-mode communication. However, the native MPIs inability to receive messages of unknown size leaves few alternatives. We hope to address this shortcoming in a future revision of the MPI standard.

A similar quote can be found regarding mpi4py. Work on the mentioned "future revision", MPI-3, has already been started. In 2008, Gregor et al. posted a paper[7] to the MPI mailing list, highlighting several areas where the MPI standard could be improved to benefit interpreted languages. As a consequence, `MPI_MPROBE` was proposed in 2009 [2] and formally accepted in April 2011[8] as part of MPI-3.0.

Gregor and Lumsdaine ported an MPI benchmark, NetPIPE, to C# to measure the runtime difference between C (the original implementation choice) and C#, the result stating the abstraction penalty, which "includes the costs associated with the .NET virtual machine, garbage collector, and interaction between managed and unmanaged code."[9] They report a "generally very small $(1-2\%)$" penalty for small message sizes, with the difference increasing with message size. Interestingly, they notice the penalty sometimes turns into a benefit, noting that "for larger messages [the results] are less obvious, with C# varying from 15% slower to 10% faster."

The results of Gregor and Lumsdaine are interestingly close to our own, despite using a completely changed software stack. The mpi.net authors obtained their test results on a 9-node cluster with each node containing a dual-core 2.13GHz Intel Xeon 3050 processor and 2GB of RAM. The paper does not contain information on the network interface cards except that the nodes were "connected by

---

[7] http://lists.mpi-forum.org/mpi-22/2008/10/0177.php
[8] http://svn.mpi-forum.org/trac2/mpi-forum-web/ticket/38
[9] All following citation can be found at [3, p. 8].

Gigabit Ethernet over a private network". Their cluster was operated using a Microsoft software stack, including Microsoft Windows Compute Cluster Server 2003, containing Microsoft's MPI implementation, MS-MPI. In contrast to our tests, Gregor and Lumsdaine used Microsoft Visual Studio .NET 2005 with full optimization to compile all tests and version 2.0.50727 of Microsoft's .NET framework to run the C# tests.

In comparison to the language issues faced by Gregor and Lumsdaine, namely the strict border between managed and unmanaged code, mpi4py benefited from the close integration of Python with C. For example, Dalcin et al. note:

> MPI for Python was improved to support direct communication of any object exporting single-segment buffer interface. This interface is a standard Python mechanism provided by some type of objects (e.g. strings and numeric arrays), allowing access in the C side to a contiguous memory buffer (i.e. address and length) containing the relevant data. [1, p. 7]

They also mention some interesting performance improvements, like persistent communication:

> Often a communication with the same argument list is repeatedly executed within an inner loop. In such a case, communication can be further optimized by using persistent communication, a particular case of non-blocking communication allowing the reduction of the overhead between processes and communication controllers. [1, p. 8]

They, too, measured the performance of their implementation with a setup very similar to ours, utilizing a Linux Beowulf cluster with Intel Pentium 4 Prescott processors with 2GB RAM, connected to a 3Com SuperStack 3 Switch and 3Com network interface cards. The software versions in their tests include GCC 3.4.4 with Python 2.4.4 and MPICH2 1.0.4p1. They also used NumPy 1.0 for their numeric array tests.

In [1, chapter 4] they find, quite in agreement with the results in this paper, that the overhead imposed by Python is negligible with direct buffers as mentioned above and significant (minimum 20%) when data needs to be serialized ("pickled").

4   Performance Evaluation

In this section, we measure and compare the performance of C#'s and Python's MPI library wrappers, that is, the additional overhead caused by interfacing between the high-level language and the underlying native MPI C-binding. To test the performance, we implemented a simple ping-pong benchmark and a slightly more complex cellular automaton in both interpreted languages and C for comparison. All measurements were taken on a local test cluster consisting of 12 nodes. Only two of them were allocated for the ping-pong test and 12 for the cellular automaton benchmark.

All involved nodes were running Linux 2.6.26 on a single Pentium 4 processor with 3GHz clock frequency and 2GB of RAM. The job scheduling was done using SGE with a shared home file system on a NFS mount. The nodes were connected via Broadcom BCM5704 Gigabit Ethernet adapters to a Netgear ProSafe GS724T switch.

The runtime software consisted of mono-2.4 system with mpi.net-1.0.0 for the C# tests and a Python-2.5.4 installation with mpi4py-1.2.1 for the Python tests. GCC 4.3.4 and openmpi-1.4.1 were used to compile the C version of the benchmarks.

First, the ping-pong benchmark was used to measure the abstraction penalty as referred to by [1]. This describes the overhead imposed by using an interpreted language — which includes garbage collection and dynamic memory allocation, setting up the interpreter if needed, just in time compiling, etc. In the first test, we were only interested in the overhead imposed by MPI library wrappers. We hence only measured the runtime of send and receive functions, providing a lower bound on the abstraction penalty. It can be argued that especially the runtime of scientific applications is highly dependent on the speed of other functions, like math and memory management. To get an idea of how big the impact on computation is, the second test deliberately includes the time taken to compute the the result. However, it should be mentioned that both C# and Python can delegate performance critical tasks to extensions written in C or even assembler, if needed.

In the first test, a byte array was created and exchanged with another node. The measured time is the average over 1000 iterations, consisting of one (buffered) send and one receive operation. In advance, 100 iterations were run without

Table 1. Runtime evaluation of *Conway's Game of Life* and a simple *Ping-Pong*. All timings in seconds. For the Ping-Pong benchmark, *Size* refers to the message size in bytes. For the Game Of Life test, *Size* refers to $\sqrt{boardsize}$ in bytes.

Game Of Life runtime				Ping-Pong runtime			
Size	C	C#	Python	Size	C	C#	Python
3	0.00004	0.00023	0.00047	8	0.000058	0.000064	0.000074
4	0.00004	0.00020	0.00047	16	0.000059	0.000064	0.000074
8	0.00005	0.00021	0.00047	32	0.000060	0.000066	0.000075
16	0.00006	0.00023	0.00049	64	0.000062	0.000067	0.000077
32	0.00009	0.00029	0.00057	128	0.000066	0.000071	0.000081
64	0.00024	0.00051	0.00080	256	0.000075	0.000080	0.000090
128	0.00079	0.00136	0.00176	512	0.000092	0.000096	0.000107
256	0.00307	0.00622	0.00718	1024	0.000121	0.000126	0.000136
512	0.01242	0.01764	0.03165	2048	0.000175	0.000179	0.000191
1024	0.05044	0.07028	0.09399	4096	0.000223	0.000228	0.000238
2048	0.20199	0.27731	0.37373	16384	0.000607	0.000617	0.000627
4096	1.27811	1.08183	1.49061	32768	0.001085	0.001105	0.001106
8192	3.27760	4.32236	5.94661	65536	0.002169	0.002225	0.002215
16384	13.12326	17.31616	24.68614	262144	0.008034	0.008223	0.008116

```c
int sum(cellState *area, size_t fieldSize, int oX, int oY) {
    unsigned sum = 0;
    for (int y = oY - 1; y < oY + 1; ++y) {
        for (int x = oX - 1; x < oX + 1; ++x) {
            if (y < 0 || y > fieldSize - 1 || x < 0 || x > fieldSize - 1) {
                continue;
            }
            sum += area[y*fieldSize + x];
        }
    }
    return sum;
}

void iterate(cellState *field, size_t fieldSize) {
    cellState *new = calloc(fieldSize * fieldSize, sizeof(cellState));
    for (int oY = 0; oY < fieldSize; ++oY) {
        for (int oX = 0; oX < fieldSize; ++oX) {
            switch (sum(field, fieldSize, oX, oY)) {
                case '2':
                    new[oY*fieldSize + oX] = (1 == field[oY*fieldSize + oX]) ? 1 : 0;
                    break;
                case '3':
                    new[oY*fieldSize + oX] = 1;
                    break;
                default:
                    new[oY*fieldSize + oX] = 0;
            }
        }
    }
    field = memcpy(field, new, fieldSize*fieldSize);
    free(new);
}
```

Fig. 2. C version to implement the update process for Conway's *Game of Life*. To reduce the nesting depth, two inner *for*-loops and a non-trivial *if*-statement have been refactored into the *sum* function.

timing to prime pipelines and caches. All three programs used heap allocation (C via `malloc()`, Python via NumPy's arrays and C# via `array` objects). Timings were obtained via `MPI_Wtime()` (C), `MPI.Environment.Time` (C#) or `MPI.Wtime()` (Python) after each run and averaged over all iterations.

The ranking with regard to transmission time was C first, then C#, then Python. Both, Python and C# compile source code into their own bytecode format and interpret it at runtime. The only difference is that C# requires this compilation to be done in a separate step. So the effectiveness of compile-time optimization as well as library instantiation time factor into the results. The mpi.net library has the advantage of building and keeping a dictionary of data types in use. For types known to the underlying C MPI library (as it is the case here), it just passes the underlying library a pointer to the data when calling send or receive functions. Mpi4py, on the other hand, has to go through an extensive function determining which datatype to pass to the C library.

The obtained timings for the first test are shown in Table 1. The actual results match the expectations. However, for message sizes larger than 2^{12} Byte, the difference between the tested libraries is insignificant. Even for small sizes, the distance between fastest and slowest is near insignificance (mind the exponential y-axis).

The second test was designed to measure one possible "real-world" situation for the use of MPI. It consisted of a cellular automaton simulation according to Conway's Game of Life with 1000 rounds. Comparable simulations are applied in natural sciences [8] to various problems and hence represent a common workload for HPC clusters. All programs used Conway's original rules: a cell (represented as an entry in a two-dimensional array) is set to *live* in an iteration iff three or four cells in its full neighbourhood (i.e. all 8 squares around a cell) were previously in the *live* state, otherwise the cell's status is set to *dead*, represented as 1 and 0. The basic steps of each program are:

1. Each node allocates a square "field" of the given size and initializes all cells with a random 0 or 1 value.
2. Communication is done to and from two "shadow" rows that represent the state of the last iteration of nodes with rank one less and one greater than the node's own. In other words, the "playing field" is vertically split among the nodes, with each node communicating its borders to its immediate neighbours.
3. Each node allocates a temporary array, computes the *live* or *dead* state of all cells and replaces the original array with the results from the temporary array.

This requires at most two send and receive operations of a single line, a one-dimensional array. The nodes with minimum and maximum rank do not exchange information directly and thus have to communicate with only one neighbour.

Table 1 shows the runtime of steps two and three for each board size. Note that the board size is the square root of the size of the array that is held by each node, not the size of the resulting "Game of Life playing field". Step one was also measured but not included, because setup time is constant and does not differ much among the tested languages. It is clearly dominated by the allocation of the two-dimensional array. In the case of C#, jagged arrays (arrays of arrays) were used instead of the more natural two-dimensional arrays, because mpi.net was not able to handle them. The C test program used a single array with continuously stored rows, so virtual 2D-array access is mapped to

```
new [ Y_offset * fieldSize + X_offset ] = 1;
```

Only Python was able to handle "real" two-dimensional arrays, which resulted in a quick neighbourhood test utilizing array slicing (see Figure 1) instead of the loops the other programs had to use (Figure 2). The ability to slice arrays also allowed the more concise notation when transmitting the "shadow" row of the array:

MPI.COMM_WORLD. Irecv ([board [−1 ,:] , MPI.CHAR] , rank+1)

Here board[-1, :] denotes the complete (with all columns) last row of the board array consisting of char objects. In the C version, it is not immediately clear that this statement receives a row of a two-dimensional array:

```
MPI_Irecv (
    &field [rowSize *(rowSize − 1)] ,
    rowSize , MPI_CHAR, myRank + 1,
    0, MPI_COMM_WORLD, &reqs [2]) ;
```

Obviously, languages like Python are more expressive. To confirm the conjecture that interpreted languages are more effective to write, the time to write the tests may be considered anecdotal evidence: while the Python version required only five hours to be written (including repeated debugging), the C version was plagued by a memory corruption, leading to 14 hours of total development time.

However, as already stated for the first test, interpreted languages do incur an abstraction overhead. When more than the MPI library performance is measured, the C test version still is the fastest, with a more distinctive difference to C#. The Python version is the slowest of the programs tested, using almost twice the time of the C program. All three programs exhibit a runtime continuously growing with the board size. The gap between C# and Python widens (note the exponential y-axis).

5 Conclusions

We gave an overview of various MPI library wrappers. All of those studied, with the exception of mpi4py, must be considered inactive. The last update to libraries like Perl's Parallel::MPI is as late as 2002, leading to build problems and hence making the wrapper unusable.

We reported the results of our experiments, a comparison between (the C interface of) Open MPI, and two of its wrappers - mpi4py and mpi.net. They were tested in a simple ping-pong benchmark and in a more complex Game of Life simulation. The results show (native C-) Open MPI leading in terms of performance with a small difference to both, mpi4py and mpi.net (C#). We also show examples from the test code to substantiate the claim that interpreted languages, their MPI adapters and Python's mpi4py in particular lower the entry barrier for users new to MPI and make general MPI programming more effective in terms of code size and development time.

As a result of our work, we strongly recommend the adoption of either mpi4py or mpi.net in cases where rapid development and early success is more important than best performance, which is still reserved for native C implementations, although (the tested) interpreted solutions came surprisingly close. Mpi4py's tight integration with the NumPy project and its clear and very readable syntax make it a good beginner's choice, while the C derived syntax and integration with the rest of the .NET environment may make mpi.net appealing to programmers with a C/C++ background.

With more user-friendly languages available, a broader audience of prospective C-illiterate MPI users will benefit from high performance computing.

References

1. Dalcin, L., Paz, R., Storti, M., Delia, J.: MPI for python: Performance improvements and MPI-2 extensions. Journal of Parallel and Distributed Computing 68(5), 655–662 (2008), http://dx.doi.org/10.1016/j.jpdc.2007.09.005
2. Gregor, D., Hoefler, T., Barrett, B., Lumsdaine, A.: Fixing Probe for Multi-Threaded MPI Applications. Tech. Rep. 674, Indiana University (January 2009)
3. Gregor, D., Lumsdaine, A.: Design and implementation of a high-performance MPI for c# and the common language infrastructure. In: PPoPP 2008: Proceedings of the 13th ACM SIGPLAN Symposium on Principles and Practice of Parallel Programming, pp. 133–142. ACM, New York (2008), http://dx.doi.org/10.1145/1345206.1345228
4. Gropp, W., Lederman, S.H., Lumsdaine, A., Lusk, E., Nitzberg, B., Saphir, W., Snir, M.: MPI - The Complete Reference: the MPI-2 Extensions, vol. 2. MIT Press, Cambridge (1998)
5. Gropp, W., Lusk, E.: Using MPI-2: A problem-based approach, p. 12 (2007), http://dx.doi.org/10.1007/978-3-540-75416-9_7
6. Ong, E.: MPI Ruby: Scripting in a parallel environment. Computing in Science & Engineering 4(4), 78–82 (2002), http://dx.doi.org/10.1109/MCISE.2002.1014983
7. Pacheco, P.: Parallel Programming With MPI. Morgan Kaufmann, San Francisco (1996), http://www.worldcat.org/isbn/1558603395
8. Packard, N.H., Wolfram, S.: Two-dimensional cellular automata. Journal of Statistical Physics 38, 901–946 (1985), http://dx.doi.org/10.1007/BF01010423, 10.1007, doi:10.1007/BF01010423
9. Snir, M., Otto, S.W., Walker, D.W., Dongarra, J., Huss-Lederman, S.: MPI: The Complete Reference. MIT Press, Cambridge (1995), http://portal.acm.org/citation.cfm?id=546703
10. The Message-Passing Interface Forum: MPI: A message-passing interface standard 2.1. Tech. rep., University of Tennessee, Knoxville, Tennessee (2008), http://www.mpi-forum.org/docs/mpi21-report.pdf
11. William Gropp, E.L., Skjellum, A.: Using MPI: Portable Parallel Programming with the Message Passing Interface, 2nd edn. MIT Press, Cambridge (1999)
12. Wilmes, J., Stevens, C.: Parallel: MPI - an MPI binding for perl (May 1999), http://cpansearch.perl.org/src/JOSH/Parallel-MPI-0.03/docs/paper.tex (retrieved September 21, 2009)

Leveraging C++ Meta-programming Capabilities to Simplify the Message Passing Programming Model

Simone Pellegrini, Radu Prodan, and Thomas Fahringer

University of Innsbruck – Distributed and Parallel Systems Group
Technikerstr. 21A, 6020 Innsbruck, Austria
{spellegrini,radu,tf}@dps.uibk.ac.at

Abstract. Message passing is the primary programming model utilized for distributed memory systems. Because it aims at performance, the level of abstraction is low, making distributed memory programming often difficult and error-prone. In this paper, we leverage the expressivity and meta-programming capabilities of the C++ language to raise the abstraction level and simplify message passing programming. We redefine the semantics of the assignment operator to work in a distributed memory fashion and leave to the compiler the burden of generating the required communication operations. By enforcing more severe checks at compile-time we are able to statically capture common programming errors without causing runtime overhead.

Keywords: Message passing, C++, Meta-programming, PGAS.

1 Introduction

The message passing paradigm is frequently used in High Performance Computing (HPC) for programming computer clusters and supercomputers. Compared to other existing parallel programming models such as OpenMP, message passing offers two basic primitives: *send* and *receive*. The burden of managing almost every aspect of the program execution including data partitioning, communication, and synchronisation between processes is left to the programmer. A low-level of abstraction is helpful in writing highly optimised programs, however, it makes distributed memory programming very difficult and error-prone.

Recently, new programming models are increasingly being used aiming at simplifying distributed programming. An example is the Partitioned Global Address Space (PGAS) model, which provides the programmer with a logically global memory address space where variables may be directly read and written by any process. Below the logical view, each variable is physically associated to a single process. Any attempt to read or write memory locations physically allocated on a different process results in a communication operation generated either by a runtime environment in the Global Array library [7]) or during the compilation process in the Co-array Fortran and UPC [8,3]. However, because

Y. Cotronis et al. (Eds.): EuroMPI 2011, LNCS 6960, pp. 302–311, 2011.

of the increased level of abstraction, the programmer loses control over the generation of communication and synchronisation operations resulting in important performance losses compared to manually written message passing applications.

The motivation for the research presented in this paper is based on the observation that sending a message from a process A to a process B is semantically equivalent to an assignment operation. The content of the memory cell owned by process B is overwritten with data residing on process A's memory space. We use the C++ *operator overloading* mechanism and *template meta-programming* techniques [4] to enable the automatic generation of low-level communication primitives by the standard C++ compiler. For example, whenever an assignment operator involving memory cells residing on different processes is encountered, the compiler generates the required communication statements. Additionally, we generate for each process rank a separate executable containing only those operations involving the assigned memory cells, which eliminates the control flow overhead incurred by the Single Program Multiple Data (SPMD) nature of the input program. The main advantage of our approach is the fact that it achieves a level of abstraction similar to PGAS-based languages by only exploiting features of the standard C++ language and compiler. Furthermore, because the underlying programming model is based on message passing, the programmer still retains full control over the resulting performance.

In Section 2, we provide an overview of our new approach of writing message passing parallel programs. In Section 3 we discuss the implementation details of the mem_wrap object that is the main abstraction behind our method. Section 4 compares our method with a UPC-based implementation for a Jacobi relaxation algorithm. Section 5 concludes the paper and highlights the future work.

2   Overview

This section gives a brief overview of our technique while further details will be given in Section 3 and 4 of the paper. Let us consider in Listing 1.1 a simple message passing program written in MPI [2], which is the *de-facto* standard for programming HPC applications. Two processes are involved in this example: process 0 computes the value of the π constant (pi) and sends it to process rank 1. The computed value is then used by both processes for further computation. One of the first characteristics of the program is the use of the SPMD technique, which generates a single executable that is spawned on multiple processors. To customize the program behaviour for a specific process rank, the programmer needs to continuously use control statements to guide the specific process flow of execution (lines 2 and 5). The use of control flow statements is in general the source of many inefficiencies and limits compiler analysis and optimizations. Additionally, miss-predicted branches cause significant performance penalties on modern pipelined CPU architectures. Because the generated executable contains code which is never executed on a particular process rank, the L1 instruction cache may be not optimally used too.

Listing 1.1. Simple message passing program in MPI

```
1   float pi;
2   if ( rank == 0 ) {
3       pi = calc_pi();
4       MPI_Send(&pi, 1, MPI_FLOAT, 1, 0, MPI_COMM_WORLD);
5   } else if (rank == 1)
6       MPI_Recv(&pi, 1, MPI_FLOAT, 0, 0, MPI_COMM_WORLD, MPI_STATUS_IGNORE);
7   use(pi);
```

A second observation is that message passing programs are often complex to read and, more importantly, to analyse. Because the programmer is forced by the programming model to describe the low-level operations (i.e. the *"how"*), the semantics of the program (i.e. the *"what"*) is mostly hidden. For example, although a connection between the send and receive operations in lines 4 and 6 exists, it is implicitly in the mind of the programmer and not made explicit in the code. This hidden knowledge could be used by the compiler to improve error checking and program performance, but it is unfortunately very complex to be captured by static analysis [5,9]. For example, the compiler could enforce the amount of received data to be not less than the amount of data sent, or use constant propagation to remove communication statements in case the transmitted value is constant (detected by compiler dataflow analysis).

Listing 1.2. Overload of assignment operator in C++

```
1   mem_wrap<float> pi; // manages memory allocation in the distributed env.
2   pi[r0] = calc_pi(); // Rank 0 executes calc_pi() and writes the returned value
3                       // into its own copy of pi
4   pi[r1] = pi[r0]; // Copies the value of pi owned by process rank 0 onto the
5                    // memory cell owned by process rank 1 (by using send/recv)
6   use(*pi);
```

In this paper, we propose a different approach which lets the programmer focus on the program semantics (the *"what"*) and lets the compiler deal with the generation of the required communication operations. The idea is not entirely new [3], however, instead of introducing a new programming model (e.g. PGAS) and an underlying language support (e.g. UPC), we exploit the capabilities of the standard C++ language and compiler. Listing 1.2 shows a simple C++ program semantically equivalent to the previous example. The first aspect is the lack of any control flow statements, which is achieved by offloading all memory operations to a new data type, i.e. mem_wrap, acting as a memory wrapper for distributed memory environments. The input program is compiled multiple times, each time for a different process rank. Keeping the value of the process rank constant at compile-time allows meta-programming techniques to be used for specializing the semantics of operations involving mem_wrap instances. For example, the initialisation of a memory cell owned by the process rank 0 results in a *no-operation* (NOP) when the program is compiled for process rank 1 (line 2).

Assignment operations involving memory cells residing on different address spaces are replaced by communication statements (line 4). Table 1 shows the codes generated at compile-time by our approach for the processes with rank 0 and 1 from program code in Listing 1.2. The SPMD input program is compiled into multiple executables (as many as the number of processes) and successively executed using the Multiple Program Multiple Data (MPMD) paradigm.

Table 1. Compiler generated codes for process rank 0 and 1

Rank 0	Rank 1

```
1  float pi;
2  pi = calc_pi();
3  MPI_Send(&pi,1,MPI_FLOAT,1,0,...);
4  use(pi);
```

```
1  float pi;
2  MPI_Recv(&pi,1,MPI_FLOAT,0,0,...);
3  use(pi);
```

Running the MPMD program generated by our technique produces very promising results. We executed both the SPMD and MPMD executables on an Intel Xeon X5570 CPU and an AMD Opteron 2435, both compiled with GCC 4.5.3 and optimization enabled (-O3). We repeated the code snippet one thousand million times and used shared memory communication (SM module of Open MPI's Modular Component Architecture) to reduce the communication overhead. The main program loop has been executed 10 times, the average value of execution time and standard deviation are depicted in Table 2. A considerable performance improvement, of around 30%, is observed for the Intel architecture, while on the AMD CPU, the improvement was of around 5%. Because the two processors have a similar L1 cache size (i.e. 64KB), we believe that the main source of performance improvement comes from the simplification of the control flow.

Table 2. Execution time for each process of the program in Listing 1.3 using SPMD and MPMD models

	SPMD		MPMD		
	Exec. time [milisec.]	Standard deviation	Exec. time [milisec.]	Standard deviation	Speedup
Intel Xeon	8180	440	6162	129	**1.32**
AMD Opteron	9638	166	9296	177	**1.04**

In order to explain the performance improvement we executed the code snippet enabling performance counters on the Intel CPU by using the PAPI library [1]. The measured values for three performance counters are depicted in Table 3. We measured the instruction cache misses for both level 1 and 2 and the total amount of conditional branch instructions. The code snippet is small to easily fit on the L2 cache, therefore no differences in terms of L2 cache misses are

Table 3. Performance counter values for the Intel architecture

Hardware counter	SPMD	MPMD
L1 Instruction Cache misses	4253718	**4246317**
L2 Instruction Cache misses	681689158	681689158
Conditional branch instructions	4260166	**4254384**

visible. However, the utilization of the L1 cache is improved for the MPMD code as we were able to reduce the amount of cache misses by a 0.5%. This is because, by removing unreachable branches, code locality is improved. Additionally, also the amount of conditional branch instructions is reduced by the same amount. This alone cannot however explain the 32% speedup which we believe to be the result of optimizations (e.g. loop unrolling and constant propagation) performed by the compiler on the MPMD code. As a matter of fact, thanks to the simplification to the control flow obtained with our meta-programming technique, we enable the compiler analysis to perform more aggressive optimizations which are not applicable on the SPMD version.

3   The `mem_wrap` Object

Meta-programming is the practice of writing a computer program that writes or manipulates other programs (or themselves) as their data. Meta-programming can be used to perform part of the computation at compile-time instead of run-time. By combining templates and meta-programming, it is possible in C++ to specialize the implementation of generic functions based on particular properties of the input parameters. For example, a generic function can have two implementations depending on whether the input parameter is a pointer or a value type. Because these checks are conducted at compile-time, it is necessary that the expressions used to select a particular implementation involve compile-time constants only.

Listing 1.3. `mem_wrap` object interface

```
1   template <class T, template <class> class Sel, class R>
2   struct mem_wrap {
3     T& operator*(); // Access to managed memory
4
5     mem_wrap<T,Sel,R>& operator=(const T&);
6     template <template <class> class Sel2, class R2>
7     mem_wrap<T,Sel,R>& operator=(const mem_wrap<T,Sel2,R2>&);
8
9     template <class R2> mem_wrap<T,Sel,R2> operator[ ](const R2&);
10  };
```

Our approach is based on a similar mechanism. The objective is to introduce an *enhanced* assignment operator which, depending on the type of the left and right hand side expressions, is specialized to implement different semantics. We introduce a new data type called mem_wrap illustrated in Listing 1.3 that manages the allocation and accesses to memory locations in the distributed memory environment. The first template parameter T represents the wrapped type which allows the management of single elements (e.g. mem_wrap<float>) or of more complex data types such as arrays (e.g. mem_wrap<vector<float>>). The second parameter Sel is the selector, which decides whether the wrapped object (of type T) has to be allocated on a particular process rank for which the input program is being compiled. For example, by using the expression Rank%2==0 as a selector, we enforce only even process ranks to allocate the memory to host the object of type T. We refer to these instances of mem_wrap as *active*. Odd ranks for which the selector is not satisfied allocate an empty wrapper instance called *shadow*. A shadow wrapper acts as a pointer to a memory location on a different machine and can be used to read data from it. To note that mem_wrap does not perform any data partitioning, the programmer is still responsible to divide the memory space among the processes. Because a mem_wrap instance can refer to memory locations on multiple address spaces, the R parameter is used to address the copy owned by a specific process rank. The mem_wrap also provides three basic methods among several others: a dereferencing operator * used to directly access the memory managed by the wrapper (line 3), an assignment operator = overloaded to work with data type instances of type T (line 5) or mem_wrap instances (line 7), and a subscript operator [] used to select a copy of the wrapped data which belongs to a particular address space.

Listing 1.4. Example of using selectors

```
1  template <class RR = mpl::int_<MY_RANK>>
2  struct even {
3      template <class Rank>
4      struct apply : public mpl::bool_<Rank::value%2==0> { };
5  };
6  mem_wrap<std::vector<float>, even> vect(100);
7  for (unsigned int i=0; i<100; ++i) { vect(i) = MY_RANK; }
8  vect[r0] = vect[r2];
```

There are two specializations of the mem_wrap class: one for active and the other for shadow wrapper instances. We define a pre-processor directive called MY_RANK as the rank of the process for which code is being generated. During the compilation process, for every instantiation of a mem_wrap, the selector is applied to the value of MY_RANK. Depending on the result, one of the two specializations is used. Furthermore, methods of the mem_wrap class have multiple specializations depending on the type of the input parameters.

To better understand how selectors work, we illustrate in Listing 1.4 a slightly more complex example of a program that allocates a vector (vect) of 100 floating point numbers on every even process rank, initialises it, and then copies its value from rank 2 to rank 0. We define the class `even` as a selector for even rank values. For simplicity, we use utilities (i.e. types and *meta-functions*) from the Boost Meta-Programming Library (MPL) [6], on which also the implementation of `mem_wrap` heavily relies. The selector is applied to a rank value using the `apply` generic inner class defined in line 2. We follow the naming convention used in MPL which enables us using existing meta-programming utilities from the MPL library. In line 4, the allocation of the variable `vect` is managed by our memory wrapper which enables the compiler to select the type of wrapper to instantiate (i.e. active or shadow) depending on the rank for which code is being generated. Accessing array elements from a wrapper instance is allowed using the () operator which, instead of returning directly the indexed value, instantiates a wrapper containing the addressed memory cell. For shadow wrappers, an assignment operator of a value of type `T` resolves to a `NOP` (e.g. loop iteration in line 5 compiled for odd processors) that compiler optimizations can easily detect and safely remove as dead code. Finally, the assignment operator in line 6 involving the two wrappers is rewritten by through send/receive, as previously shown. For odd ranks, the operation results again in a `NOP`. The `r`n constants, where n is an integer value representing the rank, are defined to easily refer to a process rank. The [] operator is used to specialize a generic wrapper to refer to a particular memory address space, method signature is shown in line 9 of Listing 1.3.

4   Jacobi Relaxation

In this section we show how an important class of HPC stencil operations can be expressed in our framework. We use as example the Jacobi relaxation method based on the nearest neighbour communication. A two-dimensional matrix is distributed among the processes, each process having a dependency to the memory cells owned by its direct neighbours. When the data is distributed in a row-wise manner, each process needs to access the memory allocated in the top `MY_RANK+1` and bottom `MY_RANK-1` neighbours. Every process allocates an equal portion `N/NPROCS+2` the matrix rows, where `N` is the matrix size. The two additional rows are used to store the first and last row received from the top and the bottom neighbors.

Listing 1.5 shows the Jacobi relaxation algorithm expressed using our method. In lines 6 and 7, two shadow wrappers are generated referring to the `top` and `bottom` neighbours. The top processor selector `top_neigh` is defined in lines 1-3. The selector for the bottom processor `bottom_neigh` (not shown because of space limit reasons) is similar with the difference that the expression `RR::value-1 == Rank::value` is used as a selector. Both `top` and `bottom` are instantiated as shadow wrappers on every processor rank because the selector expressions always evaluate to false when applied to the current rank (`MY_RANK`). Lines 10 and 11 implement the neighbor communication. In line 10, a receive operation is generated for the incoming data from the top neighbor process. Unlike

Listing 1.5. C++ Jacobi relaxation

```
1   template <class RR = mpl::int_<MY_RANK>>
2   struct top_neigh {
3       template <class Rank>
4       struct apply : mpl::bool_<RR::value+1 == Rank::value> { };
5   };
6   const size_t size = N/NUM_PROCS+2;
7   mem_wrap<carray<float>> u(size,N), tmp(size−2,N); // Active wrapper
8   mem_wrap<carray<float>, top_neigh> top( u ); // Shadow wraper
9   mem_wrap<carray<float>, bottom_neigh> bottom( u ); // Shadow wrapper
10  // Initialize matrix u...
11  for(unsigned int it=0; it<MAX_ITER; ++it) {
12      u[ slice(size−1, size, 0, N) ] = top[ slice(1, 2, 0, N) ];
13      u[ slice(0, 1, 0, N) ] = bottom[ slice(size−2, size−1, 0, N) ];
14      for (unsigned int i=1; i<size−1; ++i)
15          for (unsigned int j=1; j < N−1; ++j)
16              tmp(i−1,j−1) = 1/4 * ( *u(i−1,j) + *u(i,j+1) + *u(i,j−1) + *u(i+1,j) );
17  }
```

previous examples, the rank is not statically specified and the source rank of the message is automatically computed at compile-time in order to avoid any runtime overhead. This is done with the following procedure. The selector of the right hand side expression top_neigh is applied to a list of process ranks PL generated at compile-time as follows PL: {0,1,...,MY_RANK-1, MY_RANK+1,..., NUM_PROCS-1}, where NUM_PROCS is the total number of processes defined via a pre-processor directive, and RR (i.e. RefRank) is set to be MY_RANK. The selector is invoked several times as follows: top_neigh<MY_RANK>::apply<R>, $\forall R \in PL$. The receive operation is generated using, as a source rank, the value R which satisfies the selector, (i.e. MY_RANK+1). Speculatively, a send operation is generated towards the bottom neighbor. This requires to invert the top_neigh selector previously used to generate the receive operation. We achieve this by invoking the selector in the following way: top_neigh<R>::apply<MY_RANK>, $\forall R \in PL$. The semantics is the following, find the processes for which the top_neigh selector is satisfied when applied to the current rank value (i.e. MY_RANK). For rank values which satisfy the selector, a send operation is generated using as target rank the value of R (i.e. MY_RANK-1). The communication statements for line 11 are generated similarly but using bottom_neigh as selector. The slice function indicates the start and end rows and columns of a matrix partition which has to be either transmitted or overwritten by the incoming data.

We compared our Jacoby relaxation implementation with an UPC-based version on a shared memory machine with 10 AMD Opteron cores. The UPC implementation of Jacobi (from [10]) utilized in our experiments is depicted in Listing 1.6. The code uses a memory layout specifier (i.e. [...]) which allow the UPC runtime to distribute the u and tmp matrices assigning an equal amount of rows to each UPC thread (similar to the MPMD code). For a fair comparison, we forced UPC to use MPI as the underlying communication library

Listing 1.6. UPC based Jacobi relaxation method

```
1   shared [N*N/THREADS] float u[N][N];
2   shared [N*N/THREADS] float tmp[N][N];
3   // Initialize matrix u...
4   for( unsigned int it=0; it < MAX_ITER; ++it)
5       upc_forall(unsigned int i=1; i<N−1; i++; &tmp[i][0]) {
6           for (unsigned int j=1; j < N−1; ++j)
7               tmp[i][j] = 1/4 * ( u[i−1][j] + u[i][j+1] + u[i][j−1] + u[i+1][j] );
8       }
```

(`-network=mpi`). Furthermore, we utilized the `-T` flag which enables the UPC compiler to create an executable which runs with a fixed number of threads (i.e. `-T=10`). The Berkley UPC compiler version 2.12.2 with experimental optimization enabled (`-opt`) has been utilized. GCC version 4.5.3, with optimization flag `-O3`, has been used to compile the MPMD version of the Jacobi in Listing 1.5.

Table 4 shows that UPC performs slightly better for very small matrix sizes but, as the problem size increases, the MPMD version significantly outperforms UPC. Unfortunately we could not compile the UPC code for larger matrix sizes as the UPC compiler does not support, in the layout specifier, a block size which is greater than 1MB. We believe that the main source of inefficiency in UPC is the fact that the compiler is not able to vectorize the accesses to neighbor memory cells. Therefore every access to remote memory locations results in a separate communication operation. It is also worth noticing that compared to an SPMD-based MPI implementation of the Jacobi, the MPMD version presented here only marginally improve performance. The main advantage is indeed in the simplified programming model which, as the experiments show, do not cause any performance penalty.

Table 4. Jacobi relaxation execution time (in seconds) and speedup comparison

Matrix size	MPMD	UPC	Speedup
10x10	0.0129	0.0022	0.14
100x100	0.018	0.023	**1.28**
500x500	0.098	0.205	**1.84**
1000x1000	0.20	0.74	**3.7**
2000x2000	0.61	2.98	**4.9**

5   Conclusions and Future Work

In this paper we demonstrated with concrete examples how using advanced meta-programming capabilities of the C++ language can simplify the use of message passing and, at the same time, improve readability and performance of applications. Our approach offers three main advantages compared to traditional message passing-based programs: send/receive operations are expressed using intuitive variable assignments, it allows compile-time checking of message sizes and element types, and it facilitates compiler optimizations by generating MPMD code that eliminates harmful control flow statements.

The main drawback of our approach is the generation of a separate executable for every process, which may be not always feasible for large-scale applications. However, this is a limitation of our prototype implementation and not of the approach itself. In the future we will focus on improving the use of meta-programming techniques to statically determine groups of process with the same behavior (e.g. even and odd ranks) that reduce the number of generated executables to one per group.

References

1. PAPI: Performance Application Programming Interface, http://icl.cs.utk.edu/papi/
2. The MPI-2 Specification, http://www.mpi-forum.org/docs/docs.html
3. UPC Language Specifications, v1.2. Tech. Rep. LBNL-59208, Lawrence Berkeley National Lab Tech. Report (2005)
4. Abrahamsi, D., Gurtovoy, A.: C++ Template Metaprogramming. Addison-Wesley, Reading (2006)
5. Bronevetsky, G.: Communication-Sensitive Static Dataflow for Parallel Message Passing Applications. In: Proceedings of the 7th annual IEEE/ACM International Symposium on Code Generation and Optimization, CGO 2009, pp. 1–12. IEEE Computer Society, Washington, DC (2009)
6. Gurtovoy, A., Abrahams, D.: The Boost Metaprogramming Library, http://www.boost.org/doc/libs/1_46_1/libs/mpl/doc/index.html
7. Nieplocha, J., Palmer, B., Krishnan, M., Trease, H., Apr, E., Nieplocha, J., Palmer, B., Krishnan, M., Trease, H., Apr, E.: Advances, Applications and Performance of the Global Arrays Shared Memory Programming Toolkit. Intern. J. High Perf. Comp. Applications 20 (2005)
8. Numrich, R.W., Reid, J.: Co-array Fortran for parallel programming. SIGPLAN Fortran Forum 17, 1–31 (1998), http://doi.acm.org/10.1145/289918.289920
9. Strout, M.M., Kreaseck, B., Hovland, P.D.: Data-Flow Analysis for MPI Programs. In: International Conference on Parallel Processing, ICPP 2006, pp. 175–184 (August 2006)
10. Zheng, Y., Blagojevic, F., Bonachea, D., Hargrove, P.H., Hofmeyr, S., Iancu, C., Min, S.J., Yelick, K.: Getting Multicore Performance with UPC. In: SIAM Conference on Parallel Processing for Scientific Computing, Seattle, Washington (February 2010)

Portable and Scalable MPI Shared File Pointers

Jason Cope[1], Kamil Iskra[1], Dries Kimpe[2], and Robert Ross[1]

[1] Argonne National Laboratory, Argonne, IL 60439
[2] University of Chicago, Chicago, IL 60637
{copej,iskra,dkimpe,rross}@mcs.anl.gov

Abstract. While the I/O functions described in the MPI standard included shared file pointer support from the beginning, the performance and portability of these functions have been subpar at best. ROMIO [1], which provides the MPI-IO functionality for most MPI libraries, to this day uses a separate file to manage the shared file pointer. This file provides the shared location that holds the current value of the shared file pointer. Unfortunately, each access to the shared file pointer involves file lock management and updates to the file contents. Furthermore, support for shared file pointers is not universally available because few file systems support native shared file pointers [5] and a few file systems do not support file locks [3].

Application developers rarely use shared file pointers, even though many applications can benefit from this file I/O capability. These applications are typically loosely coupled and rarely exhibit application-wide synchronization. Examples include application tracing toolkits [8,4] and many-task computing applications [10]. Other approaches to the shared file pointer I/O models frequently used by these application classes include file-per-process, file-per-thread, and file-per-rank approaches. While these approaches work relatively well at smaller scales, they fail to scale to leadership-class computing systems because of the intense metadata loads generated they generate. Recent research identified significant improvements from using shared-file I/O instead of multifile I/O patterns on leadership-class systems [6].

In this paper, we propose integrating shared file support into the I/O forwarding layer commonly found on leadership-class computing systems. I/O forwarding middleware, such as the I/O Forwarding Scalability Layer (IOFSL) [9,2], bridges the compute and I/O subsystems of leadership-class computing systems. This middleware layer captures all file I/O requests generated by applications executing on compute nodes and forwards them to dedicated I/O nodes. These I/O nodes, a common hardware feature of leadership-class computing systems, execute the I/O requests on behalf of the application. The I/O forwarding layer on these system is best suited to provide and manage shared file pointers because it has access to all application I/O requests and can provide enhanced file I/O capabilities independent of the system and I/O software stack. By embedding this capability into the I/O forwarding layer, applications developers can utilize shared file pointers for a variety of file I/O APIs (MPI-IO, POSIX, and ZOIDFS), synchronization levels (collective and independent I/O), and computing systems (IBM Blue Gene and Cray XT systems).

Y. Cotronis et al. (Eds.): EuroMPI 2011, LNCS 6960, pp. 312–314, 2011.

We are adding several features to IOFSL and ROMIO to enable portable MPI-IO shared file pointer access. In prior work, we extended the ZOIDFS API [2] to provide a distributed atomic append capability. Our current work extends and generalizes this capability to provide shared file pointers as defined by the MPI standard. First, we created a per file shared (key,value) storage space. This capability allows users of the API to instantiate an instance of a ZOIDFS file handle and associate file state with the handle (such as the current position of a file pointer). Since a ZOIDFS file handle is a persistent, globally unique identifier linked to a specific file, this does not result in extra state for the client. To limit the amount of state stored within the I/O node and to enable recovery from faults, we are integrating purge policies for the key value store. Example policies include flushing data to other IOFSL servers or persistently storing this data in extended attribute fields of the target file.

In prior work, we implemented a distributed atomic append by essentially implementing a per file, system wide shared file pointer. In our current work, we instead require a shared file pointer per MPI file handle. This is easily implemented by storing the current value of the shared file pointer in a key uniquely derived from the MPI file handle. We modified ROMIO to generate this unique key. When a file is first opened, a sufficiently large, random identifier is generated. This identifier is subsequently used to retrieve or update the current value of the shared file pointer. To avoid collisions, we rely on the fact that the key space provided by IOFSL supports an exclusive create operation. In the unlikely event that the generated identifier already exists for the file, ROMIO simply generates another one.

By providing set, get, and atomic increment operations, the IOFSL server is responsible for shared file pointer synchronization. This precludes the need for explicit file lock management for shared file pointer support. Overall, few modifications to ROMIO were required. Before executing a shared read or write, ROMIO uses the key store to atomically increment and retrieve the shared file pointer. It then subsequently accesses the file using an ordinary independent I/O operation. To simplify fault tolerance, we plan to combine the I/O access and the key update into one operation. ROMIO's `MPI_File_close` method removes the shared file pointer key in order to limit the amount of state held by the I/O nodes. For systems such as the Cray XT series, where I/O nodes are shared among multiple jobs, we automatically purge any keys left by applications that failed to clean up the shared file pointer, for example because of unclean application termination. On systems employing a dedicated I/O node, no cleanup is necessary, since the I/O node (and the IOFSL server) is restarted between jobs.

These modifications provide a low-overhead, file-system-independent, shared file pointer implementation for MPI-IO on those systems supported by IOFSL. Unlike other solutions, our implementation does not require a progress thread or hardware-supported remote memory access functionality [7].

Acknowledgments. This work was supported by the U.S. Department of Energy, under Contract DE-AC02-06CH11357.

References

1. ROMIO: A high-performance, portable MPI-IO implementation,
   http://www.mcs.anl.gov/romio/
2. Ali, N., Carns, P., Iskra, K., Kimpe, D., Lang, S., Latham, R., Ross, R., Ward, L., Sadayappan, P.: Scalable I/O Forwarding Framework for High–Performance Computing Systems. In: IEEE International Conference on Cluster Computing 2009 (2009)
3. Carns, P., Ligon, W., Ross, R., Thakur, R.: PVFS: A parallel file system for linux clusters. In: Proceedings of the 4th Annual Linux Showcase and Conference (2000)
4. Chan, A., Gropp, W., Lusk, E.: An efficient format for nearly constant-time access to arbitrary time intervals in large trace files. Scientific Programming 16(2-3), 155–165 (2008)
5. Freedman, C., Burger, J., DeWitt, D.: SPIFFI-A Scalable Parallel File System for the Intel Paragon. IEEE Transactions on Paralllel and Distributed Systems 7(11), 1185–1200 (1996)
6. Lang, S., Carns, P., Latham, R., Ross, R., Harms, K., Allcock, W.: I/O performance challenges at leadership scale. In: Proceedings of Supercomputing (November 2009)
7. Latham, R., Ross, R., Thakur, R.: Implementing MPI-IO Atomic Mode and Shared File Pointers Using MPI One-Sided Communication. International Journal of High Performance Computing Applications 21(2), 132–143 (2007), http://hpc.sagepub.com/cgi/content/abstract/21/2/132
8. Müller, M., Knüpfer, A., Jurenz, M., Lieber, M., Brunst, H., Mix, H., Nagel, W.: Developing scalable applications with vampir, vampirserver and vampirtrace. In: PARCO 2007 (2007)
9. Ohta, K., Kimpe, D., Cope, J., Iskra, K., Ross, R., Ishikawa, Y.: Optimization Techniques at the I/O Forwarding Layer. In: IEEE International Conference on Cluster Computing 2010 (2010)
10. Raicu, I., Foster, I., Wilde, M., Zhang, Z., Iskra, K., Beckman, P., Zhao, Y., Szalay, A., Choudhary, A., Little, P., Moretti, C., Chaudhary, A., Thain, D.: Middleware support for many-task computing. Cluster Computing 13, 291–314 (2010), http://dx.doi.org/10.1007/s10586-010-0132-9

Improvement of the Bandwidth of
Cross-Site MPI Communication Using Optical Fiber

Kiril Dichev, Alexey Lastovetsky, and Vladimir Rychkov

UCD School of Computer Science and Informatics,
University College Dublin,
Belfield, Dublin 4, Ireland
Kiril.Dichev@ucdconnect.ie,
{Alexey.Lastovetsky,Vladimir.Rychkov}@ucd.ie

Abstract. We perform a set of communication experiments spanning multiple sites on the heterogeneous Grid'5000 infrastructure in France. The backbone widely employs high-bandwidth optical fiber. Experiments with point-to-point MPI communications across sites show much lower bandwidth than expected for the optical fiber connections. This work proposes and tests an alternative implementation of cross-site point-to-point communication, exploiting the observation that a higher bandwidth can be reached when transferring TCP messages in parallel. It spawns additional MPI processes for point-to-point communication and significantly improves the bandwidth for large messages. The approach comes closer to the maximum bandwidth measured without using MPI.

Keywords: Heterogeneous Communications, Optical fiber, Point-to-point, MPI.

1 Introduction and Related Work

Grid infrastructures such as Grid'5000 can have very complex communication networks involving local area networks as well as optical fiber connections between sites.

We observed that for this infrastructure several TCP connections across sites can be used in parallel with significant increase of bandwidth. We tested the bandwidth of the cross site optical fiber connection without MPI by varying the number of parallel TCP connections. We observed that the bandwidth is better when using 4 or 8 parallel connections and comes close to 1 Gbps. This observation was important for the proposed modification in point-to-point communication.

For MPI point-to-point benchmarks spanning two sites we use NetPIPE with MPI. Since MPI is used for connecting different sites, the underlying communication uses the TCP protocol. The results show a peak bandwidth of around 70 Mbps, which is a much lower bandwidth than any of the TCP benchmarks.

To improve the low bandwidth, we propose a modified MPI point-to-point communication with parallel transfer of different fragments of the same message from the sender to the receiver. Research in this direction has been done in distributed computing. In [1], a similar approach is followed by GridFTP to transfer large data

Y. Cotronis et al. (Eds.): EuroMPI 2011, LNCS 6960, pp. 315–317, 2011.

volumes in parallel over the internet by using a number of TCP data streams. The idea is less popular in the high-performance computing domain. Multi-railing is one such example, but it is mostly used for a different setting - when a number of network interfaces are available for the communicating processes at each node [2].

2   Modified MPI Point-to-Point Algorithm

We experiment with two methods of parallel transfer of different message fragments. Both cases are implemented on top of the MPI library without internal modifications.

In the first proposed implementation, we use a number of different OpenMP threads, each of which is responsible for submitting a different fragment of a message through MPI point-to-point calls. We used 2 or 4 threads per node and compared this with the original point-to-point calls. The results show no advantage of the multi-threaded implementation. We believe this is due to the internal serialization of point-to-point calls in the MPI library, which prevents true parallelization of the different communicating threads.

We then implement the same idea with MPI processes instead. At the start, a fixed number of extra MPI processes are spawned on each node (Fig. 1a). Any point-to-point communication between processes e.g. P0 and P1 is then divided into two phases – a scatter phase and a gather phase (Fig. 1b and 1c). Each phase is implemented as a linear sequence of point-to-point calls for the different message chunks of the original message. To exploit the parallelism of point-to-point calls, the scatter/gather implementation is a linear sequence of non-blocking sends and receives in MPI.

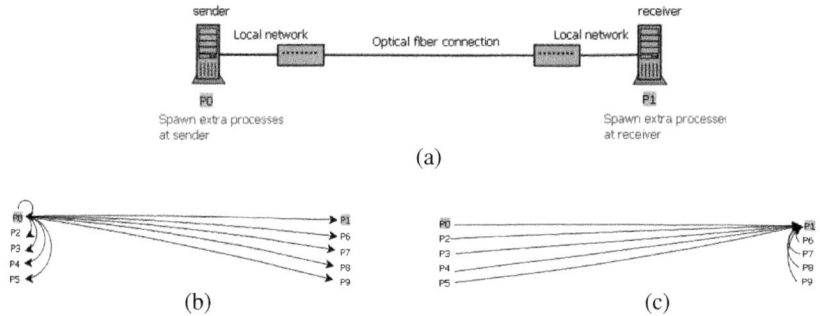

Fig. 1. Diagram of spawning extra processes at the initialization (a), and transferring a message through two-phases of point-to-point calls on the message chunks – a linear scatter (b) and a linear gather (c)

We present results of experiments with the proposed implementation for message sizes range from 100 KB to 1 MB (Fig. 2a) and from 1 MB to 10 MB (Fig. 2b). In the experiments with the proposed algorithm, we spawn 4 / 8 additional MPI processes per node and involve all of them in the modified point-to-point communication.

The runs with additional processes demonstrate increased bandwidth compared to the original runtime for all message sizes larger than 200 KB, and the improvement is relatively constant for this range. For example, for messages of 10 MB, the MPI point-to-point implementation only reaches around 80 Mbps, while the two-phase version using 16 additional processes (8 at receiver/sender) achieves 498 Mbps, which is more than 6 times increase in bandwidth. This bandwidth is still far from the peak bandwidth we could achieve with a TCP connection (nearly 1 Gbps), but is much closer to it.

The improvements are related to the optical fiber cross-site connection since for a local site run the modified algorithm does not improve the point-to-point bandwidth.

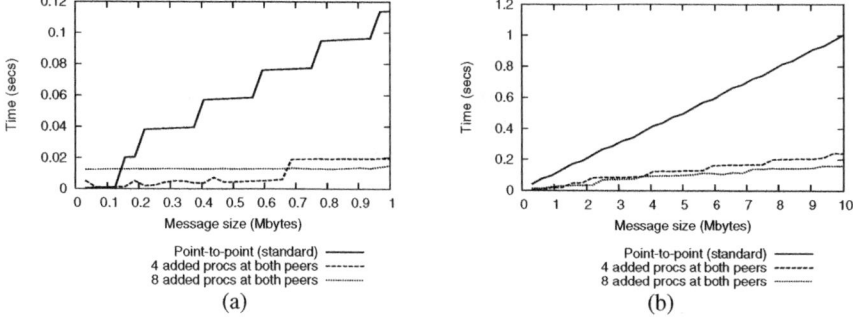

Fig. 2. Comparison of MPI point-to-point communication and the modified version for messages in the range 100 Kbytes-1 MB (a) and 1 MB-10MB (b)

Acknowledgments. This publication has emanated from research conducted with the financial support of Science Foundation Ireland under Grant Number 08/IN.1/I2054.

Experiments presented in this paper were carried out using the Grid'5000 experimental testbed, being developed under the INRIA ALADDIN development action with support from CNRS, RENATER and several Universities as well as other funding bodies (see https://www.grid5000.fr).

References

1. Allcock, B., Bester, J., Bresnahan, J., Chervenak, A.L., Foster, I., Kesselman, C., Meder, S., Nefedova, V., Quesnel, D., Tuecke, S.: Data Management and Transfer in High-Performance Computational Grid Environments. Parallel Computing 28, 749–771 (2002)
2. Moreaud, S., Goglin, B., Namyst, R.: Adaptive MPI Multirail Tuning for Non-Uniform Input/Output Access. In: Keller, R., Gabriel, E., Resch, M., Dongarra, J. (eds.) EuroMPI 2010. LNCS, vol. 6305, pp. 239–248. Springer, Heidelberg (2010)

Performance Tuning of SCC-MPICH by Means of the Proposed MPI-3.0 Tool Interface

Carsten Clauss, Stefan Lankes, and Thomas Bemmerl

Chair for Operating Systems, RWTH Aachen University, Germany
{clauss,lankes,bemmerl}@lfbs.rwth-aachen.de

Abstract. The Single-Chip Cloud Computer (SCC) experimental processor is a 48-core concept vehicle created by Intel Labs as a platform for many-core software research. Intel provides a customized programming library for the SCC, called RCCE, that allows for fast message-passing between the cores. For that purpose, RCCE offers an application programming interface (API) with a semantics that is derived from the well-established MPI standard. However, while the MPI standard offers a very broad range of functions, the RCCE API is consciously kept small and far from implementing all the features of the MPI standard. For this reason, we have implemented an SCC-customized MPI library, called SCC-MPICH, which in turn is based upon an extension to the SCC-native RCCE communication library. In this contribution, we will present SCC-MPICH and we will show how performance analysis as well as performance tuning for this library can be conducted by means of a prototype of the proposed MPI-3.0 tool information interface.

Keywords: Single-Chip Cloud Computer, SCC, MPI 3.0, Tools Support.

1 Introduction and Overview

The Single-Chip Cloud Computer (SCC) experimental processor [1] is a 48-core *concept vehicle* created by Intel Labs as a platform for many-core software research. The cores of the SCC are arranged in a 6x4 on-die mesh with two cores per network node. By means of this network, all cores can access a global shared-memory space of up to 64GByte via four on-die memory controllers. In addition to this global off-die shared memory, each core provides a chunk of 8kByte fast on-die memory that is also accessible to all other cores. These additional on-die shared-memory chunks are intended to pass messages directly between the cores, and this is why they are referred to as *Message-Passing Buffers* (MPBs). In contrast to common multi-core processors, the SCC does not provide any cache-coherency between the cores. For that reason, the global shared-memory is logically distributed in such a way that each core can boot its own Linux image. Therefore, the architecture of the SCC can be regarded as a *Cluster on the Chip* where message-passing is the programming paradigm of choice. Intel provides a customized message-passing library for the SCC, called RCCE [2], that utilizes

Y. Cotronis et al. (Eds.): EuroMPI 2011, LNCS 6960, pp. 318–320, 2011.

the fast on-die MPBs. In doing so, RCCE offers an application programming interface (API) with a semantics that is derived from the well-established MPI standard [3]. However, while the MPI standard offers a very broad range of functions, the RCCE API is intentionally kept very small [4] and far from offering all the features of the MPI standard. For this reason, we have implemented an SCC-customized MPI library, called SCC-MPICH, which in turn is based upon an extension to the SCC-native RCCE communication library. Although also Intel has recently released an MPI library for the SCC, called RCKMPI [6], we have by now applied a lot of performance tuning and additional improvements to SCC-MPICH, as for example SCC-optimized collectives. In this contribution, we will present SCC-MPICH and we will show how performance analysis as well as performance tuning can be conducted by means of a prototype of the proposed MPI-3.0 tool information interface [9]. Since this is still work in progress (effective June 2011), we will present our findings in the course of the poster session of the 18th European MPI Users' Group Meeting in September 2011.

2   A Customized MPI Library for the Intel SCC

SCC-MPICH is based on MP-MPICH [7], a multi-platform message-passing library that in turn is derived from the original MPICH [8]. In doing so, we have extended MP-MPICH by a new communication device that utilizes the fast on-die MPBs as well as the off-die shared-memory for the core-to-core communication. In turn, this new SCC-related communication device provides four different communication protocols: *Short, Eager, Rendezvous* and *SHM-Eager*. The Short protocol is low latency optimized and used for exchanging message headers as well as header-embedded short payload messages via the MPBs. Bigger messages must be sent either via one of the two Eager protocols or via the Rendezvous protocol. The main difference between Eager and Rendezvous mode is that Eager messages must be accepted on the receiver side even if the corresponding receive requests are not yet posted by the application. Therefore, a message sent via Eager mode can implicate an additional overhead by copying the message temporarily into an intermediate buffer. However, when using the SHM-Eager protocol, the off-die shared-memory is used to pass the messages between the cores. That means that this protocol does not require the receiver to copy unexpected messages into additional *private* intermediate buffers unless there is no longer enough *shared* off-die memory. The decision which of these protocols is to be used depends on the message length as well as on the ratio of expected to unexpected messages.

3   A Prototype of the MPI-3.0 Tool Information Interface

Currently, the working groups of the MPI-Forum are fostering the development of the upcoming MPI 3.0 standard. In doing so, the so-called Tools Working Group deals with the definition of additional information interfaces that should help to enhance the interaction between MPI implementations and additional

tools like debuggers, profilers and performance tuners [9]. Although the draft
for these new interfaces is still under active development and not yet mature
for standardization, we have already prototyped a major part of the proposed
functions on top of MP-MPICH for a use case evaluation [10]. Since SCC-MPICH
is based on MP-MPICH, we can already use the new information interface to
query and to tune performance and configuration parameters of the SCC-related
communication device. That way, we are able, for example, to determine optimal
threshold values between the above mentioned communication protocols.

Acknowledgments. We are not part of the MPI 3.0 Tools Working Group,
but we are observing the respective standardization process with great interest.
The presented research and development was supported by Intel Corporation.
The authors would like to thank especially Ulrich Hoffmann, Michael Konow
and Michael Riepen of Intel Braunschweig for their help and guidance.

References

1. Intel Corporation. SCC External Architecture Specification (EAS) – Revision 1.1
   (November 2010), http://communities.intel.com/docs/DOC-5852
2. Mattson, T., van der Wijngaart, R.: RCCE: a Small Library for Many-Core Communication. Intel Corporation (May 2010),
   http://communities.intel.com/docs/DOC-5628
3. Message Passing Interface Forum. MPI: A Message-Passing Interface Standard.
   High-Performance Computing Center Stuttgart (HLRS), V2.2 (September 2009)
4. Mattson, T., van der Wijngaart, R., Riepen, M., et al.: The 48-core SCC Processor:
   The Programmer's View. In: Proceedings of the 2010 ACM/IEEE Conference on
   Supercomputing (SC 2010), New Orleans, LA, USA (November 2010)
5. Clauss, C., Lankes, S., Reble, P., Bemmerl, T.: Evaluation and Improvements of
   Programming Models for the Intel SCC Many-core Processor. In: Proceedings of the
   International Conference on High Performance Computing and Simulation (HPCS
   2011), Istanbul, Turkey (July 2011)
6. Urena, I.A.C.: RCKMPI Manual. Intel Braunschweig (February 2011),
   http://communities.intel.com/docs/DOC-6628
7. Bierbaum, B., Clauss, C., Finocchiaro, R., Schuch, S., Pöppe, M., Worringen, J.:
   MP-MPICH – User Documentation and Technical Notes (2009),
   http://www.lfbs.rwth-aachen.de/users/global/mp-mpich/
   mp-mpich_manual.pdf
8. Gropp, W., Lusk, E., Doss, N., Skjellum, A.: A High-Performance, Portable Implementation of the MPI Message Passing Interface Standard. Parallel Computing 22(6), 789–828 (1996)
9. MPI 3.0 Tools Support Working Group: "Tool Interfaces", current Draft of the
   Chapter Proposal (June 2011),
   https://svn.mpi-forum.org/trac/mpi-forum-web/attachment/wiki/
   MPI3Tools/mpit/mpi-report.18.pdf
10. Clauss, C., Lankes, S., Bemmerl, T.: Use Case Evaluation of the Proposed MPIT
    Configuration and Performance Interface. In: Keller, R., Gabriel, E., Resch, M.,
    Dongarra, J. (eds.) EuroMPI 2010. LNCS, vol. 6305, pp. 285–288. Springer,
    Heidelberg (2010)

Design and Implementation of Key Proposed MPI-3 One-Sided Communication Semantics on InfiniBand

Sreeram Potluri, Sayantan Sur, Devendar Bureddy, and Dhabaleswar K. Panda

Department of Computer Science and Engineering, The Ohio State University
{potluri,surs,bureddy,panda}@cse.ohio-state.edu

Abstract. As part of the MPI-3 effort, Remote Memory Access (RMA) working group has proposed several extensions to the one-sided communication interface. These extensions promise to address several of its existing limitations. However, their performance advantages need to be clearly highlighted for widespread acceptance of the new interface. In this paper, we present design and implementation of some of the key one-sided semantics proposed for MPI-3 over InfiniBand, using the MVAPICH2 library. Our evaluation shows that the newly proposed Flush semantics allow for more efficient handling of completions and the request-based operations can help achieve close to optimal overlap in a Get-Compute-Put model.

1 Introduction and Overview

High-end computing systems have seen a tremendous growth in recent years, driven by advances in processor, network and accelerator technologies. As the capabilities of different components in a system increase, it is important for scientific applications to utilize all these components concurrently to achieve maximum performance. Programming models hold the key in enabling such usage. In the past, non-blocking message passing and one-sided communication semantics have been introduced to enable overlap between computation and communication. Earlier work has shown how one-sided communication semantics can achieve superior performance in applications than the message passing semantics [3]. However, their adaptation has been limited because of the overheads imposed by synchronization operations in MPI-2 and a mismatch with real-world use cases for one-sided communication. As part of MPI-3 effort, the Remote Memory Access(RMA) group has proposed several extensions to the existing model that promise to address many of these limitations [1].

The additions to one-sided interface include dynamic window creation, light weight synchronization (local and remote) and variety of other communication operations. However, in order for wide spread acceptance of this proposed interface, its performance advantages need to be clearly highlighted. We believe that this is a strong motivation for designing and implementing some of the key MPI-3 interfaces on a widely used commodity platform. In this work, we present the design of a key subset of the proposed MPI-3 extensions over InfiniBand. Through experimental evaluation we establish that they efficiently solve several issues faced by the MPI-2 standard. Our design of the proposed semantics is integrated into the MVAPICH2 library [2], to demonstrate a working

Y. Cotronis et al. (Eds.): EuroMPI 2011, LNCS 6960, pp. 321–324, 2011.

prototype in an open-source production MPI library. To the best of our knowledge, this is the first design and implementation of the proposed MPI-3 one-sided interface. The semantics implemented in this work are highlighted in Figure 1.

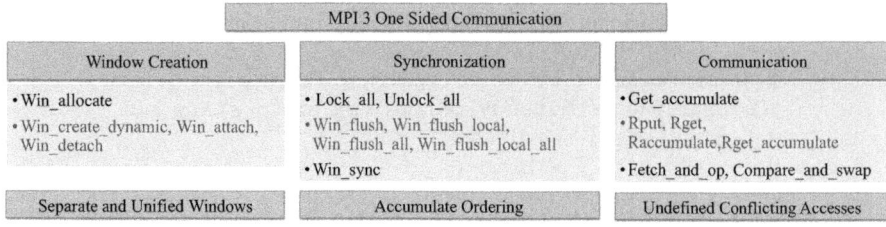

Fig. 1. Proposed MPI-3 One-Sided Communication Standard Extensions

2   Design and Evaluation

In this section, we provide a brief overview of each of the semantics addressed in this work and their implementation highlights. A detailed description about the implementation and evaluation can be found in our technical report [4].

Dynamic Windows: A window defines the memory to be used for communication in the one-sided model. In MPI-2, the location and size of memory attached to a window is specified during window creation and cannot be changed at a later point of time. This is a misfit in the case of applications and programming models with dynamic memory requirements. MPI-3 allows "dynamic" windows where each process can asynchronously attach or detach memory from a window. Implementation of one-sided communication operations over RDMA requires exchange of buffer registration information. For MPI-2, this is usually done during the window creation phase. In the case of dynamic windows, such an exchange is required each time an access happens to a newly attached buffer. However, as multiple accesses happen to each buffer, this cost can be amortized efficiently. Through micro benchmark evaluation, we show that the performance of dynamic windows is as good as that of static windows. The performance comparison of Put latency is shown in Figure 3(a). A complete set of results can be found in [4].

Flush Operations: All communication operations in MPI one-sided interface are non-blocking. In MPI-2, their completions (both local and remote) are bound to synchronization operations. This is heavy-weight. MPI-3 addresses this issue through flush operations, separating local completion from remote completion. Ensuring local and remote completions of different operations (writes, reads and atomics) in InfiniBand have different requirements and costs. The flush semantics provide flexibility to match the completion of different one-sided communication operations to the completion requirements in InfiniBand and hence provide better efficiency. A comparison of Put completion times using Lock/Unlock and Flush semantics is shown in Figure 3(b).

Request-based Operations: Request-based operations provide an easy mechanism to wait for completion of *specific* operations. This allows for finer grained overlap. Figure 2 presents pseudo-code for three versions of a Get-Compute-Put benchmark which fetches, computes and writes back N blocks of data in remote memory. Figure 2(a) shows a version without any overlap. The three phases: get, compute and put can be pipelined to overlap computation and communication. Figures 2(b) and (c) show overlapped versions using MPI-2 passive semantics and using Request-based operations respectively. The performance results are shown in Figure 3(c). We see that request-based operations provide close to optimal overlap.

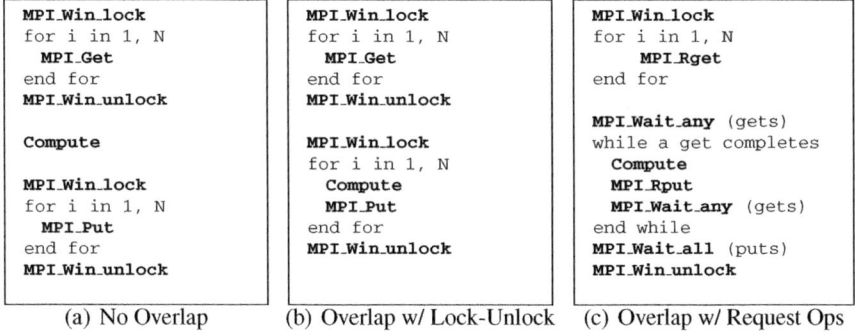

Fig. 2. Get-Compute-Put on N Blocks of Data

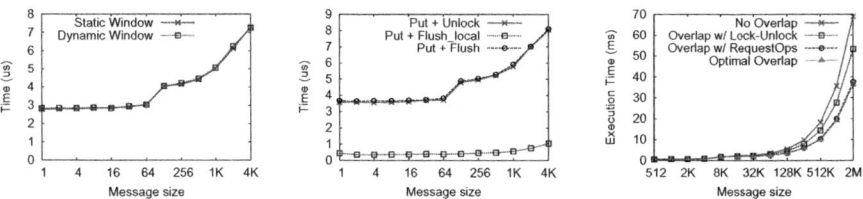

(a) Put Latency w/ Dynamic Win   (b) Put Latency w/ Flush Ops   (c) Get-Compute-Put w/ Request Ops

Fig. 3. Performance using MPI-3 One-sided Semantics

3   Conclusion

In this paper, we presented design, implementation and evaluation of a key subset of newly proposed one-sided interface. Through micro-benchmark evaluation, we have shown that the newly proposed interfaces can provide improved performance over the MPI-2 interfaces.

Acknowledgments. This research is supported in part by U.S. Department of Energy grants #DE-FC02-06ER25749 and #DE-FC02-06ER25755; National Science Foundation grants #CCF-0833169, #CCF-0916302, #OCI-0926691 and #CCF-0937842.

References

1. MPI-3 RMA,
   https://svn.mpi-forum.org/trac/mpi-forum-web/
   raw--attachment/wiki/mpi3-rma-proposal1/one-side-2.pdf
2. MVAPICH2: MPI over InfiniBand, 10GigE/iWARP and RoCE,
   http://mvapich.cse.ohio-state.edu/
3. Potluri, S., Lai, P., Tomko, K., Sur, S., Cui, Y., Tatineni, M., Schulz, K., Barth, W., Majumdar, A., Panda, D.K.: Quantifying Performance Benefits of Overlap using MPI-2 in a Seismic Modeling Application. In: International Conference on Supercomputing, ICS 2010 (2010)
4. Potluri, S., Sur, S., Bureddy, D., Panda, D.K.: Design and Implementation of Key Proposed MPI-3 One-Sided Communication Semantics on InfiniBand. Technical Report OSU-CISRC-7/11-TR19, The Ohio State University (2011)

Scalable Distributed Consensus to Support MPI Fault Tolerance*

Darius Buntinas

Argonne National Laboratory

Abstract. As system sizes increase, the amount of time in which an application can run without experiencing a failure decreases. Exascale applications will need to address fault tolerance. In order to support algorithm-based fault tolerance, communication libraries will need to provide fault-tolerance features to the application. One important fault-tolerance operation is distributed consensus. This is used, for example, to collectively decide on a set of failed processes. This paper describes a scalable, distributed consensus algorithm that is used to support new MPI fault-tolerance features proposed by the MPI 3 Forum's fault-tolerance working group. The algorithm was implemented and evaluated on a 4,096-core Blue Gene/P. The implementation was able to perform a full-scale distributed consensus in 305 μs and scaled logarithmically.

1 Introduction

As process counts in applications grow toward exascale, the length of time an application can run without experiencing a failure, known as the mean time between failures (MTBF), decreases. Applications will need to be fault tolerant in order to be useful on future exascale machines. Checkpointing can provide fault tolerance to an application without the need to modify it As the failure frequency increases, however, checkpoints will need to be taken more often, decreasing the amount of useful work the application can perform between failures.

Whereas checkpointing provides fault tolerance to an application in a transparent manner, when using algorithm-based fault tolerance (ABFT) [1][3][4], the application is aware of faults and handles them explicitly. The fault-tolerance working group of the MPI 3 Forum has been working on a proposal [5], that adds fault-tolerance features to MPI in order to support ABFT applications. The proposal defines the behavior of an MPI library if processes fail. For example, existing operations such as MPI_Comm_split are now required to either succeed at every process or return an error at every process, even if processes fail before or during the operation. The proposal also introduces new functions, such as MPI_Comm_validate_all, that require all processes to return the same list of failed processes. A distributed consensus algorithm is needed to implement these operations.

---

* This work was supported in part by the Office of Advanced Scientific Computing Research, Office of Science, U.S. Department of Energy, under Contract DE-AC02-06CH11357.

Y. Cotronis et al. (Eds.): EuroMPI 2011, LNCS 6960, pp. 325–328, 2011.

This paper presents a scalable, fault tolerant, distributed consensus algorithm used to implement the MPI_Comm_validate_all function. The MPI_Comm_validate_all implementation is evaluated on a 4,096-core IBM Blue Gene/P machine and shows $O(\log n)$ scaling.

2   Algorithm

We present the distributed consensus algorithm as it would be used in the MPI_Comm_validate_all operation. However, the algorithm could also be used in other operations requiring distributed consensus, such as MPI_Comm_split. In this section, we give a brief overview of the algorithm at a high level. A detailed description of the algorithm can be found in [2].

The MPI_Comm_validate_all function uses distributed consensus to decide on a set of failed processes, which must contain every failed process known by any participating process at the time the function is called. The same set of failed processes must be returned by the function at every process. If a process fails during the MPI_Comm_validate_all operation (i.e., after the first process calls the function and before the first process returns), the set of failed processes returned may or may not contain the failed processes.

The algorithm is similar to the three-phase commit algorithm except that, rather than sending and receiving individual messages, a reliable broadcast algorithm is used to send and collect messages. In the BALLOTING phase, after the root is chosen, the root creates a ballot containing the set of failed processes and broadcasts it to the rest of the processes. Once the processes receive the ballot, the responses to the ballot are collected back up the tree. If all the processes have accepted the ballot, the algorithm enters the COMMIT phase; otherwise a new ballot is generated, and the BALLOTING phase is repeated. In the COMMIT phase the root broadcasts a commit message. Once all processes receive the commit message, acknowledgments are collected back up to the root. The last phase is the ALL_COMMIT phase. In this phase the root broadcasts the all-commit message. Once a process receives the all-commit message, it can return from the MPI_Comm_validate_all function.

3   Performance Evaluation

To evaluate the validate-all operation, we implemented it as an MPI program. This allowed us to evaluate the operation at a large scale on a Blue Gene/P without modifying the MPI implementation. We expect the performance of the operation implemented this way to be an upper bound on the performance of the operation if it were integrated into an MPI implementation. The evaluation was performed at Argonne National Laboratory on Surveyor, a Blue Gene/P with 1,024 quad-core nodes.

Figure 1 shows the results of the evaluation. As expected, the operation scales logarithmically. For comparison, we evaluated the time taken to perform a

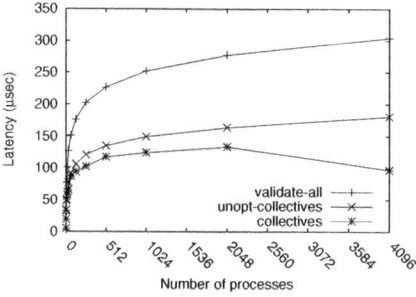
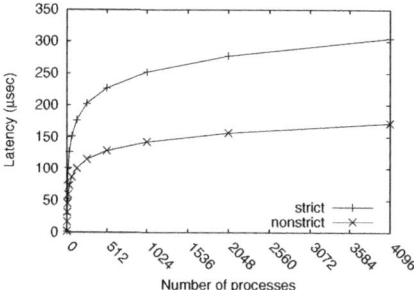

Fig. 1. Comparison of the validate-all operation with collectives operations

Fig. 2. Comparison of validate-all using strict and loose semantics

communication pattern similar to that of the validate-all operation using broadcast and reduction operations. The figure shows the results with optimized collectives using the Blue Gene/P collective tree network and with unoptimized collectives using the same torus network that the validate-all operation uses. At full scale, the validate-all implementation took 305 μs to perform the operation, which is 1.66 times slower than performing a similar communication pattern with unoptimized collectives. We expect the performance of the validate-all algorithm to improve when the operation is integrated into the MPI implementation, making the algorithm more responsive to incoming messages.

We also evaluated the performance of the operation with loose semantics, as described in the proposal [5]. Figure 2 shows the comparison. The loose implementation performs the operation 133 μs faster at full scale than does the strict implementation (which is 1.78 times as fast). Depending on the requirements of the application and the frequency at which the application calls validate-all, using the loose implementation can provide some performance improvement.

4 Conclusion

This paper presented a scalable distributed consensus algorithm used to implement the MPI_Comm_validate_all operation proposed by the MPI 3 fault-tolerance working group. The algorithm was evaluated on a 4,096-core Blue Gene/P machine and was shown to be extremely scalable. The implementation was able to perform a full-scale validate-all operation in 305 μs and scaled logarithmically.

Using the loose implementation saved only 133 μs over the strict implementation. Therefore, unless the application performs many validate-all operations, relaxing the semantics is unlikely to improve the overall performance of the application significantly.

References

1. Anfinson, J., Luk, F.T.: A linear algebraic model of algorithm-based fault tolerance. IEEE Transactions on Computing 37, 1599–1604 (1988)
2. Buntinas, D.: Scalable distributed consensus to support MPI fault tolerance. Tech. Rep. ANL/MCS-TM-314, Argonne National Laboratory (June 2011)
3. Chen, Z., Dongarra, J.: Algorithm-based fault tolerance for fail-stop failures. IEEE Transactions on Parallel and Distributed Systems 19(12) (December 2008)
4. Chen, Z., Dongarra, J.: Highly scalable self-healing algorithms for high performance scientific computing. IEEE Transactions on Computers (July 2009)
5. Fault Tolerance Working Group: Run-though stabilization proposal, http://svn.mpi-forum.org/trac/mpi-forum-web/wiki/ft/run_through_stabilization

Run-Through Stabilization:
An MPI Proposal for Process Fault Tolerance*

Joshua Hursey[1], Richard L. Graham[1], Greg Bronevetsky[2],
Darius Buntinas[3], Howard Pritchard[4], and David G. Solt[5]

[1] Oak Ridge National Laboratory
{hurseyjj,rlgraham}@ornl.gov
[2] Lawrence Livermore National Laboratory
greg@bronevetsky.com
[3] Argonne National Laboratory
buntinas@mcs.anl.gov
[4] Cray, Inc.
howardp@cray.com
[5] Hewlett-Packard
david.solt@hp.com

Abstract. The MPI standard lacks semantics and interfaces for sustained application execution in the presence of process failures. Exascale HPC systems may require scalable, fault resilient MPI applications. The mission of the MPI Forum's Fault Tolerance Working Group is to enhance the standard to enable the development of scalable, fault tolerant HPC applications. This paper presents an overview of the Run-Through Stabilization proposal. This proposal allows an application to continue execution even if MPI processes fail during execution. The discussion introduces the implications on point-to-point and collective operations over communicators, though the full proposal addresses all aspects of the MPI standard.

Keywords: MPI, Fault Tolerance, Run-through Stabilization, Algorithm Based Fault Tolerance, Fail-Stop Process Failure.

* Special thanks to the MPI Forum and Fault Tolerance Working Group members that contributed to the run-through stabilization proposal. Their comments and insights continue to help strengthen the developing proposals targeted for inclusion in the Message Passing Interface (MPI) standard. Research sponsored by the Office of Advanced Scientific Computing Research; Office of Science; Mathematical, Information, and Computational Sciences Division at Oak Ridge National Laboratory; U.S. Department of Energy, under Contract No. DE-AC05-00OR22725 with UT-Battelle, LLC; U.S. Department of Energy, under Contract No. DE-AC02-06CH11357; U.S. Department of Energy, under Contract No. DE-AC52-07NA27344 by Lawrence Livermore National Laboratory; The ARRA / DoE - Early Career Research Program; and by award #CCF-0816909 from the National Science Foundation.

Y. Cotronis et al. (Eds.): EuroMPI 2011, LNCS 6960, pp. 329–332, 2011.
© Springer-Verlag Berlin Heidelberg 2011

1   Introduction

High Performance Computing (HPC) applications, particularly those running in fault-prone environments, use fault tolerance techniques to ensure successful completion of their computational objectives. As HPC systems push toward exascale, projections indicate that these large-scale systems will become more fault-prone, posing a greater threat to the existing HPC applications [2]. In preparation for such fault-prone computing environments, applications are investigating Algorithm Based Fault Tolerance (ABFT) [5] techniques to improve the efficiency of application recovery after process failure beyond that which checkpoint/restart solutions alone can provide.

The lack of standardized fault tolerance semantics and interfaces prevents HPC applications from portably exploring ABFT techniques using the MPI standard. The MPI Forum created the Fault Tolerance Working Group in response to the growing need for portable, fault tolerant semantics and interfaces in the MPI standard to support application level fault tolerance development.

The Fault Tolerance Working Group (FTWG)'s run-through stabilization (RTS) proposal enables an MPI application to continue execution even if one or more MPI processes fail. The discussion focuses on the central themes of the proposal in the context of a communicator though all aspects of MPI are addressed in the proposal under consideration for the MPI-3.0 version of the MPI standard [4]. Various MPI implementations are currently exploring implementations of the RTS proposal. The complementary process recovery proposal is being actively developed by the FTWG.

2   Process Fault Tolerance Model

Under the RTS proposal, the primary role of the MPI implementation is to (i) inform the application of process failures, and (ii) allow the application to continue running and communicating with unaffected processes. The application is guaranteed to be eventually informed, via error handlers, of all process failures and that no process will be reported as failed before it actually fails. Therefore the MPI implementation must provide a *perfect* failure detector for *fail-stop* process failure (i.e., a process is permanently stopped, often due to a crash) [3].

From the perspective of one process, other processes can be in one of the following states (prefixed with MPI_RANK_STATE_): OK, FAILED or NULL. Processes with state OK are executing normally. Processes with state FAILED have been detected by MPI as failed-stop. Processes with state NULL are failed processes treated as if their ranks are MPI_PROC_NULL.

2.1   Validation of Process State

The RTS proposal focuses on high scalability by treating process failures differently from the perspective of point-to-point and collective communication. This is because point-to-point communication between a given pair of processes

MPI_Comm_validate{_all}(MPI_Comm c, int *newfailures)
MPI_Comm_validate_get_num_state{_all}(MPI_Comm c, int type, int *count)
MPI_Comm_validate_get_state{_all}(MPI_Comm c, int type, int incount,
                          int *outcount, MPI_Rank_info rank_infos[])
MPI_Comm_validate_get_state_rank{_all}(MPI_Comm c, int rank,
                          MPI_Rank_info *rank_info)
MPI_Comm_validate_set_state_null(MPI_Comm c, int incount,
                          MPI_Rank_info rank_infos[])

Fig. 1. Validation Interfaces for Communications (C interface shown)

is rarely affected by the failure of another process, while collective communication implies dependance upon the participation of the entire group. As such, the proposal provides two scopes of application fault recognition: *local* and *global*.

A process uses the validation functions in Figure 1 to update, access, and modify the known state of a process in a communicator. Local recognition is implemented by the variants of the MPI_Comm_validate operation, and are designed to support point-to-point communication. Global recognition is implemented by variants of the MPI_Comm_validate_all operation, and are designed to support collective communication.

A fault tolerant agreement algorithm is provided by the MPI_Comm_validate_all collective operation [1]. This operation synchronizes the fault detectors, re-enables collective operations, globally recognizes known failed processes, and provides a uniform return value across the collective group.

The failure of a process must be recognized on each communicator of which it is a member. This allows libraries, that create their own communicators, to be able to receive notification of the failure even if another library or the main application has already recognized the failure on another communicator.

2.2 Semantic Modifications

Point-to-Point. Communication between two active processes is unaffected by the failure of other non-participating processes. For example, if process A fails, process B can still send messages to process C, and vice versa. Communication with process A returns an error (MPI_ERR_RANK_FAIL_STOP) until process B *recognizes* the failed process.

Collectives. Collective operations must be *fault-aware*, meaning that they will not hang in the presence of failures. To preserve failure-free performance, collective operations are not required to provide uniform return codes. For example, using MPI_Bcast it is possible for a process to fail inside the collective such that those processes that left early returned success while the remainder will return an error. When a process fails, all collective operations are disabled in communicators that contain that process. Collective communication can be re-enabled by calling MPI_Comm_validate_all.

Communicator Management. All failed processes must be globally recognized in the participating communicator(s) before calling any communicator construction operation. If a globally recognized failed process is represented in a communicator passed to a communicator construction operation other than MPI_COMM_SPLIT, then it is represented in the new communicator. In the presence of failures, the communicator construction operations ensure uniformly consistent creation of the communicator handle and return codes.

References

1. Barborak, M., Dahbura, A., Malek, M.: The consensus problem in fault-tolerant computing. ACM Computing Surveys 25, 171–220 (1993)
2. Cappello, F., Geist, A., Gropp, B., Kale, L., Kramer, B., Snir, M.: Toward exascale resilience. International Journal of High Performance Computing Applications 23(4), 374–388 (2009)
3. Chandra, T.D., Toueg, S.: Unreliable failure detectors for reliable distributed systems. Journal of the ACM 43, 225–267 (1996)
4. Fault Tolerance Working Group: Run-though stabilization proposal,
   http://svn.mpi-forum.org/trac/mpi-forum-web/wiki/ft/
   run_through_stabilization
5. Huang, K.H., Abraham, J.A.: Algorithm-based fault tolerance for matrix operations. IEEE Transactions on Computers 33(6), 518–528 (1984)

Integrating MPI with Asynchronous Task Parallelism

Yonghong Yan, Sanjay Chatterjee, Zoran Budimlic, and Vivek Sarkar

Department of Computer Science, Rice University
{yanyh,sanjay.chatterjee,zoran,vsarkar}@rice.edu

Abstract. This paper describes a programming model that integrates intra-node asynchronous task parallelism with inter-node MPI communications to address the hybrid parallelism challenges faced by future extreme scale systems. We explore the integration of MPI's blocking and non-blocking communications with lightweight tasks. We also provide the implementation details of a non-blocking runtime execution model based on computation and communication workers.

1 Introduction

As we head towards exascale computing, it is estimated that we will need to build systems with $O(10^6)$ nodes and $O(10^3)$ cores per node [3]. This paper describes a programming model intended for such systems. We demonstrate this model (HC-MPI) by integrating intra-node asynchronous task parallelism with inter-node MPI communications in the Habanero-C (HC) research language. Our model tightly integrates MPI's blocking and non-blocking communications with HC's lightweight tasks, and allows inter-node collective MPI operations to be initiated by asynchronous tasks. This paper provides a summary of the HC-MPI model, and includes some preliminary results that show its promise.

2 HC-MPI

An HC-MPI program follows a task parallel model within a node and MPI's SPMD model across nodes. The Habanero-C language has two constructs for asynchronous task parallel programming: async and finish, which were borrowed from the X10 [1] programming language. To extend HC for distributed systems, we provide two types of message passing operations that are similar to MPI, asynchronous message passing (async MP), async_send and async_recv , and synchronous message passing (sync MP), sync_send and sync_recv . The async MP calls use asynchronous tasks to pass messages, and do not block the current execution flow. The enclosing finish statement of async MPs serves as a synchronization scope that guarantees that all the asynchronous operations (tasks and MPs) will be completed before proceeding to the code after the finish scope. The sync MP operations, sync_send and sync_recv , synchronize the sender and receiver in a message passing transaction. At the intra-node level, the sync calls

Y. Cotronis et al. (Eds.): EuroMPI 2011, LNCS 6960, pp. 333–336, 2011.

only suspend the current execution context, and do not block the CPU on either the sender or the receiver side. The compiler transforms HC-MPI constructs to standard C code that handles context suspension, resumption, and interaction with the runtime system.

The HC-MPI runtime contains several computation workers and a single communication worker. Computation workers perform computation tasks, and the communication worker performs MP and global collective operations delegated to it by computation workers.

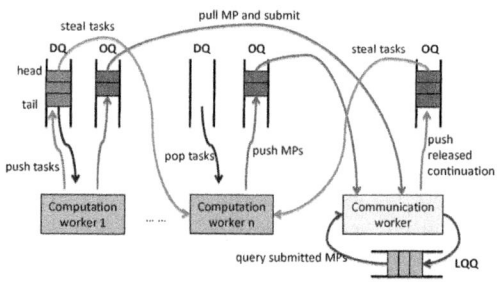

DQ: double-ended queue (deque), OQ: outbox queue, LQQ: local query queue

Fig. 1. HC-MPI work-stealing runtime with MPI integration

A **computation worker** has two queues, a double-ended queue (deque) to manage computation tasks, and a regular queue that acts as an outbox for MP operations. A computation worker pushes or pops asynchronous computation tasks from the tail end of the deque when it creates tasks (through async) or picks up new work. When the deque is empty, it steals tasks from the head of the deque of other workers, or from the communication worker's outbox.

A **communication worker** has an outbox queue and a local query queue. It retrieves and pulls MP operations from the outbox queues of computation workers, and processes each MP operation in turn. After submitting an MP operation, the worker moves it to the query queue. The worker alternates between the retrieving/submission and query cycles. When the communication worker detects that an MP operation is complete and needs to release a previously suspended task, it pushes the task on its outbox deque so that a computation worker can steal it and continue its execution. When the communication worker performs a collective MPI operation, it suspends untill the operation is completed. The computation workers can continue executing tasks during this time.

3   Experimental Results

We used the test suite developed for evaluating multi-threaded MPI communications [4]. The experiment setup is a cluster with quad-core Intel Xeon CPU's, 4GB memory per node, OpenMPI 1.4.3 compiled with support for multi-threaded MPI. The MPI environment was initialized with MPI_THREAD_SINGLE for HC-MPI experiments since HC-MPI uses a single communication worker per

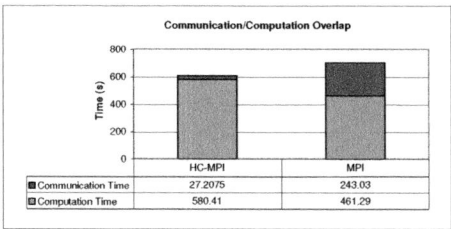

Fig. 2. Computation/Communication Overlap

MPI process. "xP-yT" indicates a configuration of x MPI processes, and each process has y threads in MPI cases, or y tasks in HC-MPI case.

As shown in Figure 2, HC-MPI is able to reduce the communication time by a factor of 10x as compared to MPI. In this experiment, communications are performed with MPI_Isend and MPI_Irecv in the MPI test, and by async_send and async_recv in HC-MPI test, thus to allowing overlap with computation. The trade-off between communication and computation time is paid off as we see that the total time is reduced by about 15% by HC-MPI.

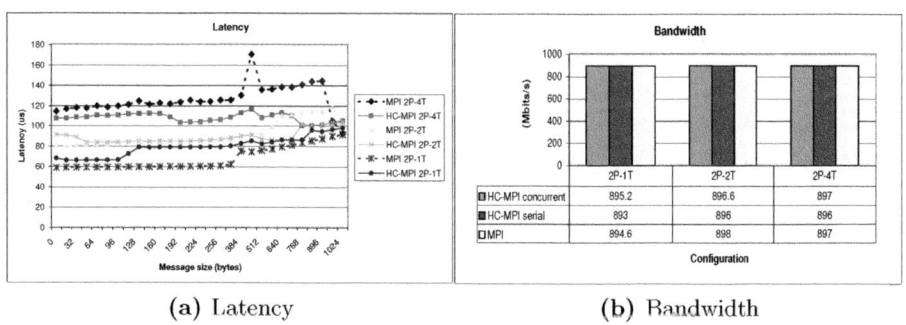

(a) Latency   (b) Bandwidth

Fig. 3. Latency and Bandwidth

Figure 3a shows the latency between two processes of both HC-MPI and pure MPI implementation of the benchmark. On average, HC-MPI is able to reduce the latency by about 5-10% compared to the corresponding MPI examples for the "2P-2T" and "2P-4T" cases. The bandwidth test is performed by repeatedly sending 8Mbyte messages and then measuring the bandwidth [4]. From the test results shown in Figure 3b, the overhead incurred by HC-MPI implementation poses no noticeable impact on the bandwidth.

4 Related Work

A similar approach was taken in a recent MPI/SMP effort [2] in which MPI message passing operations are wrapped as tasks and blocking communication operations are annotated as restartable to allow runtime to switch to other tasks when the current task is blocked on messaging operations. HC-MPI uses a scalable work-stealing approach, while MPI/SMP used a shared list for all tasks among computation workers which may easily become a scalable bottleneck.

5   Conclusions and Future Work

In this paper, we presented our work on integrating asynchronous task parallelism with MPI in the Habanero-C programming language, runtime and compiler for latency hiding and communication/computation overlap. The HC-MPI compiler support and a scalable workstealing runtime guarantees no CPU is blocked when an execution context needs to be suspended. The non-blocking collective operations for MPI is to be investigated in future.

References

1. Charles, P., et al.: X10: an Object-Oriented Approach to Non-Uniform Cluster Computing. In: Proceedings of OOPSLA, pp. 519–538. ACM, New York (2005)
2. Marjanović, V., Labarta, J., Ayguadé, E., Valero, M.: Overlapping communication and computation by using a hybrid mpi/smpss approach. In: Proceedings of the 24th ACM International Conference on Supercomputing, New York, NY, USA, pp. 5–16 (2010)
3. Sarkar, V., Harrod, W., Snavely, A.E.: Software challenges in extreme scale systems. Journal of Physics: Conference Series 180(1), 012045 (2009)
4. Thakur, R., Gropp, W.: Test suite for evaluating performance of multithreaded mpi communication. Parallel Computing 35(12), 608–617 (2009)

Performance Evaluation of Thread-Based MPI in Shared Memory

Juan-Antonio Rico-Gallego and Juan-Carlos Díaz-Martín

University of Extremadura, Spain
{jarico,juancarl}@unex.es

Although Unified Parallel C and OpenMP are being proposed for supporting more efficiently multicore architectures, the fact is that MPI is still used as a useful model on shared memory machines. Traditional mainstream MPI implementations as MPICH2 and OpenMPI build each MPI task as a process, an approach that presents some disadvantages in shared memory because message passing between processes usually requires two copies. One-copy communication can be achieved with operating system support, through kernel modules like KNEM or Limic, techniques like SMARTMAP, or special system calls like *vmsplice*, with disadvantages mainly in portability and limited improvements in performance. It is also possible building each MPI task as a thread. This is not a new concept. Implementations such as TOMPI, TMPI, or the newer MPI Actor or MPC-MPI run an MPI node as a thread, each one stressing different goals. AzequiaMPI is a thread-based but still a full conformant open source implementation of the MPI-1.3 standard. AzequiaMPI shows that MPI performance can be significantly improved by fully exploiting a single shared address space.

1 System Design and Performance

AzequiaMPI has a simple design based on three queues per MPI task. LFQ is the communication interface to the rest of tasks, a lock-free queue supporting a single receiver (its owner node) and many senders. Once dequeued, a message goes to the internal ordinary double-linked MBX queue if no matching is found in the also private PRQ queue, where the node posts its receiving requests.

Some performance improvement techniques have been applied to AzequiaMPI. Firstly, using the two-copy fastbox mechanism (taken from MPICH2, [1]) reduces the latency for messages up to 1 KB, gaining 50% over MPICH2in messages up to 32 bytes. Secondly, we use a new technique called *split copy* in synchronous communication. In a rendezvous, sender just copies the first half of its user buffer while receiver gets charged of the second half. Split copy increases the bandwidth up to 90% around 128 Kbyte messages.

We compare the overall performance of AzequiaMPI versus that of MPICH2-Nemesis in an Intel Xeon E5620 Nehalem machine. All tests were run with ranks bound to cores. Figure 1 illustrates the point to point performance using the Netpipe benchmark, while Figure 2 shows the results from the IMB benchmark on collective operations of eight ranks. HPL and NPB SP Multizone results are hardly influenced by AzequiaMPI improvements because they are computation intensive benchmarks.

Y. Cotronis et al. (Eds.): EuroMPI 2011, LNCS 6960, pp. 337–338, 2011.

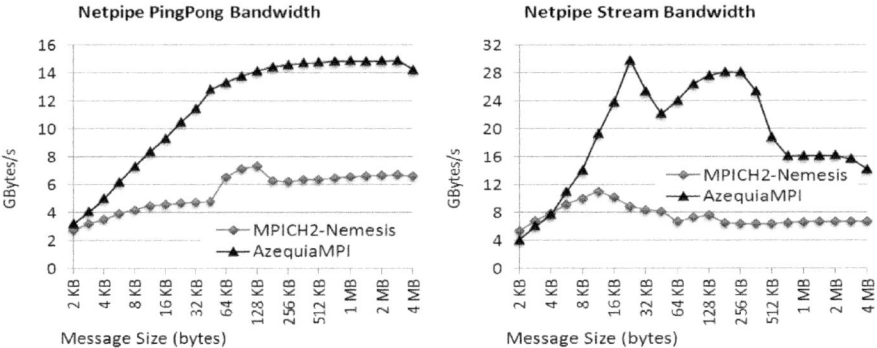

Fig. 1. Bandwidth of AzequiaMPI versus that of MPICH2-Nemesis in Netpipe benchmark

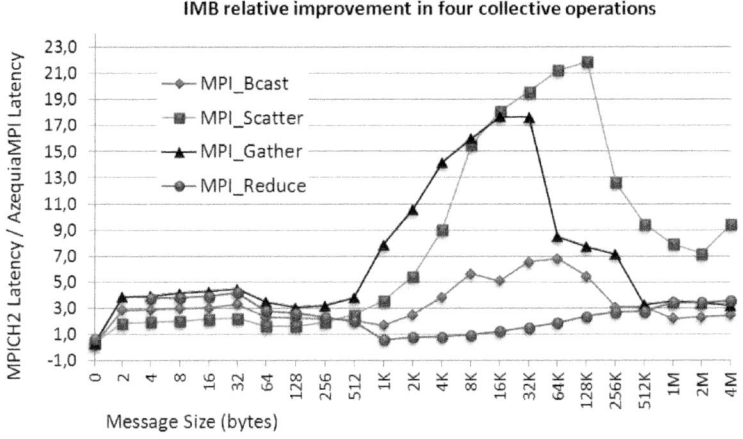

Fig. 2. IMB performance of AzequiaMPI versus MPICH2-Nemesis in four collectives

As conclusions, the thread-based design, together with the lock-free communication and further optimizations that exploit the properties of a common address space makes of AzequiaMPI to outperform other MPI distributions in a significant manner.

Acknowledgments. This work has been supported by Spanish National Research program "Ingenio 2010, subprogram CENIT-2005" ("Hesperia") and the Spanish project TIN2008-03063. Thanks to the HLRS laboratory in Stuttgart and the HPC-EUROPA2 program (project number: 228398), with the support of the European Commission - Capacities Area - Research infrastructures.

Reference

1. Buntinas, D., Mercier, G., Gropp, W.: Design and Evaluation of Nemesis, a Scalable, Low-Latency, Message-Passing Communication Subsystem. In: Proc. of the Sixth IEEE International Symposium on Cluster Computing and the Grid, CCGRID 2006 (2006)

MPI-DB, A Parallel Database Services Software Library for Scientific Computing

Edward Givelberg[1], Alexander Szalay[2,3], Kalin Kanov[3], and Randal Burns[3]

[1] Department of Physics and Astronomy, The Johns Hopkins University
givelberg@jhu.edu
[2] Department of Physics and Astronomy, The Johns Hopkins University
[3] Computer Science Department, The Johns Hopkins University

Abstract. Large-scale scientific simulations generate petascale data sets subsequently analyzed by groups of researchers, often in databases. We developed a software library, MPI-DB, to provide database services to scientific computing applications. As a bridge between CPU-intensive and data-intensive computations, MPI-DB exploits massive parallelism within large databases to provide scalable, fast service. It is built as a client-server framework, using MPI, with MPI-DB server acting as an intermediary between the user application running an MPI-DB client and the database servers. MPI-DB provides high-level objects such as multi-dimensional arrays, acting as an abstraction layer that effectively hides the database from the end user.

Keywords: Databases, data-intensive computing, software library.

Introduction. In virtually every field of science very large data sets are generated by measurement instruments (telescopes, high-energy particle accelerators, gene sequencing machines, etc.), as well as by computer simulations, changing the way we do science. In astrophysics, following the example of the SDSS Sky-Server, the Millennium simulation [3] created a remotely accessible database with a collaborative environment [2], which drew hundreds, if not thousands of astronomers into analyzing simulations.

The emerging challenge in processing these data sets is scalability and parallelism. Scientists at the JHU Institute of Data-Intensive Engineering and Science (IDIES) have been developing technology to integrate parallel distributed databases with traditional high-performance parallel scientific computing. MPI-DB is planned as the foundational software component in this strategy, enabling direct parallel data ingest and retrieval between HPC nodes and a database cluster, the equivalent of ODBC for high perfornce computing.

The NSF has recently awarded our group to build a 5PB cluster, the Datas-cope, for extreme data-intensive computations. The driving goal behind the Data-Scope design is to maximize stream processing throughput over 100TB-size datasets. MPI-DB will enable us to establish peer-to-peer connections between nodes on the HPC cluster and the Data-Scope I/O nodes – both for the on-the-fly ingest of data generated by an MPI application, or for the parallel compute-intensive analysis of large data sets read from the parallel database.

Y. Cotronis et al. (Eds.): EuroMPI 2011, LNCS 6960, pp. 339–341, 2011.
© Springer-Verlag Berlin Heidelberg 2011

MPI-DB: Description of the Software Library. The MPI-DB software library provides database services to scientific computing processes. The user client application is compiled and linked against the MPI-DB client library. At run time it connects to an MPI-DB server at a specified network address, which services clients' requests by querying the database and sending the results back to the clients. The library uses MPI for client-server communications, is supported on both Linux and Windows and supplied with C, C++ and Fortran language bindings. We plan to provide support for major databases, such as SQL Server, MySQL and PostgreSQL.

As illustrated in Figure 1, MPI-DB is *a distributed scheduler for the database* addressing scalability for both data-intensive and CPU-intensive applications.

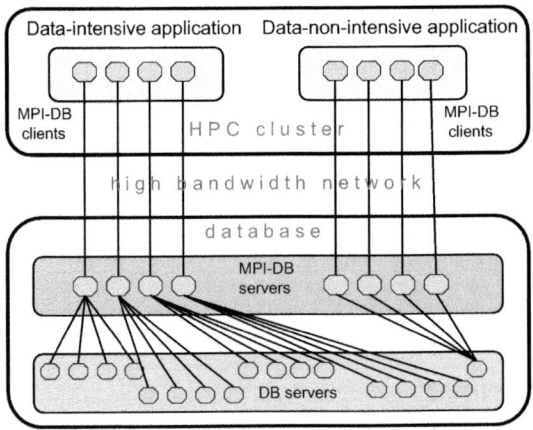

Fig. 1. Scalable architecture of the MPI-DB software library: The HPC cluster and the database are connected by a high bandwidth network. MPI-DB acts as a scheduler for the database: it allocates multiple database server connections for each process of a data-intensive applications, while fewer database server connections are allocated for an application with lower data requirements.

The MPI-DB software is built as a collection of layers (See Figure 2). The user is no longer required to write SQL queries: MPI-DB provides a set of high-level *programming objects*, with methods to store and retrieve from the database. These include a rich set of multi-dimensional array operations, incorporating the basic features proposed in [1]. as well as basic functionality to store and manipulate strings and binary obejcts, such as images or checkpointing files.

Remarks on Implementation. MPI-DB is being developed as object-oriented software in C++ using MPI-2 functionality. Presently, for MPI-DB the Intel MPI library is the only satisfactory MPI implementation, while other MPI implementations contain run-time bugs, do not provide adequate hardware drivers or do not support MPI-2 standard altogether.

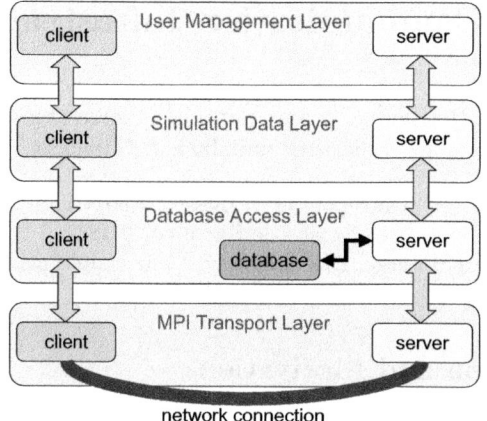

Fig. 2. Multi-layered software architecture for MPI-DB. Vertical arrows denote software dependencies. The client and the server of the MPI Transport layer communicate over the high bandwidth network using the MPI protocol. The client and the server in a higher layer communicate with each other using the client and the server of the layer below, respectively. The server in the Database Access Layer acts as a client to a database server.

We tested MPI-DB in parallel data ingestion over a 10 Gigabit/sec Ethernet. Client data was sent to the server and then ingested into mysql database. We were able to realize scalable data ingestion with maximum aggregate throughput of up to 700 Megabytes per second when all processor cores were utilized.

References

1. Dobos, L., Csabai, I., Milovanovic, M., Budavari, T., Szalay, A.S., Tintor, M., Blakeley, J., Jovanovic, A., Tomic, D.: Array requirements for scientific applications and an implementation for microsoft sql server. In: Baumann, P. (ed.) Proc. of the EDBT/ICDT 2011 Workshop on Array Databases, Uppsala, Sweden (2011)
2. Lemson, G., The Virgo Consortium: Halo and galaxy formation histories from the millennium simulation: Public release of a vo-oriented database (2006)
3. Springel, V., White, S.D.M., Jenkins, A., Frenk, C.S., Yoshida, N., Gao, L., Navarro, J., Thacker, R., Croton, D., Helly, J., Peacock, J.A., Cole, S., Thomas, P., Couchman, H., Evrard, A., Colberg, J., Pearce, F.: Simulations of the formation, evolution and clustering of galaxies and quasars, vol. 435, pp. 629–636 (2005)

Scalable Runtime for MPI: Efficiently Building the Communication Infrastructure

George Bosilca[1], Thomas Herault[1], Pierre Lemarinier[2],
Ala Rezmerita[3], and Jack J. Dongarra[1]

[1] University of Tennessee, Knoxville
[2] Université de Rennes 1, IRISA
[3] Grand-Large, INRIA Saclay – Université Paris-Sud

1   Introduction and Motivation

The runtime environment of MPI implementations plays a key role to launch the application, to provide out-of-band communications, enabling I/O forwarding and bootstrapping of the connections of high-speed networks, and to control the correct termination of the parallel application. In order to enable all these roles on a exascale parallel machine, which features hundreds of thousands of computing nodes (each of them featuring thousands of cores), scalability of the runtime environment must be a primary goal.

In this work, we focus on an intermediate level of the deployment / communication infrastructure bootstrapping process. We present two algorithms: the first to share the contact information of all runtime processes, enabling an arbitrary set of connections, and the second to distribute only the information needed to build a binomial graph. We implemented these two algorithms in ORTE, the runtime environment of Open MPI, and we compare their efficiency, one with the other, and with the runtime systems of other popular MPI implementations.

2   Evaluation

We use the underlying launching tree to exchange contact information at the runtime level, and let the runtime system build for itself a communication infrastructure mapping a binomial graph [1] topology. This topology has several interesting properties such as redundant links to support fault tolerance and binomial tree shape connectivity rooted in each peer. A precedent work [3] proposes a self-stabilizing algorithm to build such an overlay on top of a tree. Such an algorithm provides two main features: 1) inherent fault-tolerance and 2) self-adaptation to the underlying tree topology, which negates the need for initializing the construction of the binomial topology.

We compare our implementation with three other setups: the implementation of ORTE described in [4] (prsh with improved flooding), MPICH2 [5] version 1.3.2p1 using Hydra [6,2] with rsh and MVAPICH2 version 1.7a using the ScELA [7] launcher. All versions are compiled in optimized mode and the experiments based on rsh were using ssh as a remote shell system.

Y. Cotronis et al. (Eds.): EuroMPI 2011, LNCS 6960, pp. 342–344, 2011.

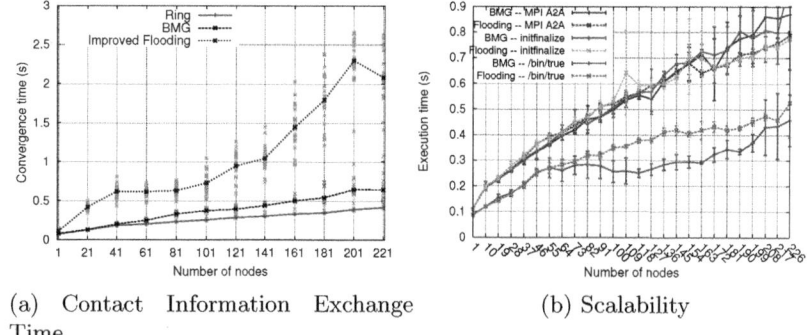

(a) Contact Information Exchange Time

(b) Scalability

Fig. 1. Comparison of ORTE prsh with BMG and ORTE prsh with Flooding

First, we compare the two implementations in ORTE together, in the figures presented in Fig. 1. The first micro benchmark, presented in Fig. 1a measures the time taken to solely exchange the Contact Information, following the Improved Flooding Strategy, or the BMG Building strategy, as functions of the number of nodes. The latter consists in two phases: first the building of the ring, then of the BMG, and the two phases are represented in the figure. Individual measurements are represented with light points, and mean values are connected with a line.

The BMG algorithm presents a better convergence time, in practice, than the Improved Flooding algorithm. This is expected, since it exchanges much less information (each node receives only the contact information of $O(\log_2(n))$ nodes) than the Flooding algorithm $(O(n))$. The ring construction time occupies a major part of this time, but the system appears to scale faster than linearly.

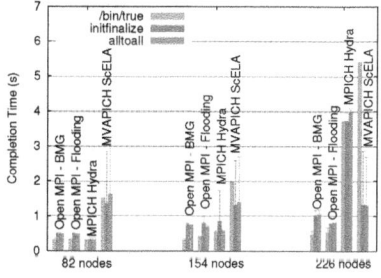

Fig. 2. Comparison with other MPI runtime systems

This is also demonstrated in Fig. 1b, which presents how both implementations perform when increasing the number of nodes. On the /bin/true benchmark, the BMG construction algorithm demonstrates a better scalability than the Flooding Algorithm, with noticeable steps that characterize the logarithmic behavior of the algorithm. This logarithmic behavior disappears, when launching a communicating MPI application, like a2a, or even a simple empty MPI application, like initfinalize. This is due to the third phase of the launching in ORTE, the modex, that introduces a linear component to the performance.

Fig. 2 compares the two ORTE implementations with Hydra (MPICH2), and ScELA (MVAPICH) for the three benchmarks, and various number of nodes. Although Hydra performs slightly better than both ORTE implementations at a small scale, ORTE reaches the same performance for 154 nodes and above.

After about 166 nodes, both Hydra and ScELA for the `/bin/true` benchmark suffer from connections storms, that impact the performance by introducing a delay of 3s, due to TCP SYN packets retransmission.

3   Conclusion

In this paper, we presented two strategies for the construction of a runtime communication infrastructure running in parallel with the deployment of the runtime processes and the deployment of the parallel application. The first strategy uses an improved flooding algorithm, that enables any runtime process to communicate with any other directly, thus providing an arbitrary routing topology for the runtime. The second strategy uses an ad-hoc self-adapting algorithm, that transforms the initial spawning tree into a binomial graph, not only sharing the needed contact information (and only this information), but also establishing at the same time the corresponding links. We implemented both algorithms in ORTE, the runtime system of Open MPI, and compared the implementations with the state of the art runtime environments for MPI. Experiments demonstrated an improved scalability, highlighting the importance of tight integration between launching and communication infrastructure construction, and the advantages of a flexible routing topology at the runtime level.

References

1. Angskun, T., Bosilca, G., Vander Zanden, B., Dongarra, J.: Optimal routing in binomial graph networks. In: Eighth International Conference on Parallel and Distributed Computing, Applications and Technologies, PDCAT 2007, pp. 363–370 (December 2007)
2. Balaji, P., Buntinas, D., Goodell, D., Gropp, W., Krishna, J., Lusk, E., Thakur, R.: PMI: A scalable parallel process-management interface for extreme-scale systems. In: Keller, R., Gabriel, E., Resch, M., Dongarra, J. (eds.) EuroMPI 2010. LNCS, vol. 6305, pp. 31–41. Springer, Heidelberg (2010)
3. Bosilca, G., Coti, C., Herault, T., Lemarinier, P., Dongarra, J.: Constructing resiliant communication infrastructure for runtime environments. Advances in Parallel Computing 19, 441–451 (2010), doi:10.3233/978-1-60750-530-3-441
4. Bosilca, G., Herault, T., Rezmerita, A., Dongarra, J.: On scalability for mpi runtime systems. In: IEEE International Conference on Cluster Computing (to appear, 2011)
5. Mathematics, and Computer Science Division, A. N. L. MPICH-2, implementation of MPI 2 standard (2006), http://www-unix.mcs.anl.gov/mpi/mpich2/
6. Mathematics, and Computer Science Division, A. N. L. Hydra process management framework (2009),
http://wiki.mcs.anl.gov/mpich2/index.php/HydraProcessManagementFramework
7. Sridhar, J., Koop, M., Perkins, J., Panda, D.: ScELA: Scalable and extensible launching architecture for clusters. In: Sridhar, J., Koop, M., Perkins, J., Panda, D. (eds.) HiPC 2008. LNCS, vol. 5374, pp. 323–335. Springer, Heidelberg (2008)

Writing Parallel Libraries with MPI - Common Practice, Issues, and Extensions

Torsten Hoefler and Marc Snir

University of Illinois, Urbana, IL 61801, USA
{htor,snir}@illinois.edu

Abstract. Modular programming is an important software design concept. We discuss principles for programming parallel libraries, show several successful library implementations, and introduce a taxonomy for existing parallel libraries. We derive common requirements that parallel libraries pose on the programming framework. We then show how those requirements are supported in the Message Passing Interface (MPI) standard. We also note several potential pitfalls for library implementers using MPI. Finally, we conclude with a discussion of state-of-the art of parallel library programming and we provide some guidelines for library designers.

1 Introduction

Modular and abstract structured programming is an important software-development concept. Libraries are commonly used to implement those techniques in practice. They are designed to be called from a general purpose language and provide certain functions. Libraries can be used to simplify the software development process by hiding the complexity of designing an efficient and reusable collection of (parallel) algorithms. High-performance libraries often provide performance portability and hide the complexity of architecture-specific optimization details. Library reuse has been found to improve productivity and reduce bugs [2,17].

In this work, we discuss principles for designing and developing parallel libraries in the context of the Message Passing Interface (MPI) [18]. We recapitulate the features that have been introduced 18 years ago [21,6], add newly found principles and interface issues, and discuss lessons learned. We also analyze current practice and how state of the art libraries use the provided abstractions.

We show that the key concepts are widely used while other concepts, such as process topologies and datatypes, did not find very wide adoption yet. In the following sections, we describe principles for modular distributed memory programming, introduce a taxonomy for parallel libraries, discuss several example libraries, derive common requirements to support modular programming, show how MPI supports those requirements, discuss common pitfalls in MPI library programming, and close with a discussion of common practice.

Y. Cotronis et al. (Eds.): EuroMPI 2011, LNCS 6960, pp. 345–355, 2011.

2   Modular Distributed Memory Programming

Modular programming plays an important role in "Component-based software engineering (CBSE)", which suggests to program by composing large-scale components. Korson and McGregor [12] identify ten generally desirable attributes that each serial and parallel library should bear: wide domain coverage, consistency, easy-to-learn, easy-to-use, component efficiency, extensibility, integrability, intuitive, robust, well-supported.

Those principles are also valid in distributed memory programming. The main difference is that distributed and parallel programming requires to control multiple resources (processing elements). Several major language techniques to support the development of distributed libraries have been identified in the development of the Eiffel language [16]. The list includes (among others) several items that are relevant for the development of parallel libraries:

classes reusable components should be organized around data structures rather than action structures

information hiding libraries may use each others facilities, but internal structures remain hidden and "cross talk" between modules is avoided

assertions characterize semantic properties of a library by assertions

inheritance can serve as module inclusion facility and subtyping mechanism

composability especially performance composability and functional orthogonality. This requires to query relevant state of some objects.

Writing distributed libraries offers a large number of possibilities because spatial resource sharing can be used in addition to temporal sharing with multiple actors. We identify the following main library types that are commonly used today: (1) spatial libraries use a subset of processes to implement a certain functionality and leave the remaining processes to the user (e.g., ADLB [15]), (2) collective loosely-synchronous libraries are called "in order" (but not synchronously) from a statically specified process group, and (3) collective asynchronous libraries are called by a static group of processes but perform their functions asynchronously.

2.1   A Taxonomy for Parallel Libraries

We classify existing parallel application libraries into four groups:

Computational Libraries provide full computations to the user, for example the solution of PDEs, n-body interactions, or linear algebra problems. Example libraries include PETSc [1], ScaLAPACK, PMTL [14], PBGL [4], PPM [20].

Communication Libraries offer new high-level communication functions such as new collective communications (e.g., LibNBC [11]), domain-specific communication patterns (e.g., the MPI Process Group in PBGL), or Active Messages (e.g., AM++ [23]).

Programming Model Libraries offer a different (often limited) programming model such as the master/slave model (e.g., ADLB [15]) or fine-grained
objects (e.g., AP [24]).

System and Utility Libraries offer helper functionality to interface different architectural subsystems that are often outside the scope of MPI (e.g.,
LibTopoMap [10], HDF5 [3]) or language bindings (e.g., Boost.MPI, C# [5]).

Some of the example libraries and their used abstractions are discussed in the
next section.

2.2   Example of Libraries

We now describe some example libraries that we categorized in our taxonomy
that either utilize MPI to implement parallelism or integrate with MPI to provide
additional functionality. This collection is not supposed to be a complete listing
of all important parallel libraries. It merely illustrates one example for each type
of parallel library and shows which abstractions have been chosen to implement
parallel libraries with MPI.

PETSc. The PETSc library [1] offers algorithms and data structures for the
efficient parallel solution of PDEs. PETSc provides abstractions of distributed
datatypes (vectors and matrices) that are scoped by MPI communicators and
hides the communication details from the user. The passed communicator is
copied and cached as attribute to ensure isolation. PETSc uses advanced MPI
features such as nonblocking communication and persistent specification of communication patterns while hiding the message passing details and data distribution from the user. It also offers asynchronous interfaces to communication calls,
such as VectScatterBegin() and VecScatterEnd() to expose the overlap to the
user.

PBGL. The Parallel Boost Graph Library [4] is a generic C++ library to
implement graph algorithms on distributed memory. The implementation bases
on *lifting* the requirements of a serial code to base a parallel implementation
on it. The main abstractions are the process group to organize communications,
the distributed property map to implement a communication abstraction, and a
distributed queue to manage computation and detect termination. The library
offers a generic interface to the user and uses Boost.MPI to interface MPI.

PMTL. The Parallel Matrix Template Library [14] is, like the PBGL, a generic
C++ library. It uses distributed vectors and matrices to express parallel linear
algebra computations. As for PBGL, the concepts completely hide the underlying
communication and enable optimized implementations.

PPM. The Parallel Particle Mesh Library [20] provides domain decomposition
and automatic communication support for the simulation of continuous systems.
The library offers a high-level application oriented interface which is close to a
domain-specific language for such simulations. It offers support for advanced
functions of MPI.

ADLB. The Asynchronous Dynamic Load Balancing Library, developed at Argonne [15], offers a simplified programming model to the user. The master/slave model consists of essentially three function calls and the scalable distribution of work and parallelization of the server process is done by the library. The library expects an initialized MPI and uses communicators during its init call to divide the job into master and slave groups. The master group "stays" within ADLB while the user has full access to the slave group to issue ADLB calls to the master. The library has been used with multi-threaded processes on BlueGene/P to achieve 95% on-node speedup.

LibNBC. LibNBC [11] is an asynchronous communication library that provides a portable implementation of all MPI-specified collective operations with a non-blocking interface. It uses advanced MPI features to provide a high-performing interface to those functions and faces several of the issues discussed in Sections 3.3 and 3.4. LibNBC also offers an InfiniBand-optimized transport layer [9] that faces problems described in Section 3.5.

LibTopoMap. LibTopoMap [10] provides portable topology mapping functionality for the distributed graph topology interface specified in MPI-2.2 [18]. It replaces the distributed graph topology interface on top of MPI and caches the new communicator and new numbering as attribute of the old communicator. This shows that a complete modular implementation of the Topology functionality in MPI is possible.

HDF5. The HDF5 library [3] offers an abstract data model for storing and managing scientific data. It offers support to specify arbitrary data layouts and its parallel version relies heavily on MPI datatypes and MPI-IO. As a system library, it faces several problems as discussed in Section 3.4.

2.3   Common Requirements of Parallel Libraries

Based on the survey of libraries, we distill the common requirements for a parallel runtime environment such as MPI. Parallel libraries require performance, scalability, usability, error handling, isolation (a "safe and private" communication space) for point-to-point and collective communication, and virtualized process naming (e.g., topologies or a virtual one-dimensional namespace). In addition, high-quality programming frameworks may offer topology mapping, fault-tolerance, and data management support to libraries.

3   The Loosely Synchronous Model in MPI

We now discuss how MPI supports parallel libraries by providing many of the required features listed in Section 2.3. The loosely synchronous model for parallel libraries is specified in Section 6.9 of the MPI standard [18]. In this model, all processes in a communicator invoke parallel subroutines in the same order. Those processes do not have to synchronize before the invocation.

We now discuss the main concepts in MPI that support the development of parallel libraries.

Communication Contexts in the form of MPI communicators are the most important concept for libraries. Communicators offer spatial and temporal isolation because they can specify disjoint process groups and isolate communication on overlapping process groups. This "communication privatization" is similar to the important "data privatization" in object oriented languages. Section 3.2 discusses potential issues with reentrant libraries. Each communicator has an associated **Process Group** that offers a virtual one-dimensional namespace for processes. Communicators also provide a scope for collective communication and support the concept of "functional and spatial composability" [16].

Virtual Topologies allow for domain-specific process naming schemes that can be passed to and queried by libraries. This extends the simple one-dimensional naming of process groups to arbitrary Cartesian naming schemes or general graph topologies (which can be enumerated by graph traversals, such as Breadth First Search).

Attribute Caching can be used to associate state to communicator objects. MPI allows to attach arbitrary data to communicators, windows, and datatypes in order to pass context or state information between library calls. MPI guarantees that this information can be quickly retrieved and is consistent. It also offers functionality to strictly control the inheritance of attributes in communicator copy functions. This allows to mimic the concept of "inheritance" [16] of general object oriented programming.

Datatypes defines an interface to exchange the layout of data structures for communications between libraries and user applications. MPI offers the required functions to create private copies of datatypes (MPI_Type_dup) and manipulate them. It also offers functions to query the composition of existing datatypes and serialize or deserialize (MPI_Pack/MPI_Unpack) them into/from buffers. Those abstractions support the abstract definition of datatypes in libraries.

MPI's Modular Design allows to implement full sections of the MPI standard as separate libraries (e.g., Sections 5 (Collective Communication), 7 (Process Topologies), and 13 (I/O) can solely be implemented with the core functionality of MPI 1). This supports and encourages the implementation of external communication libraries, such as LibTopoMap or LibNBC.

3.1   Where It Breaks

MPI's support for parallel libraries is comprehensive. However, library writers have to exercise care when using several functionalities in MPI and define external contracts with the users of the library.

The most prominent example is multi-threading. If a library requires a certain thread level, e.g., MPI_THREAD_MULTIPLE, then the user must ensure that MPI is initialized with this thread level. This can be tricky if multiple parallel libraries are used in a single program and can lead to performance degradation if the thread support is only needed for small parts of the code.

A second limitation applies to the MPI_Info values that can be specified during the creation of several objects. Specified values cannot be queried or reset by libraries and need to be communicated out-of-band or enforced via external "contracts". This can influence performance, or even correctness if specified info arguments change the object's semantics (e.g., no_locks).

In the following sections, we discuss several application-specific issues and limitations that library-writers may be confronted with.

3.2   Reentrant Libraries

A parallel library invocation will be passed a communicator argument that indicates the group of processes performing the call. A well designed library will pass this communicator as an explicit argument. The library needs a communication context that is distinct from the communication contexts of the invoking code. This is usually done by creating a communicator private to the library that is a duplicate of the argument communicator shows how this private communicator can be cached with the communicator argument, so that the private communicator is created only at the first invocation.

This method provides a static communication context, shared by all library instances. It ensures that sends inside the library cannot match receives outside the library, and vice-versa; but it does not ensure that a send performed by one instance of the library be matched by a receive in another instance.. Such a library is *nonreentrant*: it requires that only one invocation instance be active on a communicator at the same time (no recursion, no new invocation before a previous one completed at all processes). : One can build reentrant libraries in various ways: E.g., by having a barrier, either at entry or at exit (which may have severe impact in performance due to unnecessary synchronization), by creating a new communicator instance for each invocation (several communicators could be pre-dup'd and managed in a stack-like manner as attributes), or by imposing a communication discipline that avoids out-of-order message matching: no wildcard source receives (MPI_ANY_SOURCE), no cancel operations and messages produced within a dynamic scope are consumed within the same dynamic scope.

3.3   Nonblocking Libraries

Nonblocking or asynchronous libraries pose the challenge of progress and control transfer. We differentiate between "manual" progress (the user periodically transfer control to the library, for the library to progress, cf. coroutines) and "asynchronous" progress (the library spawns an asynchronous activity, e.g., a thread) [8]. Manual progress is required on some HPC systems because of limitations on multithreading, limitations on signaling between the communication hardware and (user) threads, and lack of an appropriate scheduling policy.

MPI-2.2 offers generalized requests to integrate completion checks of operations in nonblocking library routines with the usual MPI completion calls (e.g., MPI_Test). However, the specification requires asynchronous progress and does not work on systems where manual progress is needed. A simple fix for this, which adds manual progress facilities to generalized requests, has been proposed for MPI-3 [13].

The parallel invocation method described in Section 3.3 requires that the communicator argument be duplicated, at least at the first library invocation on the communicator. However, MPI_Comm_dup is a blocking collective routine and may require synchronization. This makes it impossible to implement pure nonblocking collective libraries. The alternative of Initializing each communicator before using it with a library is an unnecessary burden for library users even though it is common practice (e.g., in ScaLAPACK). A nonblocking MPI_Comm_dup call would solve this problem.

3.4   Complex Communication Operations

Libraries are often used to implement new, higher-level communication operations. We already discussed issues with nonblocking interfaces of such libraries, however, implementers need to consider two more potential hurdles.

If the library on top of MPI-2.1 was to perform a reduction with either a predefined or user-defined MPI operation, then the library needed to implement the reduction operation itself (since the new library cannot access the function pointer associated with an MPI_Op). MPI-2.2 introduces MPI_Reduce_local to solve this problem. MPI_Reduce_local performs a single binary reduction with an MPI_Op handle as it would be performed by a collective reduction operation. It is recommended to use this functionality to implement reduction communication operations, such as nonblocking MPI_Reduce on top of MPI.

3.5   Process Synchronization Outside of MPI

The MPI standard does not specify the interaction of MPI with other, potentially synchronizing, communication mechanisms outside of MPI. This can pose problems when such operations are mixed, e.g., if communication libraries are tuned for low-level transport interfaces [9]. The implementer has to ensure that all communication interfaces make progress. However, MPI may require manual progress but does not offer an explicit progress call. This may be emulated (rather inelegantly) by calling MPI_Iprobe in a progress loop.

4   Hybrid Programming

Hybrid Programming mixes MPI with other programming models such as Pthreads, OpenMP, or PGAS models. The implementations of runtimes for those models often use external communication layers and may suffer from issues discussed in Section 3.5. However, the interaction between different parallelization schemes can have more complex effects. We discuss two issues with the interaction of MPI and threads. We remark that the discussion is not limited to threads and applies to other models, such as PGAS, or languages, such as C#.

4.1 Thread-Safe Message Probing

MPI offers a mechanism (MPI_Probe/MPI_Iprobe) to peek into the receive queue and query the size of found messages before posting the receive. This enables the reception of dynamically-sized messages. However, this also creates problems in the context of multiple threads [5] since one thread can query the message and another thread can receive it (the queue is a global shared object). A matched probe call that removes the message from the queue while peeking has been proposed to MPI-3 to solve this problem [7]. This addition enables low-overhead probing for threaded libraries and languages.

4.2 Control Transfer and Threading

Threaded libraries pose additional problems for the interfaces. This is because threaded libraries encapsulate resource requirements in addition to functionality. For single-threaded libraries, the control is handed from a single thread running on a single processing element (PE) to a single thread. In multi-threaded environments, we differentiate four scenarios:

1. A single application thread calls a single-threaded library.
2. A single application thread calls a multi-threaded library.
3. Multiple application threads call a single-threaded library.
4. Multiple application threads call a multi-threaded library.

Scenario 1 is identical to the single-threaded case while all other scenarios require some kind of resource management. Scenario 2 is simple because the library is the only consumer of PE resources, while Scenario 3 can solved by synchronizing all threads before the library is called (this is commonly used today, e.g., in [15]). Scenario 4 is most tricky and requires advanced resource management.

Resource management can either be performed by the operating system (time multiplexing) or explicitly by the user with ad-hoc mechanisms such as querying the number of available cores and thread-pinning. A promising OS-based space multiplexing (core allocation) approach is proposed in [19].

4.3 Communication Endpoints

Special care has to be taken if the communication layer requires multiple client threads per node in order to achieve full performance. This has to be addressed in hybrid programming by either using multiple threaded MPI processes per node, or a scheme similar to Scenario 4. A proposal for MPI-3.0 [22] shows an extension for MPI to provide multiple logical network endpoints in a threaded hybrid MPI application.

5   Guidelines for Library Designers

We now conclude this work by providing some hints and guidelines for MPI library developers. All those guidelines are in addition to the well-known serial library design rules, such as privatization and abstraction. In general, libraries should utilize the features provided by MPI while paying attention to the pitfalls discussed above. In particular, libraries should use communicators to specify spatial decomposition of the process space and to present safe communication contexts for temporal decomposition. Created communicators and library internal state and data-structures should be cached with the user communicator (which then becomes the central communication object that needs to be passed to every library call, special care has to be taken for reentrant libraries, cf. Section 3.2). Libraries should take advantage of virtual topologies to specify process topologies and possibly perform topology mapping (this may conflict with the user program or other stacked libraries). If library-specific structures are passed to communication functions and from or to the user, then those should be specified with MPI datatypes. Parallel libraries should also handle errors internally and provide library-specific error messages to the user. This can be achieved by attaching a library-specific error handler to the library's private communicator.

Library and communicator initialization can either be done explicitly or implicitly (at first invocation). Communicator initialization must be done collectively and we discuss issues with nonblocking communication in Section 3.3.

5.1 What to Avoid!

Libraries should never use the passed communicators directly (just attached attributes); this includes the global communicator MPI_COMM_WORLD. Synchronization or draining messages at entry or exit from a library call may impose unnecessary overheads and should be avoided. Libraries also don't need to limit themselves to disjoint process groups. Overlapping communicators are managed well within MPI.

5.2 Progress

There is no generally good strategy for highly-performance library progress: The use of asynchronous progress may be too inefficient or even impossible, while the use of manual progress breaks isolation and may lead to deadlock when multiple libraries are composed, with no systematic use of manual progress at each interface. Thus, progress should be ensured for each library separately. Also, repeated library invocations for manual progress add a superfluous overhead on systems with asynchronous progress. The cleaner solution would be to provide adequate asynchronous progress on all systems. Baring this, it is very desirable to provide manual progress calls that are macro-expanded into noops on systems that do not need them.

6 Summary and Conclusions

In this paper, we showed principles for designing parallel libraries, described a taxonomy of existing libraries and several library examples. We then derived general requirements for parallel libraries and described how they are supported in MPI. Furthermore, we show issues with the current MPI specification that may present pitfalls to developers. Finally, we summarize current practice and good practices for designing parallel libraries.

We conclude that MPI is very well suited to support the development and use of parallel libraries. It offers mechanisms for space- and time-multiplexing processes and an object-oriented interface. It is crucial that other parallel programming environments, such as upcoming PGAS languages, provide a similar level of support for library development.

Acknowledgments. This work was supported by the Blue Waters sustained-petascale computing project, which is supported by the National Science Foundation (award number OCI 07-25070) and the state of Illinois.

References

1. Balay, S., Gropp, W.D., McInnes, L.C., Smith, B.F.: Efficient management of parallelism in object-oriented numerical software libraries, pp. 163–202 (1997)
2. Basili, V.R., Briand, L.C., Melo, W.L.: How reuse influences productivity in object-oriented systems. Commun. ACM 39, 104–116 (1996)
3. Folk, M., Heber, G., Koziol, Q., Pourmal, E., Robinson, D.: An overview of the HDF5 technology suite and its applications. In: Proceedings of the EDBT/ICDT 2011 Workshop on Array Databases, AD 2011, pp. 36–47. ACM, New York (2011)
4. Gregor, D., Lumsdaine, A.: Lifting sequential graph algorithms for distributed-memory parallel computation. In: Proceedings of OOPSLA 2005, pp. 423–437 (2005)
5. Gregor, D., Lumsdaine, A.: Design and implementation of a high-performance MPI for C# and the common language infrastructure. In: Proceedings of PPoPP 2008, pp. 133–142. ACM, New York (2008)
6. Gropp, W., Lusk, E., Skjellum, A.: Using MPI: portable parallel programming with the message-passing interface. MIT Press, Cambridge (1994)
7. Hoefler, T., Bronevetsky, G., Barrett, B., de Supinski, B.R., Lumsdaine, A.: Efficient MPI Support for Advanced Hybrid Programming Models. In: Keller, R., Gabriel, E., Resch, M., Dongarra, J. (eds.) EuroMPI 2010. LNCS, vol. 6305, pp. 50–61. Springer, Heidelberg (2010)
8. Hoefler, T., Lumsdaine, A.: Message Progression in Parallel Computing - To Thread or not to Thread? (September 2008); accepted at the Cluster 2008 Conference
9. Hoefler, T., Lumsdaine, A.: Optimizing non-blocking Collective Operations for InfiniBand. In: Proceedings of IEEE IPDPS 2008 (2008)
10. Hoefler, T., Snir, M.: Generic Topology Mapping Strategies for Large-scale Parallel Architectures. In: Proceedings of ICS 2011, pp. 75–85. ACM, New York (2011)
11. Hoefler, T., Lumsdaine, A., Rehm, W.: Implementation and Performance Analysis of Non-Blocking Collective Operations for MPI. In: Lumpe, M., Vanderperren, W. (eds.) SC 2007. LNCS, vol. 4829. Springer, Heidelberg (2007)
12. Korson, T., McGregor, J.D.: Technical criteria for the specification and evaluation of object-oriented libraries. Softw. Eng. J. 7, 85–94 (1992)

13. Latham, R., Gropp, W., Ross, R.B., Thakur, R.: Extending the MPI-2 Generalized Request Interface. In: Cappello, F., Herault, T., Dongarra, J. (eds.) PVM/MPI 2007. LNCS, vol. 4757, pp. 223–232. Springer, Heidelberg (2007)

14. Lumsdaine, A., Mccandless, B.C.: Parallel extensions to the matrix template library. In: Parallel Processing for Scientific Computing (1997)

15. Lusk, E.L., Pieper, S.C., Butler, R.M.: More scalability, less pain: A simple programming model and its implementation for extreme computing. In: SciDAC Rev., vol. 17, pp. 30–37 (2010)

16. Meyer, B.: Lessons from the Design of the Eiffel Libraries. Commun. ACM 33(9), 68–88 (1990)

17. Mohagheghi, P., Conradi, R., Killi, O.M., Schwarz, H.: An empirical study of software reuse vs. defect-density and stability. In: Proc. of ICSE 2004, pp. 282–292 (2004)

18. MPI Forum: MPI: A Message-Passing Interface Standard. Version 2.2 (September 4, 2009), http://www.mpi-forum.org/docs/mpi-2.2/mpi22-report.pdf

19. Pan, H., Hindman, B., Asanović, K.: Lithe: enabling efficient composition of parallel libraries. In: HotPar 2009, p. 11 (2009)

20. Sbalzarini, I.F., Walther, J.H., Bergdorf, M., Hieber, S.E., Kotsalis, E.M., Koumoutsakos, P.: Ppm: a highly efficient parallel particle-mesh library for the simulation of continuum systems. J. Comput. Phys. 215, 566–588 (2006)

21. Skjellum, A., Doss, N.E., Bangalore, P.V.: Writing libraries in mpi. In: Proceedings of the Scalable Parallel Libraries Conference, pp. 166–173 (October 1993)

22. Snir, M.: Endpoint proposal for mpi-3.0. Tech. rep. (2010)

23. Willcock, J., Hoefler, T., Edmonds, N., Lumsdaine, A.: AM++: A Generalized Active Message Framework. In: Proccedings of ACM PACT 2010 (2010)

24. Willcock, J., Hoefler, T., Edmonds, N., Lumsdaine, A.: Active Pebbles: Parallel Programming for Data-Driven Applications. In: Proc. of ACM ICS 2011, pp. 235–245 (2011)

Author Index

GPSR Compliance

*The European Union's (EU) General Product Safety Regulation (GPSR)
is a set of rules that requires consumer products to be safe and our
obligations to ensure this.*

*If you have any concerns about our products, you can contact us on
ProductSafety@springernature.com*

In case Publisher is established outside the EU, the EU authorized
representative is:

Springer Nature Customer Service Center GmbH
Europaplatz 3
69115 Heidelberg, Germany

Batch number: 09490872

Printed by Printforce, the Netherlands